To So Few

The Verdict

Books by Cap Parlier:

———

<u>Anod series</u>
The Phoenix Seduction (1995)
Anod's Seduction (2004) [reprint of The Phoenix Seduction]
Anod's Redemption (2004)

———

Sacrifice (2000)
The Clarity of Hindsight (2016)

———

<u>To So Few series</u>
To So Few – In the Beginning (2014)
To So Few – The Prelude (2014)
To So Few – Explosion (2015)
To So Few – The Trial (2016)
To So Few – The Verdict (2017)

———

<u>and with **Kevin E. Ready:**</u>
TWA 800 - Accident or Incident? (1998)

———

Coming soon from Cap Parlier, **To So Few – Defliction**,
the sixth book of the series novel of flight and a warrior's life.

———

To So Few

The Verdict

by
Cap Parlier

SAINT GAUDENS PRESS
Wichita, Kansas & Santa Barbara, California

Saint Gaudens Press
Post Office Box 405
Solvang, CA 93464-0405

Http://www.SaintGaudensPress.com

Saint Gaudens, Saint Gaudens Press
and the Winged Liberty colophon
are trademarks of Saint Gaudens Press

Cover quotation from speech by Prime Minister Winston L.S. Churchill, CH, TD, MP, 20.August. 1940

Print edition ISBN: 978-0-943039-41-1

Library of Congress Catalog Number - 2016921242

Printed in the United States of America

The TO SO FEW series books are works of fiction. Any reference to real people, objects, events, organizations, or locales is intended only to give the fiction a sense of reality and authenticity. Other names, characters and incidents are the products of the author's imagination and bear no relationship to past events, or persons living or deceased.

Dedication

—

To all those who have gone before us, and given their last full measure of devotion to the cause of freedom and the defense of our liberty.

Acknowledgments

—

To John Richard and Roger Benefiel for research assistance.

To my wife, spouse, partner, sponsor and cheerleader, Jeanne, who tolerated the hours, days, weeks, months and years of research, writing and discussion. She has and continues to tolerate my love of flight and the need to tell a story about the greatest event in human flight.

To my reviewers: my wife, Jeanne; John Richard; and Leta Buresh, for their patience, reflection, opinions and suggestions. I believe they made it a stronger story.

Credit and gratitude go to John Richard for the concept of the unique portion of the front cover artwork.

A special recognition must be offered to numerous individuals who provided their knowledge, experience and precious time to assist my historical research.

-- Imperial War Museum – Dr. Neil Young, Research and Information Office

-- Royal Air Force Museum – Mr. Mungo Chapman, Research & Information Services

-- Duxford Aerodrome Museum

-- R.J. Mitchell Memorial Museum – Mr. D.G. Upward, Director

-- National Railway Museum – Mr. Philip Atkins, BSc, Librarian

-- Churchill Archives Centre – Ms. Carolyn Lye

-- House of Lords – Mr. D.L. Prior, Record Office

If there are errors in the representation of historical details, the responsibility rests solely with me and must in no way reflect upon the experts acknowledged above.

Prologue

In the first year of war in Europe, the German armed forces executed an aggressive, masterful and overwhelmning series of campaigns, and along with Italy, the Axis now dominated virtually all of Europe except for the British Isles. With the continent securely under their control, the Germans turned their undefeated attention on the British. On 10.July.1940, the German Air Force began its operation to establish air superiority over the English Channel and Southeast England – a prerequisite for Operation SEALION – the German cross-Channel invasion of Great Britain. While the *Luftwaffe* pounded the Royal Air Force, the German Army gathered its forces and the necessary transport for the pending amphibious operation.

During the summer of 1940, the British faced the prospect of their first invasion since 1066 AD, this time at the hands of *der Wehrmacht* – the German armed forces. The British Army had been seriously diminished in the Battle of France and stripped of a large portion of its armor and artillery during the evacuation known as the Miracle of Dunkirk. The Royal Navy fought desperately the defend the vital Atlantic supply convoys and kept the hammer cocked should the German Navy move warships to support a crossing of the English Channel. The defense of Great Britain during the tenuous summer of 1940 fell upon the skills of several hundred young fighter pilots in the Royal Air Force, Fighter Command.

Prime Minister the Right Honorable and Gallant Winston Leonard Spencer Churchill, CH, TD, Member of Parliament for Epping, had formed a coalition War Cabinet and government comprised of all political parties – no small task by itself. He bolstered the British people, and freedom-loving citizens around the world, with his moving, poignant and inspiring words. Against the counsel of his colleagues and advisors, Churchill also made several trips to France in the weeks after the German invasion in a Herculean, personal and risky effort to sustain the fight, right up to the moment the French government collapsed and signed the dictated armistice with the Germans on 22.June.1940.

While the British stood alone and struggled to defend the Home Islands against the German invasion that appeared inevitable, Churchill delicately worked his personal relationship with President of the United States Franklin Delano Roosevelt to find vital support to resupply its decimated army and makeup the essential goods shortfall created by the successes of the German U-boat offensive in the Battle of the Atlantic.

The Americans were dominated by an isolationist mood and the slow, plodding recovery from the Great Depression. President Roosevelt had already served two terms as president and had just accepted his party's nomination for an unprecedented third consecutive term as president with the election scheduled for the approaching fall. The President brought several influential Republican politicians to fill key cabinet posts, as he quietly prepared the government for war. He also walked a rather thin, misty line in providing material support to the British. Roosevelt sent several civilian and military emissaries to Great Britain during the summer of 1940, with one question: can the British repel a German invasion and remain free? The answer to that question became vital to overcoming the risks of supporting the British.

—

By the end of August 1940, the surviving pilots of Fighter Command were near exhaustion. They had been flying three, four, five combat sorties each day for nearly a month, against mind-numbing odds with an overwhelming and highly experienced adversary. Two friends and brothers-in-arms had completed their qualification training together and had been assigned to No.609 (West Riding of Yorkshire) Auxiliary Fighter Squadron at RAF Drem since December 1939. The squadron moved their Supermarine Spitfires south as the Battle of France opened in May 1940, and settled at RAF Middle Wallop as part of No.10 Group for the duration of the Battle of Britain.

Pilot Officer Brian Arthur 'Hunter' Drummond was the first and youngest of seven, American volunteer pilots in Fighter Command. Immediately after graduating from high school and turning 18 years of age, he left the only home he had ever known on the Great Plains of Kansas, defied U.S. federal law, and crossed the border into Canada to join the Royal Air Force. Brian had tallied 14 aerial combat victories and had three aircraft shot out from under him. In the last of those shootdowns, he landed unconscious under his parachute in a large pond on a Hampshire farm. The farm owner, a war widow herself, risked her life to save him from drowning. For her bravery and risking her life to save Brian, King George VI personally awarded the George Cross to Charlotte Grace Palmer née Tamerlin, and at the same ceremony, the King also awarded Brian the Distinguished Flying Cross. From that shared event, Charlotte and Brian began a tenuous relationship.

Brian's best friend and squadron mate, Pilot Officer Jonathan Andrew Xavier 'Harness' Kensington came from an upper middle class family near Newcastle-upon-Tyne. He graduated from Newcastle University with a degree in mechanical engineering in 1939, and joined the Royal Air Force. Jonathan and Brian became friends during their time in the advanced flight-training unit.

Fortunately, they were both assigned to the same operational fighter squadron. Early into The Battle, Jonathan had been handpicked to join a dozen operational pilots in support of the exploitation team evaluating captured and refurbished German aircraft, including fighters and bombers.

By the start of the Battle of Britain, Pilot Officer Samuel Oreland had transferred to Coastal Command, having lost his nerve to fly single seat fighter airplanes. By the end of the first month of The Battle, the puts and takes of desperate aerial combat took their toll on No.609 Squadron, as it did every other squadron in Fighter Command that summer of 1940. Of the original pilots that began The Battle, they lost:

– Pilot Officer Stephen 'Mongo' Strickland, KIA: 9.July.1940,

– Flying Officer George 'Angle' Ashcroft, KIA: 11.July.1940, and

– Flying Sergeant James 'Junior' Carrolton, KIA: 11.July.1940.

A wave of pilots evaded capture after the fall of France and made their way to Great Britain over the next few weeks after the French-German armistice. Of seven American volunteer fighter pilots who served with le Armée de l'Air during the Battle of France, three American pilots were assigned to and joined Drummond with No.609 Squadron on 12.July.1940:

+ Pilot Officer Frank Oscar 'Red' Burns,

+ Pilot Officer Henry 'Hank' Maxwell – the first of the squadron's American volunteer pilots to fall, KIA: 27.July.1940, just a couple of weeks after joining the British,

+ Pilot Officer Stanley Jordan 'Slim' Koenig – KIA: 31.August.1940.

Another original pilot perished as Fighter Command struggled with makeup the losses:

– Pilot Officer Reginald 'Organ' Foxworth, KIA: 18.July.1940.

Also, among the continental escapees were two former Czechoslovak Air Force fighter pilots, who joined the squadron on 22.July.1940:

+ Pilot Officer Kormer 'Curly' Mansek, and

+ Pilot Officer Janus 'Crazy' Kradilcek

The surviving fighter pilots of No.609 Squadron and their flight positions were:

'A' Flight, Blue Section:

Commanding Officer Squadron Leader Horatio Michael 'Spike' Darling (English),

Pilot Officer Roland 'Boxer' Stockard (English),

Mansek,

'A' Flight, Green Section:

Flight Lieutenant Roger 'Jackstay' Beamish (Scottish),

Kradilcek,
Drummond,
'B' Flight, Red Section:
Flight Lieutenant Robert Gates 'Sparky' Morrow (English),
Burns,
Kensington,
'B' Flight, Yellow Section:
Flight Lieutenant John 'Waggle' Davies (Welsh),
Flying Sergeant Miles 'Fog' Johnson (English),
Open seat, not yet filled.

The combat losses for Fighter Command mounted sharply, as the German Air Force focused it aerial might on domination of skies above Southeast England. The net loss rate had become so precipitous that the leaders of Fighter Command implimented a desperate stablization scheme to keep the seats of their frontline fighter aircraft filled. They cancel all rest leaves for fighter pilots, all transfers, and reassignment of experienced fighter pilots from Training Command back to combat seats. Despite the extraordinary efforts, the loss continued to push more negative by the day. The War Cabinet agreed to bomb Berlin to provoke Hitler to change his focus.

More than a dozen countries from the corners of the British Empire and many of the occupied countries of Europe now flew with their British brethren, including more than a handful of American volunteers. Despite the welcome additions, the challenge for the leadership of Fighter Command, the entire Royal Air Force and ultimately for His Majesty's Government came down to managing the net equation for both pilots and aircraft as the Germans used their advantage to bleed the British.

And so, here begins our story.

———

Chapter 1

Sleep after toil,
port after stormy seas, ease after war,
death after life does greatly please.
-- Edmund Spencer

Friday, 6.September.1940
RAF Middle Wallop
Middle Wallop, Hampshire, England
14:30 hours

Week 9

The pilots of No.609 Squadron collapsed into the lawn chairs or fell onto the grass outside the Dispersal Hut after their third combat mission of this long day. 'Hunter' Drummond was no different. His body landed in one of the lawn chairs. The sun was near its local zenith due to the two hour advanced time shift of British Double Summer Time implemented for the duration of the war. The warmth of the sun felt so . . . good . . . just . . . a . . . . few . . . moments . . . .

The clang of the scramble bells startled Brian back to consciousness. He bolted upright like a Pavlovian dog, groped for his flight equipment, only to realized he was still wearing everything – his leather helmet with earphones still in place, flight goggles on his forehead, and oxygen mask dangling from the left side of his helmet. Even his flight gloves still covered his hands. Brian jumped to his feet and ran toward his 'PR-F' fighter. By the time he reached his aircraft, the propeller was turning and the engine was running smoothly at fast idle. His crew chief Leading Aircraftman Bernard 'Bernie' Gordon stood on the left wing root, forward of the cockpit access hatch, braced against the air stream from the propeller.

Brian bounded onto the wing and into the cockpit. While he strapped into his parachute harness and latched his shoulder and lap belts, 'Bernie' attached his earphone and microphone chords, and plugged in his oxygen mask. No words were exchanged. Both men knew the routine and procedures. Once satisfied with his readiness, Brian gave Gordon a short salute. 'Bernie' closed and latched the access hatch, and then jiggled it to make sure it was closed and secure. Gordon saluted Brian and jumped off the wing.

Brian quickly scanned his instruments. His attitude gyro was uncaged and erect, thanks to 'Bernie' preparatory efforts. His heading was properly set into the directional gyro and he set the field elevation in his altimeter. Brian looked across the line of Spitfires just in time to see 'Spike' Darling's hand signal to taxi. He waited for his section leader 'Jackstay' Beamish to move before he advanced his throttle to roll forward.

The squadron was airborne in less than two minutes and climbing at full power toward the south. The radio calls between 'Spike' and sector control confirmed they were headed south to join a big fight already in progress over the Isle of Wight and defending Southampton and the Supermarine plant at Woolston. The initial assigned altitude was Angels Two Zero – 20,000 feet. Chain Home, the radar surveillance network, apparently indicated an enemy fighter force inbound from France to assist the attack.

Brian checked his guns were armed and ready, the gun camera enabled, and his gunsight was on and set. He attached his oxygen mask over his nose, mouth and chin, switched on his supplemental oxygen, and then checked the indicator on the system control panel to make sure he was receiving the oxygen he would soon need to keep his mind clear and alert. Brian adjusted his flight goggles over the top of the oxygen mask. All was ready for the fight.

They could see the melee below them, but that part of the fight belonged to other squadrons. Middle Wallop Sector Control transferred their control to Tangmere Sector Control, callsign 'Horse.' Tangmere adjusted their vector and altitude twice. They were now into the English Channel and searching for their targets down sun.

"Tally ho," called 'Sparky.' "Bandits, ten o'clock, slightly below the horizon."

Brian opened his position just a little and searched across the squadron. There they were . . . perhaps twenty plus 109s. *Two to one odds . . . better than we've seen lately.*

"'Sparky,' you take the lead. We'll take the backside."

"Wilco."

"OK, lads, here we go," 'Spike' broadcast. He kept the squadron high and turned to intercept the inbound enemy fighters.

The Germans wheeled their two squadrons and climbed to face No.609 Squadron. Brian quickly rechecked all his combat switches were up and ready. He instinctively scanned the sky behind both wings and above them to make sure they were not being baited into an ambush. The sky was clear.

Then, for some unexplainable reason, the Germans all rolled sharply to the south and dove away from the Spitfires. Both flights rolled to give chase. These encounters made Brian very unsettled. He scanned the sky around them continuously. He felt a trap. No one else saw anything either. There were no calls. They chased the Germans for several miles farther out into the English

Channel. Brian's head remained on an incessant swivel, searching for what he suspected were other German fighters.

"'Sorbo' . . . 'Sorbo Leader,' break it off."

Brian throttled back as the other pilots did and stayed with 'Jackstay,' as they leveled off, turned back north, and reformed as a squadron. *Maybe they'll let us have at the retreating attackers.* Tangmere gave them several more vectors but no action.

"'Sorbo Leader,' this is 'Horse.' We're done for this one. RTB. Switch 'Bandy.'"

"'Horse,' this is 'Sorbo Leader.' Wilco. Switching 'Bandy.' Good day."

Darling did not need to issue instructions. All the No.609 Squadron pilots switched radio frequencies to the assigned Middle Wallop Sector Control frequency. They were returning to base and all they had to do was follow their leader.

They landed at Middle Wallop with the gun port tapes still in tact. The ground crews would know they had not fired their guns and would not need ammunition. They would need fuel and oxygen. They parked, shutdown their engines, briefly conferred with their crew chiefs, and then returned to their spots so abruptly vacated 80 minutes earlier. 'Red' and 'Fog' dragged their lawn chairs into the sun. 'Hunter' left his chair where it was, now in the advancing shade of the Dispersal Hut. They had another four hours of daylight ahead of them and the potential of another one or two combat missions. Not a single cloud offered any hope of a respite.

'Spike' said, "I'll take the intelligence debriefing. You lads had better grab whatever rest you can. We are not done for this day." He did not wait for a response and disappeared into the Dispersal Hut with Flying Officer James Royster, the squadron's assigned intelligence officer, in tow.

Not a word was heard among the pilots. They needed no encouragement to catch whatever sleep they could while they remained on Available status. The ground crews would report their aircraft fully ready for take off in a few minutes. Sleep claimed them all.

———

Friday, 6.September.1940
House of Commons
Westminster, London, England
16:00 hours

The Right Honorable Winston Churchill stood before the Prime Minister's Despatch Box as the House became silent and the Speaker announced the Prime Minister's report. He scanned the great hall to see virtually every set of eyes directly on him – many faces familiar to him from his years of service as a Member of Parliament. He took pride in his accomplishments and did not shrink from his failures, but the past paled in comparison to the steadily darkening situation faced by the United Kingdom. All the signs continued to point down with no relief in sight, and yet he knew he had to find the words to reflect hope to a thirsty nation.

He cleared his throat and lowered his head as he prepared to speak. "My honorable colleagues," he began in a low, mumbled voice, "I stand before you with humility and encouragement. We are near the watershed for our beloved island. As our ancestors so many times in history stood with dark storm clouds about them, they found the vigor and fortitude to withstand the natural tendency for despair, and the courage to face the daunting task ahead. So, shall we. So, shall we.

"I am happy to report to the House, our American brothers are giving us the tools to achieve victory over that madman on the Continent. Just yesterday, we took delivery and manned the first handful of American destroyers. The first two vessels set sail this morning to join the Fleet in the Battle of the Atlantic. They shall ferret out the vermin that lurk under the waves and in the cover of darkness, and then they shall destroy them. The Royal Navy, as it has done so admirably for centuries, shall maintain the sea lanes open to commerce with our Empire, with the Commonwealth, and with our brothers who now stand with us against the common foe. While we do not yet enjoy their comradeship in arms, they are with us nonetheless."

He paused to absorb the scene frozen before him. They wanted his words of encouragement. The normal raucousness of the House had vanished. "As all of us are painfully aware, our young pilots are engaged in the greatest air battle in the history above our venerable nation, and perhaps the world. They continue to rise repeatedly each day to face daunting odds, deadly fast engagements against a determined enemy, and mind numbing fatigue. They have kept the enemy at bay for more than a month, and they shall continue to keep the enemy from our shores.

Churchill paused again. His thoughts flashed upon Air Chief Marshal Sir Hugh Caswall Tremenheere Dowding, GCVO, KCB, CMG, known affectionately as 'Stuffy' by his men – his chicks as he called them. Dowding had served as Air Officer Commanding-in-Chief, Fighter Command, since its formation in July 1936.

"The King and Queen, this day, visited the Headquarters of Fighter Command. They met with Air Chief Marshal Sir Hugh Dowding as well as his staff and group commanders. His Royal Highness told me about their experience, watching the progress of the battle as some of us have done from the gallery of the Fighter Command Operations Room. He asked me to convey how moved he was observing the calm, cool, professionalism of the staff, controllers, technicians and personnel as they dealt with such staggering conditions as we experienced earlier this afternoon. He believes morale remains high and wishes me to state that Fighter Command is up to the task at hand." Cheers interrupted his speech for the first time. Churchill nodded his head slightly in acknowledgment of the spontaneous applause like the whistle of a pressure cooker venting steam. He allowed the cheers to continue for a minute or so before he raised his right hand to request quiet.

"At this afternoon's War Cabinet meeting, we reviewed the most recent report of the Combined Intelligence Committee prepared just this morning. We have examined a broad range of intelligence sources including extensive aerial photographic reconnaissance of the French, Belgian and Dutch coastlines. It is the collective and considered opinion of the Combined Intelligence Committee as well as the War Cabinet that the heinous Nawzee juggernaut may yet attempt a crossing of the English Channel. Later this evening, if the afternoon reconnaissance information remains the same, we shall issue Invasion Alert Number Two to our armed forces. Alert Number Two means we believe that an invasion attack is probable within the next three days."

Gasps, grunts and disparaging remarks stopped his words and communicated the reception of the disappointing news by the Members. Winston knew many of the Members not in or close to the War Cabinet would be surprised by the announcement. As was usually the case, it was better to deal with any rancor inside the Commons Chamber rather than in the media or public. He decided to wait patiently for the hall to calm.

"Order," said the Speaker of the House the Right Honorable and Gallant Edward Algernon FitzRoy, DL, Member of Parliament for Daventry. "Order . . . order . . . order," he repeated in progressively louder voice. "The House will come to order." The Speaker tried to regain control.

Winston Churchill continued to wait for quiet and the attention of Members. This was not a particularly pleasant speech, but it was one the War Cabinet wanted to be delivered. Any debate that might arise had to be fully addressed in the House. The objective was unanimity of position, although in this situation none was required.

As the words and sounds diminished, Winston continued. "The Prime Minister and the War Cabinet wish to inform the Honorable Members that all possible precautions have been taken. While our armed forces are not at the desired strength and preparedness, every sailor, soldier and airman stands ready to defend our island. If invasion should come, the protection of our way of life may and indeed most likely will fall upon all our citizens. We must defend every house, every farm, every street, every bridge in our land. We must show no quarter to the enemy."

Cheers mixed with disagreement. The mood in the air told Churchill the objective was not close at hand. He held up one hand, and then both hands as if two would do better than one. The effect was not appreciably better. For the first time in years, Churchill found himself losing patience with the House of Commons. The often-petty comments along with the still lingering few pacifists who wanted to sue for peace despite the horrors of the last few months defied the Prime Minister's tolerance. He had more important things to do with an invasion even closer.

He shook his hand in the air. Before either the Prime Minister or the Speaker of the House could speak the droning tones of the air raid siren took possession of the Chamber. Members began to move before any words were spoken.

"The House stands adjourned until the conclusion of the present situation," the Speaker announced.

Members moved quickly but calmly in an orderly manner through the main entrance to the hall. The Prime Minister joined his colleagues although the interruption further stretched his patience. Although bombs had fallen on the outer reaches of the city, most likely unintentionally based on German targeting history. Winston felt no need to seek shelter. If he could stand in the middle of a primary target during a raid, the leaders of His Majesty's Government could certainly maintain their concentration in the as yet unscathed capital city. He wanted to ask his colleagues to remain so he could finish his speech and return to the business of leading the nation's defense.

The bomb shelter designated for Parliament was actually a portion of the Underground rail complex, Westminster Station to be exact. The dim lighting added to the sinister character of the enemy's actions above the city. It did not stop the small pools of continued conversation, discussion and debate from proceeding just as though they were in the Member's lounge.

Winston Churchill brooded and paced not wanting to join any of the conversations. He considered several times returning to Westminster Palace and the House of Commons Chamber to finish his speech without the Members and audience. There was more important business to be conducted rather than skulking around in a dreary, damp, Underground station platform area. He contained his frustration as best he could although he began pacing with his head down more like a battering ram than a brooding politician.

"What seems to be the problem, Winston?" asked Lord Privy Seal the Right Honorable Clement Richard Attlee, Member of Parliament for Limehouse and Leader of the Labour Party.

"There is no use having these banshee howlings of the siren two or three times a day, simply because hostile aircraft are flying to or from some target. We must prepare for heavier fighting in the month of September. The need of the enemy to obtain a decision before the autumnal inclement weather is very great. The whole nation will take its example from our airmen and will be proud to share part of the danger with them."

"Surely you do not mean we should ignore the warnings."

"Yes, I most certainly do. While bombs may soon fall on London itself, they have not yet, and so here we sit underground like gophers when we have a nation to lead."

"Your courage is renowned, laudable and a beacon for us all, Winston, but we don't know when the siren's wail will announce the bombardment of Westminster and Whitehall. We cannot afford to lose you, Winston. While you may feel the urge to defy the enemy, I am afraid the War Cabinet and His Majesty's Government must insist upon your safety. You have kept us together these last few months, and you shall see us through this affair."

Winston Churchill stopped his pacing to look into the concerned eyes of his political rival. He recognized the validity of Attlee's words. They eased his frustration to a tolerable level. The all-clear signal sounded, and Winston checked his watch. Thirty-seven minutes passed, and no thuds from exploding bombs were heard.

The clear skies and late afternoon sun strained their eyes as they walked across Parliament Bridge Road to the ornate Gothic palace that represented the seat of British government for centuries. The Prime Minister finished his speech, answered several questions and they adjourned. The evening War Cabinet meeting commenced 45 minutes later in the Cabinet Conference Room at No.10 Downing Street.

21:00 hours

Invasion Alert No.2 was issued. The United Kingdom moved one step closer to the final conflagration when His Majesty's Government informed all military and government agencies that the impending German invasion was now probable.

———

Saturday, 7.September.1940
Headquarters, No.11 Group
Uxbridge, Middlesex, England
15:00 hours

"**W**hat the bloody hell is going on?" protested Air Commodore John Henry Randolph Spencer, CMG, DFC, the newly appointed senior controller for No.11 Group.

The small group of controllers stared at the empty map board for their area of operations. All morning, there had only been three comparatively very small raids of no consequence. Only the high altitude, solo, enemy reconnaissance flights seemed to provide some touch of routine. The rather feeble effort so far defied comprehension. The weather was near perfect. The Chain Home Radio Direction Finding network had numerous holes from bomb damage and equipment failure. The pilots of Fighter Command were beyond the limits of human endurance. Even high-octane fuel and the ammunition supplies approached the levels of insufficiency where knowledgeable officers questioned the viability of the organization.

Why, after months of concerted assault and three weeks of relentless, punishing bombardment of the Southeast fighter bases, would they let up now? The question plagued all of them as they agonized over reasons and possible explanations. Such a dramatic departure from the *modus operandi* of the enemy made everyone uneasy. Even the talkers waiting for the information from the Filter Room of Fighter Command fidgeted and shuffled from lack of action. It was like an obscene form of withdrawal symptoms from a powerful, perverse narcotic. The enormous onslaught repeated everyday over the previous weeks had become the routine, the norm, the commonplace, for the personnel of Fighter Command. The change was too obvious, too dramatic.

"The War Cabinet issued Invasion Alert Number Two last night and nothing happens," John said, more to himself than the others.

"Does Two mean an invasion within three days?" asked one of his fellow junior controllers.

"Probable, within three days," answered John. "Not certain."

"This is a strange way to begin an invasion."

"Maybe it is the quiet before the cyclone," the senior controller added.

One of the many telephones rang. "Sir, Biggin Hill is wondering what is happening as well. They asked if there has been communications or equipment failures?"

John stared at the board without responding. His mind raced through the endless possibilities. Had they won the battle? Had the Germans decided victory was beyond their reach and diverted their attention elsewhere?

"Sir?"

"Yes, yes. Tell Biggin Hill as well as the other sector stations the network is working properly. We have no appreciable activity."

"Yes, sir."

The muffled conversations drifted into the background as the communications system carried the message to the sector stations, and then to the aerodromes, squadrons and pilots. The situation made John nervous. What was he missing? What were they missing? Had the Germans figured out how to defeat the Chain Home early warning system? The Observer Corps reported nothing. Unless the enemy had also found a way to make bombers invisible and silent, they could not have accomplished that feat. Even the group commander, Air Vice-Marshal Keith Rodney Park, MC+Bar, DFC, stopped by the controller's booth to ensure all elements of the system were functional.

John lifted the direct telephone line to Air Marshal Geoffrey Leonard, CBE, DFC, Chief of Operations, Fighter Command. Something had to be happening outside the Southeast area of operations. Maybe an invasion had begun in the North?

"Air Marshal Leonard, here," he answered.

"Air Commodore Spencer, sir. What do you make of this?" he asked, dispensing with the usual amenities of polite discourse.

"The prevailing thought here is, preparation for something big."

"Does Chain Home show anything?" John asked, knowing full well what the answer would be – the Radio Direction Finding network was operating correctly, and there were no targets to reflect the radio energy back to the receivers.

"You know perfectly well the answer, John. No. The scopes are blank. We have queried each of the Observer Corps sector stations. They report clear skies, perfect visibility, and no targets or sounds."

"I just felt the urge to ask. The sector stations are asking us."

"There is nothing we can say," added Leonard without conviction. "You know the Cabinet issued Alert Two last night?"

"Yes, sir."

"Maybe the Huns are taking the day or two off to rest up for the big day."

John Spencer surprised himself with his laughter. "Sorry, sir. I simply can't imagine the Germans taking a day off to rest."

Leonard chuckled as well. "I think you have a point there, John."

"I suppose we can only wait for them to make their move."

"Precisely, John. I might add, we are checking other sources to see if we can turn up some further clues."

The No.11 Group Senior Controller knew that meant intelligence sources. They probably had the Air Ministry pumping SIS and other intelligence collection agencies for information of any kind. He glanced at the large wall clock opposite his balcony position.

15:25 hours

There were only four more hours of daylight. Maybe they were shifting to night operations? If they did, it would be a major departure. All their night operations so far had been very limited in size except for the Liverpool raid a week ago. The thought of maneuvering a formation of hundreds of aircraft in the sky at night still baffled Air Commodore Spencer.

"'Stuffy' just arrived."

"Talk to you later, sir. Thank you for the update."

"Cheers, John," he said and hung up the telephone.

The face of all the controllers turned to him. John smiled. "Nothing."

"So, we wait for nothing," a junior controller said out of frustration.

Air Commodore Spencer considered chastising the squadron leader for wasted words, but chose not to admonish the younger officer. He felt the same frustration. "Let's call our sector stations, again. Tell them we are checking every possible source of information. We will let them know as soon as we have anything."

As the task was being carried out, John returned his attention to the empty map board and the range of possibilities. The thought that they had won the battle seemed too attractive to be real. They were nearly beaten, or maybe they had already been beaten. How could the Germans not know how thin the fighter forces were, and how vulnerable Great Britain was at this very moment? They had a barely effective Fighter Command and virtually no Army in the Home Forces. If they landed, they could probably take all of Great Britain without much of a serious fight. It was impossible to believe they might have seriously overestimated the strength of Fighter Command and were hesitating short of invasion because of a few remaining Spitfires and Hurricanes.

Saturday, 7.September.1940
RAF Middle Wallop
Middle Wallop, Hampshire, England
17:15 hours

"This is a pleasant change," said Pilot Officer Jonathan 'Harness' Kensington. "Damn Germans giving us a day off."

"Maybe they decided we were too difficult for them," said Flying Sergeant Miles 'Fog' Johnson.

"Maybe."

"I doubt it," responded Squadron Leader Horatio 'Spike' Darling. "Why?"

He held up a piece of paper. "You read the last message. Invasion Alert Number Two means the War Cabinet believes that invasion is probable within three days. They would not issue such a warning if there were not substantial evidence indicating an invasion. I suspect we are about to experience a major shift in German tactics. Maybe they are making final preparations. Whatever the reason, we should prepare ourselves for hard times ahead."

Several pilots laughed and said in unison, "Hard times," as if they had not experienced hard times, yet. They laughed, but they knew without saying it that Squadron Leader Darling was most probably correct. Pilot Officer Brian 'Hunter' Drummond sensed the electricity in the air like the smell of rain before it arrives. Something was indeed going to happen.

"Maybe they start invasion," said Pilot Officer Janus 'Crazy' Kradilcek.

"We would have heard something, if that were true," answered Darling.

"Maybe it start, and telephone lines broke."

"The radio works, and sector would have told us if something was wrong. The sole purpose of the last telephone call was to tell us everything was normal."

"Normal is a sky full of Krauts," interjected Pilot Officer Frank 'Red' Burns, the only other surviving American volunteer pilot in the squadron.

The comments sparked off another round of laughter and derogatory remarks at the parentage and morals of their enemy. Brian always found it refreshing that the pilots could laugh at adversity. It kept them sane.

The telephone rang at the dispersal tent. All the pilots sat up and turned toward the entrance but did not speak.

"Squadron to Readiness," announced squadron operations clerk Corporal Jennifer Warren.

"That answers that."

"Just getting a late start on the day, are we now."

"Overslept, I guess."

"And so it goes."

The pilots became more animated. Several retrieved their flight equipment. Kradilcek and Burns walked to their aircraft. Even the ground crews began to fidget with the Spitfires standing at full readiness. The tension of waiting affected them all.

"It is half past five. Bloody idiots only have another few hours of daylight," observed Flight Lieutenant Robert 'Sparky' Morrow. "What do they hope to accomplish before dark?"

"Make us land in the dark," said 'Boxer' Stockard.

"Hell of a thought, if I do say so," 'Waggle' Davies said.

They did not have to wait long for action. The telephone rang, again, barely ten minutes later. Johnson and Stockard started walking toward their aircraft before the announcement.

"Scramble the squadron."

Brian joined the others as they ran to their fighters. Leading Aircraftman Bernard Gordon started the Merlin engine in the 'PR-F' Spitfire. Most of the engines were turning over at idle as the pilots arrived. Bernie stood on the wing braced against the propwash from the large, 3 blade, deHavilland propeller. The smell and bite of half burnt, high octane, gasoline greeted Brian as he jumped into his cockpit. He took his shoulder straps from Gordon to buckle himself tightly into his parachute and seat while Gordon made the connections for his oxygen mask and microphone as well as his helmet earphones. When everything was ready, Bernie Gordon saluted his pilot quickly and jumped from the wing. Brian waited for his section leader, 'Jackstay' Beamish, and right wingman, 'Crazy' Kradilcek, to initiate their taxi before he nudged his throttle forward. They were airborne within two minutes.

"Bandy, this is Sorbo Leader calling," Darling radioed, as they passed through 500 feet.

"Roger, Sorbo. Vector Oh Nine Oh, angels Two Six."

"We have it, Bandy."

'Brooklands patrol or back up for Kenley,' Brian said to himself, as he rechecked his cockpit ready for combat. Off his right wing, he saw Winchester in the distance. He could see the area of Standing Oak Farm although he could not see any of the details that would identify Charlotte Palmer's house. He violated one of his rules, allowing his thoughts to create her image. He wondered when he would see her again. Brian wanted to be with her. He shook his head, scanned the sky around him, and then checked his instruments to ensure they matched his senses.

As they climbed above the top of the haze layer at 8,000 feet, features of the scene became evident and began to describe the situation for the pilots. Dissipating straight contrails from the south changed to a disorderly, twisted jumble of spaghetti in the sky. An enormous cloud of black smoke rose for thousands of feet from what had to be London.

"Sorbo Leader, this is Bandy calling," the Middle Wallop controller radioed in a calm, professional voice. The radio remained silent for many moments. Brian could only imagine Squadron Leader Darling in his cockpit absorbing the scene in front of them. "Sorbo Leader, Bandy here," the call came with a strain of query.

"Bandy, Sorbo Leader, go ahead."

"Roger, Sorbo Leader, thought we might have lost you there for a moment. Short Jack advises, vector one oh five, angels three one."

Darling turned the squadron slightly to the right before he responded to the direction passed from Biggin Hill Sector Control through Middle Wallop Sector Control. They continued their climb at full power passing through 20,000 feet on their way to 31,000 feet. The direction change skipped over both Tangmere and Kenley sectors that lay between Middle Wallop and Biggin Hill. As No.609 Squadron pressed on with their throttles against the emergency gates, more detail popped out of the confusing scene. The sky was filled with airplanes across many thousands of feet of altitude. Brian had never seen so many aircraft at one time on the ground or in the air. There had to be thousands of aircraft scattered across the entire Southeast of England.

"Sorbo Leader, Bandy."

"Sorbo Leader."

"Switch to Short Jack." All of the No.609 Squadron pilots switched to the correct radio frequency.

"Roger. Short Jack, this is Sorbo Leader."

"Sorbo Leader, this is Short Jack. Target is a very large raid."

"Maybe an understatement, Short Jack. Tally-ho. The sky is filled with bandits."

The radio fell silent as everyone continued to assimilate the details and create an impression of what was happening. They leveled off at 31,000 feet. Their straight condensation trails streamed behind them. Darling adjusted their heading to take the best entry point into the battle. He called for line abreast. They were all going into the fight together. They still had several minutes before they reached the battle.

Brian could not avoid the growing black cloud. As they moved closer, the characteristic curves of the River Thames and the Eastern districts of the capital city came into view. He thought he could see the bright orange and yellow flickers of the flames at the base of the black cloud, which told them the Germans had struck the Docklands area of central London. Other dockyards along the Thames to the East had to be involved as well. The Tower Bridge could be seen at the edge of the huge fire. Whitehall and Westminster did not appear to have been hit.

"My God," someone radioed, reflecting the staggering scene before them.

"Fucking bastards!"

"Jesus Christ," someone else broadcast.

"Keep it down, lads," said Darling. "Concentrate. We have work to do."

Brian rechecked his switches, again. His guns were armed and live. The 'PR-F' Spitfire was ready. Brian's eyes picked up the intensity, focus and concentration of the hunter he was. The sun behind them clearly illuminated the green and brown, curved lines of other Spitfires, and more importantly, the gray, sharp-edged, squared lines of the Bf109s. Most of the fight remained slightly below them. Fighters dove and climbed across the condensation line at 28,500 feet allowing the moisture of the engine exhaust to condense into a cloud stream. The sporadic blooming of the condensation trails, or contrails as the pilots called them, added to the surreal scene around them. Brian found his first few targets.

"Let's get to it," 'Spike' Darling radioed, as he lowered his nose gently and only slightly to pick up more speed and avoid coughing his engine.

The pilots of No.609 (West Riding of Yorkshire) Squadron followed their leader into the largest single air battle in history.

Brian pushed his throttle through the emergency gate wire, as the others did as well. He held his first target, a mottled, black and gray, Bf109 with a white propeller spinner arching over the top of what could loosely be described as a loop. His target was chasing and closing on another Spitfire. Brian quickly scanned the sky. He was not yet a target. Rolling, Brian chose to follow the German using a good pursuit angle to close the distance between them. He worked to fill the Bf109 in his sight, and then squeezed the trigger. Flashes spotted the German. Pieces and chunks of his aircraft flew off as Brian kept his bullet stream on his prey. Brian released the firing button as his target disintegrated. The Spitfire's nose rose well above the horizon as Brian rolled the aircraft to scan the sky for another target.

Brian realized in an instant there was no question of finding another target. The sky blossomed with dark gray targets like dandelion seeds cast into the wind. He wasted no time trying to pick one. He moved his gunsight reticle onto the closest German fighter. Brian worked hard in the cold, thin air. As was always the case in large engagements, the hunter on occasion becomes the hunted. He pulled off his target to counter his attacker. The pilots had to constantly protect themselves and each other in the twisting turning fights. Although Brian never took a mental tally of his engagement ratios, he closed within firing range on more than a quarter of his attempts without having to take evasive action. He fired at least a short burst on most of the valid targets. He could observe the distinctive flashes of bullets vaporizing metal on about a third of the targets he engaged. His instincts and defensive techniques enabled him to sort out good targets from bait.

Changes in the character of this monster engagement did not escape Brian's awareness. The white noses of the Bf109s became briefly green, and then red. They were now among the elite of German fighter squadrons, JG1, *Jagdgeschwader Freiherr von Richthofen Nr. 1*, Baron von Richthofen Fighter Wing No.1, named in honor of the highest scoring fighter pilot in the history of any air force and the progenitor of modern fighter tactics. Brian did not alter his technique and keen skills, although the difficulty of finding good targets increased substantially. The first few white noses began to reappear when Brian exhausted his ammunition.

He called out his condition for the others. Brian heard others begin to call their ammunition depletion. Darling called for the disengagement. As they dove away from the fight, the clear lines of contrails from other Spitfires merged with the fight. The timing was perfect. The fresh fighters kept the Germans occupied allowing No.609 Squadron to disengage.

As the squadron rejoined to the east of the flooding stream of aircraft between London and the Kent and Sussex coastlines, the Sun touched the Western horizon. They would be landing in twilight. Darling asked each pilot to report his fuel state. Brian thought he had enough fuel to make it back to RAF Middle Wallop. Their leader had to consider whether they should divert for fuel. In the end, he decided not to divert, and the waning daylight meant it was their last sortie for this momentous day.

Brian looked back at the still growing black cloud of smoke above London. The fires were spreading probably from the bombing rather than a lack of fire fighting services. They had seen so much damage, destruction and death in the months of war, but London burning had the most sickening feeling for Brian. This had not been a mistake. The bombing of London on this Saturday evening was a deliberate act by a despicable Nazi leader.

No.609 Squadron landed safely at RAF Middle Wallop. The debriefings took much longer than usual, as everyone including the intelligence folks dealt with the dramatic change in German tactics. They missed the evening meal at the Mess, plus several of the pilots wanted to know what was happening on the ground in London. Brian and Jonathan joined Beamish, Davies, Stockard and Johnson at the Operations building. A civil radio was tuned to the BBC, as the news of the bombing continued to flow. The broadcast reminded Brian of some natural disaster like the Hindenburg crash, or a California earthquake, or a major fire in some city. The broadcaster tried to stay with the facts presented to him and caught himself several times adding adjectives to the facts. The Operations personnel filled in some of the details.

Estimates of the initial raid that began shortly after 19:00 hours ranged upward to several thousand German bombers and fighters. Additional waves of bombers and fighters appeared as the previous waves delivered their deadly loads on London. The darkened sky had not stopped the flood of bombers. While both German and British day fighters were grounded by the darkness, the bombers continued their missions unabated. The night fighters of No.604 Squadron took to the air as the sunlight faded. They would have a very long night.

By all reports, London had been the sole target of the German bombers. There were no reports of any attacks on other cities, ports, airfields or other military facilities. The attack which continued well into the night was a purposeful, concentrated abuse of the civil population of the capital city of Great Britain, seat of the British Empire and one of the most influential metropolitan centers in the world.

As they listened with riveted attention to the reports from London, the clatter of the teletype broke their silent vigil. All eyes remained on the machine. Wing Commander Herbert Worly, the Station Operations Officer, tore off the paper. He read the short message and life appeared to drain from his face. He passed the cable signal to Flight Lieutenant Beamish. Brian was the fifth person to read the message.

```
ZZZZ/7221GGI4931/HFHQ-ALLSE/8771136/ZZZZ
DATE:  07.09.40, 2007 HOURS
FROM:  GHQHF
TO:  ALL SOUTHERN AND EASTERN COMMANDS
SUBJECT:  NOTICE
BREAK
CROMWELL.  ALL INFORMED.
END
ZZZZ/7221GGI4931/HFHQ-ALLSE/8771136/ZZZZ
```

Brian did not know what it meant, but knew from Worly's reaction it had to be important and not good. He passed the paper to Jonathan. It was 'Boxer' Stockard who eventually asked the question.

"What does Cromwell mean?"

Wing Commander Worly looked up from his feet. He stared at Stockard for several long moments without the slightest change in his dire expression. "Cromwell means the invasion has begun."

"No, God damn it!" shouted Beamish.

The pilots and Operations personnel looked at one another in disbelief. Brian looked to Jonathan. His eyes must have said what his mind was thinking. *Had they fought so hard for nothing? If the Germans had truly landed, everyone knew there was not much left of the Army to stop them. Groups of old men with wooden parade rifles and pitchforks would not stop the German Panzers.*

"We can still stop this," Jonathan whispered to his friend. "They have not won, yet."

Brian just shook his head, not in disagreement with Jonathan, but disbelieved that the invasion had actually begun. *If it had begun, why were they going after bombers attacking London instead of the invasion forces landing on the beaches?*

Brian wanted to call Mary Elizabeth Ann Spencer née Armstrong, wife of Air Commodore Spencer, to make sure she and their unborn baby were all right. As far as he knew, Charlotte Palmer and Rosemary Alice Kensington, Jonathan's younger sister, were not close to London. All three were women Brian cared about and would be in some form of danger. Mary lived toward the northern edge of London, so she probably did not get bombed, but he had no way to know. He hoped she was not in Central London shopping or something. Many of the pilots in the quiet of the Hampshire airfield operations building felt the need to call friends, loved ones or lovers. They all knew they could not use the telephone lines at such a critical time for non-official business. They would just have to wait for news.

The raids on London began to dwindle in size toward midnight. The pilots managed to scrounge up some light food at the local public house. Brian went to sleep wondering if he would be awakened with the *Wehrmacht* approaching Middle Wallop. Tomorrow would undoubtedly be a full and eventful day.

Chapter 2

Anger is a momentary madness,
so control your passion or it will control you.

-- Horace

Sunday, 8.September.1940
Headquarters, No.11 Group
Uxbridge, Middlesex, England
08:00 hours

Week 10

"**W**ake up, sir."

Only the glare of lights greeted Air Commodore Spencer. The stiff neck, sore spots on his forehead and both forearms along with the distinctive, leadened, numb, tingling resulting from insufficient blood supply to his hands made the lack of restful sleep all the more debilitating. He had remained at his post even after the last of the night raiders left the map board.

"Sir, Air Vice-Marshal Park has called a morning staff meeting," said the duty junior controller.

"What time is it?"

"Oh eight hundred, sir."

John looked at the blank map board with only two talkers awaiting the first intruders. "How is the weather outside?"

"A fine specimen of a late summer's day."

"Damn. No rest for the weary."

"Yes, sir."

"What time did the boss call the meeting?"

"Half eight, sir."

"I should freshen up then, or my colleagues are likely to give me the boot," John said with a feeble smile that was the best he could conjure up, as the tingling in his arms began to dissipate.

He found a fresh shirt in his office desk drawer along with the last clean pair of underwear he had stashed away. The hot shower in the officer's changing room felt good to his aching body. He allowed the heavy spray to envelop his body and drain away the pain. As he began to feel like a human being again, John turned the hot water handle nearly off. The not quite cold water invigorated his mind and tightened his muscles. The refreshment would last him at least an hour or two.

Air Commodore Spencer entered the conference room buzzing with the professional conversation about the extraordinary events of the previous evening. He took a chair along the wall behind the conference table. He still did not remember the names of most of the staff partially due to being a new arrival, but more significantly because of the mounting mental fatigue. Everyone seemed to be preoccupied with their collective analysis of the change in German tactics.

Air Officer Commanding-in-Chief, No.11 Group, Air Vice-Marshal Keith Park entered the room with a hard grim expression on his face and purpose in his stride. He stood to the head of the long conference table and looked each man in the eyes. For some strange reason, John Spencer's attention was drawn to the top row of ribbons stacked above his left breast pocket of his tunic and beneath his pilot's wings. The first decoration was a ribbon of white-purple-white vertical stripes of equal width identifying him as a holder of the Military Cross awarded to junior officers of the Army for gallant and distinguished services in action. John remembered part of the story about Park's actions as a New Zealander captain in the Royal Flying Corps, at the time it was part of the Army. In Park's case, his Military Cross was with Bar, which signified a second award. The ribbon next to the Military Cross was identical to his own, the alternating violet and white horizontal stripes of the Distinguished Flying Cross awarded at the end of the Great War for acts of valor, courage and devotion for duty performed by the officers of the flying services. The DFC ribbons awarded today differed from the awards during the Great War in that the alternating violet and white stripes were diagonal rather than horizontal. Air Vice-Marshal Park was an accomplished fighter pilot decorated for his skill and achievements.

"Before we begin," Park said without changing his expression, "I might tell you that I took my Hurricane up last night after the last of the raiders had departed. I circled the city for what seemed to be an eternity. The sight of London on fire . . . well," he said and cleared his throat, as he fought to control his emotions.

John remembered the 'OK-1' Hawker Hurricane held at RAF Northolt just for Park. It was the only fighter in the RAF with a number as a designator. The pilots told stories about the 'OK-1' Hurricane arriving at airfields at almost anytime, day or night, even in France after the real war had started, and the tall, slim figure jumping from the fighter wearing beige coveralls like a test pilot. Park held a very special place in the eyes, minds and thoughts of the fighter pilots.

"If I may say, it was nauseating," Park said, practically spitting out the last word. "The entire Southeast was ink well dark except for the hideous flames of the conflagration caused by the enemy bombs in the center of the city. Last night's raid is a horrific insult to our people and our nation, but it is also the sign of our salvation." Numerous heads including John's jerked toward the group commander. "Aloft last night, I distinctly remember thanking the Lord Almighty for our deliverance." Everyone remained frozen with a flood of thoughts and images in their minds. "I should point out that not one Fighter Command aerodrome was attacked yesterday. Göring and his henchmen appear to have delivered to us the very break we have all prayed for these last few weeks. If he continues to focus his attention on London, we shall regain our strength."

"What about civil morale?" someone asked.

Park smiled. "Leave that to the Prime Minister. He shall take care of morale, as he has done so admirably to date."

"Won't they see us as having failed to protect the city?"

"Maybe, but as long as Mister Churchill remains Prime Minister, he will ensure the people understand the importance of their sacrifice. I talked with Sir Hugh a short time ago. He, in turn, conferred with Sir Cyril, Sir Archibald and the Prime Minister himself earlier this morning. Everyone recognizes the terrible damage London and the other major cities will most likely suffer over the next days and weeks, and maybe months, but the enemy has given us the opportunity to rebuild our nearly depleted fighter forces and facilities. Please do not misunderstand me, we have many hard days ahead, however this could very well be the watershed we have hoped for. We must not lose this moment. We must redouble our efforts to replace our lost pilots, and where possible, give them some much-needed rest. Now, let us proceed, before Gerry decides to have at us again."

The group intelligence officer began his briefing as Park sat in his chair. "The Air Ministry is still collecting all the reports, however the current assessment confirms Air Vice-Marshal Park's observations. There were four distinct waves, which tended to commingle during the engagement and may have exceeded 3,000 enemy bombers and fighters. Hardest hit areas were the Docklands and Canning Town districts. Nine fires were reported by the Fire Brigade as 100 pumps each." Several mumbled expressions of surprise and anger paused the report. "In addition, 19 were reported at the 30 pump level, 40 at 10, and there were more than 1,000 lesser fires. All in all, a very difficult night for the Fire Brigade and other emergency services." He looked at the wall clock.

08:45 hours

"As of three quarters of an hour ago, only two of the major fires had been contained and brought under control. The newspapers have reported approximately 400 killed and in excess of 1,400 injured."

"What of Cromwell?"

"The Cromwell alert issued by General Headquarters, Home Forces, last night, is still being investigated."

"Have there been any landings, as yet?"

"None that have been reported to the intelligence bureaus."

"What is the prognosis for today?" asked John with impatience that surprised him.

The briefer stared at the Senior Controller for a few moments as if he needed time to understand what John said. "Until we know more, I should think, confused is the best we can predict." He paused, waiting for a reaction as his patience began to wear thin. When none came, he continued. "According to Fighter Command and the Air Ministry, we are likely to see more of the same, I'm afraid. Attacks will probably continue on London and may very well spread to other major cities. A concerted effort is being mounted to determine the status of the invasion fleet and forces. The last report we have reflected the situation yesterday afternoon, and there were no overt signs of initiation. We shall have to wait and see what the enemy does next to ascertain his immediate objectives."

"Gentlemen," interjected Park, "while the results of yesterday's insult are dramatic and significant to our people, I must remind you that it is our pilots who are stopping the Germans. Lord knows, it is not the Army. The Fleet has not pressed the Channel in many weeks now. It is our command of the skies that is holding the wolves at bay. Rather than cogitate over damages to civil infrastructure and population, I suggest we find ways to bolster our pilots and keep the Huns on their side of the moat."

"Hear, hear," someone said.

"Let us seize this fortuitous opportunity to redeploy and invigorate our forces. Air Commodore Killigon," the group operations officer, "I would be most appreciative of a draft instruction to focus our efforts. We must create a layered response that keeps the pressure on the incoming raids both to and from their targets. We should consider relocating our squadrons to give them the best occasion for intercept. I will take your recommendations, however it would appear to me that we should amass the Spitfires against the high fighter screen, keep them occupied, and let the Hurricanes concentrate on the bombers. Please coordinate your proposal with Air Marshal Leonard at HQ, and seek their concurrence to withdraw the remaining Defiants and Blenheims from our area of operations. We need whatever Spitfire squadrons can be spared. Under Sir Hugh's Stabilization Scheme, we should be able to get them."

"Not likely from One Two Group," John said and instantly regretted his fatigued quip.

Air Vice-Marshal Park smiled at John. "You may very well be correct, Air Commodore Spencer, however it is consistent." He looked away to scan the other staff officers. "Now, if there are no other questions, we have much to do and many long days ahead. Let us put our shoulder to the wheel and bolster our boys in those cockpits." The shuffle of papers, chairs and light conversation marked the conclusion of the short staff meeting. "Air Commodore Spencer, if you would be so kind . . . ," said Park above the sounds of exit.

John knew what was coming. His sharp tongue, less controlled by the nearly perpetual headache and misty concentration brought on by the accumulation of inadequate sleep, was about to gain him a quiet reprimand from his new commander. He waited for the others to leave the room and did not respond to the few raised eyebrows of other officers.

"John, you simply cannot carry this war on your shoulders. Your weariness is clouding your usual clear judgment. I am one step short of seeking the flight surgeon's counsel to remove you for medical recuperation. You must get some proper rest, John. I need your brain, your experience, your leadership, not your sarcasm."

John felt the concern of the highly decorated, fighter pilot. "My humblest apologies, sir. I shall heed your advice."

"Go home to Mary. Let her remove some of your pain and frustration."

"She feels her own."

"As all our families do, John, but let her help."

"I'll try, sir."

"You can do better."

"Yes, sir."

"Do you want to leave now?"

John considered his commander's question. The thought of feeling Mary's soothing hands stroking away the pain in his head had a real attraction for him. "No, sir. Too many things happening. I shall try to go home tonight."

Park stared into John Spencer's eyes as though he did not believe the Senior Controller and might find some definitive clues inside his eyes. "As you say, then. Please mind my words."

"Yes, sir."

"Now, get back in your hole," Park said with a joking air to his words.

John nodded his head, left the conference room, walked down the short hall out into the large garden, and entered the very small, thick concrete building that established the entrance to the No. 11 Group underground operations bunker.

He sat in the commander's gallery and stared at the nearly blank map board of southern England and Wales. Two reconnaissance flights were being tracked across the Midlands with a pair of fighters after each intruder. The intruders were represented on the map board as "h" blocks for hostile, while the "F" blocks for Friendly on the board marked the location of the defenders. No signs of any major raids could be seen. In the fog of his fatigue, his thoughts turned to Mary and their unborn child.

John had not talked to Mary in nearly two days. All the reports he saw told him no bombs had fallen in the vicinity of Bushey Heath, so she should be all right. He knew he could not leave the Operations Room to visit his wife, which made the urge to call her all the greater. A non-official call of a personal nature was strictly prohibited at this time of national emergency. He needed to feel her touch, but maybe her voice would ease his anxiety. John checked the map board. No activity. He scanned the Operations Room. Everyone waited for action. He lifted the outside telephone and asked the switchboard operator for his home number.

"Bushey Heath Two Four Seven One," she said in a soft, clear voice.

"Mary, it's me."

"Dear God, John. Are you all right?"

"Yes, yes. I was calling to ask you the same thing."

"Quite all right, although rather shaken, I suppose. The bombs, the fires, the grotesque smoke rising from the center of the city . . . what happened, John?"

"The Germans have apparently turned their fury on London."

"How dreadful."

"Yes, quite, however we see it as our salvation."

"You what!" she protested.

"Mary, listen. I don't have much time. I should not be calling anyway. Not one bomb fell on any of our fighter bases for the first time in a month. The respite gives us the very opportunity we have prayed for, namely, if this keeps up, it gives us time to strengthen our fighter defenses. We may very well have turned the corner. It is far too early to be jubilant, however this is the first ray of light among the storm clouds."

"John, that is ghastly, simply ghastly."

"If it helps us save Britain from invasion, it shall be worth it."

The telephone line remained silent for several moments as Mary absorbed John's words. "How is Brian?"

John did not know the answer. He looked around the commander's gallery for a casualty list, and then motioned for a junior officer to come to him. "Please wait a moment."

"Yes, sir."

"Do we have yesterday's casualty list?"

"Yes, sir."

"All of Fighter Command?"

"Yes, sir."

"Will you be so kind to fetch it for me?"

"As you wish, sir."

"Just a moment, Mary, I shall find out for both of us."

The junior officer handed John the mimeographed listing. John scanned each listing of killed, wounded and missing. Brian Drummond's name did not appear on any of the listings.

"He appears to be all right."

"Thank you, John."

One of the assistant controllers walked toward him with a yellow piece of paper that meant an important signal. "I must go, Mary."

"I love you, John."

"The same for me. I shall talk to you later." John replaced the telephone handset in its cradle. He waited for the officer to extend his hand with the message. John took it.

```
ZZZZ/1171ADR5469/HFHQ-ALLSE/5556715/ZZZZ
DATE:  08.09.40, 1045 HOURS
FROM:  GHQHF
TO:  ALL SOUTHERN AND EASTERN COMMANDS
SUBJECT:  NOTICE
BREAK
CROMWELL ISSUED LAST NIGHT FOR INFORMATION
ONLY, NOT FOR ACTION.  REPEAT.  CROMWELL NOT
FOR ACTION.
END
ZZZZ/1171ADR5469/HFHQ-ALLSE/5556715/ZZZZ
```

"Bloody hell. False alarm."

"So, it appears."

"Please make sure everyone knows, and send a runner up to the boss."

"On the way."

A young pilot officer accepted the message and carried it out of the Operations Room. As the assistant controller used the address system to inform the duty personnel that the invasion warning was for information purposes only and that there were no signs of enemy invasion as yet, the first h blocks began to appear on the far side of the English Channel. The h blocks remained stationary as the talkers began to change the size numbers on the blocks. The raids were beginning to build up over the airfields in Northern France and Belgium. The lights on the squadron tote boards began to step down as the RAF No.11 Group fighter squadrons were brought to higher alert in anticipation of the day's attacks from the South. The long day was just beginning. At least there has been no invasion, John told himself. Thank God for small blessings.

———

Sunday, 8.September.1940
Headquarters, Secret Intelligence Service
No.54 Broadway
Westminster, London, England
10:00 hours

"**W**hat is the news from the outside?" asked the Director-General Colonel Stewart Graham 'C' Menzies, DSO, MC, as the door closed behind the three other men in his modest office – Director of Naval Intelligence Vice Admiral Sir Geoffrey Ian 'Jumper' Pike, KCB, DSC; Head of SIS Operations Carl Ambrose Acton; and, Commander Alastair Ignatius Denniston, CMG CBE, Head of the Government Code and Cypher School at Bletchley Park, the keepers of the Enigma machine and the source of ULTRA.

"Hell of a way to spend the Lord's day," complained the venerable naval officer.

"Quite."

"I checked with the Air Ministry just before coming in, 'C.' The initial raids are approaching. London appears to be the target, once again," said Acton.

"Winston appears to have finally managed to coax Hitler to take the bait. If that squirt of a corporal presses these attacks on London, it shall be his undoing," said Pike.

"In what way, Admiral?" asked Acton.

"Fighter Command shall have just the break they need to replenish. Masterful stroke on Winston's part . . . bombing Berlin . . . masterful. Whatever it took to get their bombsights off our fighter aerodromes may have very well saved us all from humiliation under the treads of German tanks."

"Yes, right then," interjected 'C.' "Sir Geoffrey, as I indicated on the telephone, Denniston here, apparently has some interesting material we should share promptly. Alastair, if you please," he said, motioning toward the chief codebreaker.

Denniston opened the locked case manacled to his left wrist to retrieve several pieces of paper. He placed the case on the floor next to his chair accompanied by the tinkling of the hardened steel chain links, and rearranged the papers on his lap. When he was ready, he cleared his throat and shifted his position nervously.

"I have here several quite recent ULTRA intercepts. The volume under the Red key remains quite high and most productive, although the contents are predominately logistical or routine. Three days ago, we encountered two new keys. We have designated them, Green and Black. Based on the information I am about to show you, the Black key belongs to the SD, *Sicherheitsdienst*, Heydrich's security service; however, more significantly, Himmler's Gestapo, the *Geheime Staatspolizei*, the Nazi Secret State Police, have apparently adopted the same key. This is rather unusual, but quite fortuitous."

"What about the Green key?" asked Acton.

"Yes, well, the Green key represents a new turn as well. Normally," Denniston chuckled softly with his hand over his mouth, as if he were about to belch, "if there ever is a normal in this business, each service command has it's own key. Red for the *Luftwaffe*, our most prolific. Brown for *Oberkommando der Heeres*, OKH, the Army, and Blue for *Oberkommando der Kriegsmarine*, OKM, the Navy, Raeder's jewel, which we unfortunately have not yet reliably broken. I might add we think we are making progress on Blue. They appear to use a most unique trigger to roll the wheel sequence. We occasionally find the sequence in a day or two, however no yield on Blue code messages in nearly two weeks. It is not time, date, weather or anything else we have been able to determine."

"Green," Acton said with some impatience.

"Sorry. In this business, we find the mechanics so fascinating. Yes, as I was saying, Green is a new key instituted by OKW, the German Armed Forces High Command, *Oberkommando der Wehrmacht*. Normally, as I said, each service has its own key, which OKW uses in communications with that respective headquarters – their own form of isolation, I suppose. The Green key is the first time a common key has been used across all services." He smiled and looked to each man, as if he told a joke and waited for the laughter.

"Alastair," said 'C' with the patience of a teacher, "if you will notice, your audience is quite aware of the German military and police apparatus. I would suggest we get down to the messages you have so carefully arranged on your lap."

Denniston lost his smile and shifted in his chair. "So sorry, 'C.' The first message is an SD order to all Gestapo and SS units. It is rather grim."

Menzies extended his right hand signaling his desire for the message. Denniston took the two steps necessary to hand the message to the leader of the SIS.

MOST SECRET - ULTRA

```
SECRET
DATE 5TH SEPTEMBER 1940
TO SECRET STATE POLICE GROUP COMMANDS
FROM HEADQUARTERS SECURITY SERVICE EMPIRE
CENTRAL SECURITY OFFICE
OPERATION SEALION
BEGIN
TO AMPLIFY INSTRUCTIONS PROVIDED BY TELEGRAPHY
2ND SEPTEMBER ARMY COMMANDERS HAVE BEEN
INSTRUCTED BY OKW TO ASSIST SS AND GESTAPO
UNITS BREAK DESIGNATED SS UNITS WITH
INVASION FORCES ARE TO EXECUTE INSTRUCTIONS
DEFINED IN CURRENT OPERATIONS ORDER BREAK
FURTHER ALL MALES 17 TO 45 YEARS OF AGE ARE
TO BE IMMEDIATELY INTERNED IN APPROPRIATE
TEMPORARY FACILITIES AND PREPARED FOR
MOVEMENT TO CONTINENT BREAK NO EXCEPTIONS
ARE AUTHORISED BREAK MALES OUTSIDE ABOVE AGE
RANGE WHO ARE NOT APPROPRIATELY SUBDUED MAY
BE ADDED TO INTERNMENT LISTS AT DISCRETION
OF LOCAL COMMANDERS BREAK FEMALES OFFERING
RESISTANCE SHOULD BE DEALT WITH ACCORDING TO
STANDING PROCEDURES BREAK LOCAL COMMANDERS
MUST PRIORITISE INDIVIDUALS FOR TRANSFER
ON AVAILABLE SPACE RETURN TRANSPORT TO
BE DESPATCHED AND ULTIMATELY INTERNED AT
CONTINENTAL FACILITY TO BE IDENTIFIED LATER
BREAK TARGET LIST INDIVIDUALS DEEMED CANDIDATES
FOR FURTHER EXPLOITATION OR MORE EXTENDED
INTERROGATION SHOULD BE GIVEN PRIORITY ONE
POSITION FOR TRANSPORT TO THE CONTINENT BREAK
QUERIES OR ADDITIONAL GUIDANCE SHOULD BE
DIRECTED TO SD HQGB OR THIS HEADQUARTERS
```

```
END
SECRET
```

MOST SECRET - ULTRA

Colonel Menzies showed no reaction to the content of the message. He handed it to Admiral Pike. 'C' waited for Sir Geoffrey and his head of operations to complete their reading of the intercept. "It is indeed rather grim."

"It does raise one's anger to think these monsters might actually get to do this," said Admiral Pike.

"What in God's name is standing procedures for dissident females?" asked Acton, as he looked to Denniston.

"We do not know, I'm afraid."

"You don't want to know," interjected Admiral Pike. "Our mutual friend," he said, nodding to Menzies, "has sent several reports that give us a clue. Suffice it to say, there is nothing pleasant, gentlemanly, honorable or decent in anything we have been able to learn about their so-called, 'standing procedures.'"

"What should we do with this?" asked Acton.

"At least the PM should be aware of the contents. The implications cannot be ignored." The others nodded. "What else do you have?" 'C' asked.

"The next three are of interest – two from OKL, and one from OKW," Denniston answered, as he repeated the sequence.

MOST SECRET - ULTRA

```
SECRET
DATE 6TH SEPTEMBER 1940
TO ALL AIR FLEET COMMANDERS
FROM HEADQUARTERS AIR FORCE HIGH COMMAND
OPERATION SEALION
BEGIN
EMPIRE MARSHAL GORING HAS DECIDED TO TAKE
COMMAND OF THE AIR FORCE AND PERSONALLY DIRECT
THE FINAL STAGES OF OPERATION SEALION BREAK AIR
FLEET COMMANDERS ARE TO REPORT TO HIM DIRECTLY
BREAK FURTHER THE ENEMY SHALL BE SEVERELY
PUNISHED FOR THEIR UNFORTUNATE AND ILL ADVISED
RESISTANCE TO THE UNIFYING FORCES OF EUROPEAN
UNION BREAK TARGET LOGE HAS BEEN DESIGNATED
```

```
AS SOLE AND PRIMARY BREAK TARGET LOGE SHALL
REMAIN SO UNTIL OPERATION SEALION IS COMPLETE
OR RESISTANCE CEASES BREAK DECISION EXECUTION
DATES REMAIN SAME
END
SECRET
```

MOST SECRET - ULTRA

As 'C' handed the message to 'Jumper,' he said, "So, the fat man has decided he wants the glory."

"It appears so," answered Denniston.

"I presume Target Loge is London?"

"Yes, sir. Given yesterday's events, I think we can confirm that assumption."

"European union," 'C' snorted. "Sugar in Caster Oil does not make it taste better."

"With professional detachment," said Pike, "you must admire these brutes for their audacity and arrogance. The bloody bastards are maniacal .. . certifiably insane. They shall not have our island."

"Many share your sentiments, Sir Geoffrey," said Acton.

"The second Air Force message is the primary, confirmatory, essential element of information. You should note, London is now referred to in text. The objectives of the current aerial bombardment are quite clear. Several key words and phrases match various other documents and intercepts that establish the accuracy of the message."

"Have they given us any strategic objectives?" asked Pike. "The bombing of London certainly cannot be considered to be strategic, vengeful maybe, but certainly not strategic."

"Not yet."

"No, Admiral Pike," interjected Acton, "other than they must believe the subjugation of London is an important element of their invasion plans. Everything we have seen, know about, or suspect points toward a singular strategic objective – the forceful occupation of Great Britain and completion of their dominance of Europe."

"And maybe the world," whispered Denniston.

"And maybe the world," repeated Acton.

Menzies turned his attention to the second Air Force intercept.

MOST SECRET - ULTRA

```
SECRET
DATE 8TH SEPTEMBER 1940
TO ALL AIR FLEET COMMANDERS
FROM HEADQUARTERS AIR FORCE HIGH COMMAND
OPERATION SEALION
BEGIN
THE COMMANDER WISHES TO REINFORCE THE CURRENT
OPERATING ORDERS BREAK THE MAINTAINING OF THE
ATTACK AGAINST LONDON IS INTENDED TO TAKE PLACE
BY DAY THROUGH AIR FLEET 2 WITH STRONG FIGHTER
AND DESTROYER UNITS BREAK BY NIGHT AIR FLEET
3 WILL CARRY OUT ATTACKS WITH THE OBJECT OF
DESTROYING HARBOUR AREAS THE SUPPLY AND POWER
SOURCES OF THE CITY BREAK THE CITY IS DIVIDED
INTO TWO TARGET AREAS THE EASTERN PART OF
LONDON IS TARGET A WITH ITS WIDELY STRETCHED
OUT HARBOUR INSTALLATIONS TARGET AREA B IS
THE WEST OF LONDON WHICH CONTAINS THE POWER
SUPPLIES AND THE PROVISION INSTALLATION OF THE
CITY BREAK ALONG WITH THIS MAJOR ATTACK ON
LONDON THE DESTRUCTION RAIDS WILL BE CARRIED ON
AS MUCH AS POSSIBLE AGAINST MANY SECTORS OF THE
ARMAMENT INDUSTRY AND HARBOUR AREAS IN ENGLAND
IN THEIR PRECIOUS SCOPE
END
SECRET
```

MOST SECRET - ULTRA

'C' waited for the other intelligence officers to complete their reading of the second Air Force message. "It would appear the enemy intends to remain focused on our capital city." Everyone nodded their heads. "At least there is no pretense with this Target Loge rubbish."

"It would appear so," said Admiral Pike. "The Prime Minister will be most pleased. He believes, as many of us do, that London can take the pounding. Relief for Fighter Command and strengthening of the air defense system are the highest priorities for all of us at the moment."

"Yes, yes. It is sad, nonetheless, that our beautiful city shall have to suffer this atrocity to save England."

MOST SECRET - ULTRA

```
SECRET
DATE 8TH SEPTEMBER 1940
TO OKH OKM OKL
FROM OKW
OPERATION SEALION
BEGIN
THE LEADER HAS ESTABLISHED THE FOLLOWING KEY
DATES BREAK THE DECISION TO EXECUTE OPERATION
SEALION IS SET FOR 11TH SEPTEMBER BREAK
FINAL OPERATIONS ORDERS SHALL BE DISTRIBUTED
BY SPECIAL COURIER ON 18TH SEPTEMBER BREAK
EMBARKATION OF SEABORNE FORCES SHALL COMMENCE
ON 19TH SEPTEMBER AS WELL AS EXECUTION OF
AUTUMN CRUISE BREAK INVASION FLEET SAILS ON
20TH SEPTEMBER BREAK THE GLORIOUS DAY OF THE
LANDING ASSAULT IS SET FOR 21ST SEPTEMBER BREAK
THE LEADER DESIRES PHASE I TO BE COMPLETE
BY 27TH SEPTEMBER AND PHASE II COMPLETE BY
10TH OCTOBER BREAK OKH COMMANDERS DIRECTED TO
SUPPORT SD OBJECTIVES AS HIGHEST PRIORITY AFTER
COMPLETION OF PHASE II BREAK SUBSEQUENT ORDERS
SHALL ESTABLISH OBJECTIVES OF PHASE III AND BE
ISSUED NO LATER THAN 30TH SEPTEMBER BREAK HAIL
THE LEADER
END
SECRET
```

MOST SECRET - ULTRA

"This intercept certainly does not leave much doubt," observed Acton, as the last person to read the message.

"Thank the good Lord above us all for Enigma," Admiral Pike said, as he looked to the heavens. "The value of that little box . . . well."

"And must be protected."

"At all costs," added Pike.

"Please convey our gratitude to your codebreakers, Alastair," said Menzies. "I might add Sir Geoffrey, the box would just be a jumble of wires, if it were not for the wizards at Bletchley who somehow figured out the code wheel sequences for the box."

"Without question," stated Pike.

"Well, then, that appears to be quite a set of intercepts, Alastair. Once again, congratulations to you and your team."

"Thank you, sir. If you will permit me, I am compelled to offer a word of caution. We were lucky with these. We are still not to a point of reliable decryption."

"Anything else?" 'C' asked, sweeping his hand across all three men.

"There are a few other items, if you are interested. Some snippets from American newspapers." None of the others reacted other than to stare at Carl Acton. "We managed to collect up several interesting journalistic pieces. The first is from the *New York Times*, which cites a quote, authoritative German source, unquote, as saying that only a fraction of the German air army has been utilized in the attacks on us. The article goes on to say that the attacks will continue unrelenting and with increasing intensity until their objectives are met."

"Rather inflammatory, don't you think," said 'C.'

"Quite. The article also states the German government has made high level diplomatic contacts with the Americans to assure them the Nazi intentions are local, meaning they do not involve any American interests, and they have only peaceful, humanitarian intentions."

All three listeners grunted or exhaled a single, ha. Pike shook his head in disbelief. Denniston waited for their attention to return.

"I must add we have confirmation from the U.S. State Department, however, apparently their president has directed them to ignore the contacts, and they are prohibited from responding."

"Wouldn't they love to see those SD intercepts? The bloody heathen bastards!" spat Pike, causing the others to look with surprise at the rare profanity from the old Admiral and veteran intelligence professional.

"Yes, well," said Menzies with a slight smear of uneasiness. "What else?" he asked Acton.

"We also have a transcript of a recent broadcast report by the American correspondent, Edward R. Murrow," he said handing a copy to each man. "I would encourage you to distribute these within your organizations. I find his words calm, melodic, maybe even a bit poetic, informative and actually quite buoyant. My contacts and counterparts in America say that this Murrow fellow may be the key to changing American public opinion more our way."

"About bloody time," snorted Pike.

"I hope not too late," added Menzies. "Anything else?"

"No, sir," said Denniston and Acton in unison.

"Then, we stand adjourned. I shall talk to the PM later this evening about this information. Thank you, gentlemen."

Monday, 9.September.1940
Shoreham Hotel
2500 Calvert Street NW
Washington, District of Columbia
United States of America
09:00 hours

"Good morning. Welcome to the capital of the United States of America," announced Chief of the British Technical and Scientific Mission to the United States of America Sir Henry Thomas Tizard, KCB, FRS.

This was the first assembly of the entire team since their 10.August meeting in London. After months of uncertainty, negotiations, indecision and finally commitment, they arrived safely at their primary destination to begin their important mission for His Majesty's Government. British Ambassador to the United States Lord Lothian – Philip Henry Kerr, Kt, CH, PC, DL, had been ambassador to the U.S. since the beginning of the war in Europe and had become the 11th Marquess of Lothian upon the passing of his father in 1930 – attended the meeting, but sat in the back of the Diplomat Conference Room, not wanting to interfere with or disturb Sir Henry's initial team meeting.

"I trust everyone has settled into their rooms nicely. For those not overlooking Rock Creek Park, I offer my apologies. It was our intention to give everyone a pleasant view during his stay in the city. I am quite certain everyone has noted the lack of blackout conditions here in America. I must say, if we do our mission well, we can return our cities to the lights. If you have any difficulties whatsoever during your assignment on this mission, please contact Arthur Woodward-Nutt or myself immediately. Do not let any obstacle affect the performance of your duties.

"I am sorry you had to witness the bombing of Liverpool on the evening of your departure. I also imagine all of you are aware of events in London two nights ago. The Germans carried out their first, concentrated, aerial attack on the capital. Major fires were started, primarily in the Dockland and Canning Town districts. Some fires are still burning. And, they attacked again last night. Civilian casualties have been rather high, but the city is holding up. I say this not to add anguish, but only to punctuate the vital importance of our mission here.

"For the past nearly four weeks, Group Captain Pearce and I have conducted the final coordination with the Americans, and completed the necessary preliminary arrangements with each of the sub-teams. We are scheduled to meet for the first time beginning tomorrow morning with our American counterparts at a secure facility across town at Naval Station Anacostia. We have arranged for two buses to transport us from the hotel, which will be our base of operations for now, to the meeting site. Doctor Vannevar Bush, the leader of the American contingent, who you will meet tomorrow, and I have agreed to the agenda for the plenary session, which we expect to last two days. We have decided to organize the general session around the eight sub-teams. Doctor Bush has agreed to our sub-team structure. Thus, we will introduce each sub-team in sequence along with a general overview of the content by our sub-team leader, as we discussed in London. Each of you should be prepared to give that summary of the items and information in your sub-group. Doctor Bush will cover the administrative details for operations in this country. Then, we expect to break up into our sub-teams to open the next level of detail, and agree on the timetable and travel arrangements for the real work to begin. As we discussed in London, we expect the sub-team work to be completed by the end of this year, and both Doctor Bush and I further expect some teams may continue their work for the duration of the war. After this initial exchange, the leaders of the sub-teams will decide on subsequent activities with concurrence by my office. Now, lastly, the base office for our mission has been temporarily established in Room 705 of this hotel. Once we get through this week, I hope to find a proper office for us in short order. Are there any questions?"

"Sir Henry, if it does not violate any of your preparations or agreements, could you give us an idea of the next stage site locations?" asked Doctor John Douglas Cockcroft, a member of the Committee for the Scientific Study of Air Warfare, and more importantly, a member of the Military Application of Uranium Detonation (MAUD) Committee.

"Yes, certainly. We will cover the answer in sufficient detail tomorrow. Doctor Bush and I have agreed upon the following locations: Electronics at the Radiation Laboratory, Massachusetts Institute of Technology, Cambridge, Massachusetts; Explosives at the National Bureau of Standards, Washington, DC; Chemical Warfare at Fort Detrick, Frederick, Maryland; Naval Ordnance at Naval Proving Ground, Dahlgren, Virginia; Aeronautical Ordnance at Naval Air Station Anacostia, Washington, DC; Aeronautical at Wright Field, Dayton, Ohio; Ground Air Defense Systems at Aberdeen Proving Ground, Aberdeen, Maryland; and, Support Systems at the Engineer Research and Development Laboratory, Fort Belvoir, Virginia."

"Thank you, Sir Henry."

"We will cover a lot more detail tomorrow and should answer all your questions. I must say all of our cargo has been accounted for, secured in a building at Anacostia, and guarded by a squad of United States Marines. When the appropriate time arrives, which I might say we expect next week or so, each sub-team leader should contact Mister Woodward-Nutt to coordinate final transfer details. By the plan we set in motion last month, the Americans actually took custody when the crates entered the Anacostia building; yet, we feel a professional responsibility to ensure each respective item safely arrives at its final destination. Those details will be finalized later this week. I just asked that you keep Arthur informed and involved until the process is complete. From that point, you will work through the testing and replication sequences with your sub-teams. I must remind everyone, our exchange cargo now belongs to the Americans. Do not let any sense of possession enter your deliberations. There must not be any . . . any sense of *quid pro quo*. If the Americans share their research and technology, please accept their offerings in the spirit in which they are given and take copious notes. Make absolutely certain you protect any material provided or spoken, regardless of whether it is classified. When you are able please provide your notes and any other materials provided to Mister Woodward-Nutt for registration, safekeeping and sharing appropriately with our brethren back home. Our task now is to help the Americans fully and completely understand each and every item on our list.

"Now, when we are done here, we will go on a little tour, so everyone can familiarize themselves with our office here. Now, I would like Arthur Woodward-Nutt to address the administrative support information – contact numbers, payroll, communications home, and such. Arthur?"

Woodward-Nutt discussed all of the known administrative details to ensure all of the team could keep their minds on the mission. He answered numerous questions. Then, Arthur led them up to the 7th floor 'office' of the mission. Lastly, before they broke up for the afternoon, they set the time to meet in the Lobby for their transportation to first day's meetings. They were ready.

***

Monday, 9.September.1940
RAF Middle Wallop
Middle Wallop, Hampshire, England
16:30 hours

The frustration of an uneventful patrol, when everyone knew there was serious fighting to the East, brought his nerves and patience to their limits. Pilot Officer Brian Drummond did not want to patrol. He wanted the satisfaction of the successful hunt. He craved the rush of victory in his gunsight. Brian knew he was not alone with these feelings. Standing guard over Brooklands Aerodrome and the Hawker Hurricane factory for the second time in one day along with the mad dash back to Southampton to chase off a futile, small raid attempt on RAF Middle Wallop did not ease the frustration.

They reached their patrol altitude of 26,000 feet, entered an elongated racetrack patrol pattern, and throttled way back to maximum endurance power. At maximum endurance, the nine still serviceable Spitfires of No.609 Squadron could stay airborne for more than two and a half hours. Sitting in a slow orbit with his Merlin engine protesting against the nearly idle power setting seemed about the furthest from the top of Brian's things-to-do list. He rechecked his armament switches in the proper position and ready to go as a means of breaking the monotonous boredom of an orbiting patrol.

"Sorbo Leader, this is Horse calling."

"Go ahead, Horse," 'Spike' Darling answered Tangmere Sector Control. They spent most of their airborne time in the Tangmere, Kenley and Biggin Hill Sectors. Brooklands was the northernmost aerodrome in the Tangmere Sector.

"Very large raid approaching from Southeast. Suggest you climb short of contrails."

"Roger, Horse." Darling added some power, but not much, to save fuel for what had to be a slight altitude change. "'Jackstay,' call my contrail."

"Roger," answered Flight Lieutenant Roger 'Jackstay' Beamish.

Brian, as well as most of the other pilots, watched the air behind Darling's tail. As they climbed slowly, Brian scanned the sky around them several times to make sure no one sneaked up on them. Wispy clouds in the distance made sighting of the large raid difficult. When Brian saw the telltale white stream behind the 'PR-A' Spitfire, he checked his altimeter. The engine exhaust vapor condensation level started at 27,850 feet on this late summer day.

"You have it, 'Spike,'" radioed 'Jackstay.'

Darling did not acknowledge the contrail call other than he began a gradual descent and leveled off at 27,500 feet. Brian kept searching the horizon to the southeast. During the eastern turn to head back west, they were over the southwest reaches of London. Parts of the city sprang smoke clouds tailing off toward the east. The fire brigades were still fighting the enormous fires from the previous evening. A sense of unfairness crept into Brian's thoughts. They did not have the fires extinguished from the previous day's brutality, and another very large raid was heading in the direction of London.

As Brian looked back to the southeast, the wispy clouds had grown and separated into distinct strings – the contrails of the high fighter screen for the bombers. A dark cloud of many dots stretched out beneath the contrails. They did not appear to be headed toward them. The extension of the contrails as well as the movement of the black cloud of dots pointed toward the center of London.

They watched as the contrails began to curve. The fighter engagement had begun. Even the orderly spacing of the bombers began to vary as the Hurricanes waded into the sea of attacking bombers. Brian rolled his aircraft toward the attackers and pushed his throttle forward.

"'Hunter,'" radioed someone to pull him back.

Brian returned to his position off the left wing of the 'PR-D' Spitfire of 'Jackstay' Beamish. They had been told to remain in patrol position over Brooklands and to engage any enemy aircraft that might threaten the Hawker Works. The temptation of the approaching bombers made the protection of an aircraft factory seem far less important than protecting the capital city of Great Britain. The details of all the aircraft came into clear focus as the horde passed to the east of No.609 Squadron. Watching the comparatively small numbers of brown and green British fighters among the grays of the German bombers and fighters produced a horrendous allurement of so many targets so close and yet beyond their reach until they either made a move toward Brooklands, or Tangmere Sector Control released them from the shackles of their patrol duties.

The attraction of the attackers became more pronounced, as they witnessed the tiny black specks falling from the bellies of the bombers onto the center of London past the light gray barrage balloons strung out at various heights above the city. The inaudible bursts of detonating bombs were clearly seen from their position overhead Brooklands. The orange tongues of ensuing flames erupted from the broken buildings, and more black smoke joined the still rising smoke trails from earlier insults.

They continued to watch the slow motion tragic ballet every time they turned onto the eastbound leg of the patrol orbit. The urge to continue straight just to watch became unbearable. Each time they had to turn away toward the west approached impossibility.

The black-gray puffs of anti-aircraft shells bursting among the bombers separated the Hurricanes from their prey. The slower cousins to the Spitfires rose to join in with the fighters still above and behind the bombers. As the bombers turned south after dropping their bombs, they moved out of the anti-aircraft cannonade, and the Hurricanes descended upon them again. The fighter engagement continued unabated as the cloud moved back to the south. Aircraft of all shapes and colors fell from the sky usually with black streamers marking their path to oblivion. Occasionally, the white hemisphere of a parachute blossomed into the confusing sky.

No.609 Squadron was forced to observe the enormous air battle and renewal of the conflagration on the ground. Before the first wave disappeared from sight, the first indications of the second wave appeared from a position farther to the east. Brian considered leaving his squadron over Brooklands rather than watch another raid strike London. He yelled into his mask to relieve the mounting anger he felt. They had been aloft not quite two hours. Low fuel would soon render them ineffective and cause a descent from their perch. The sun moved lower on the Western horizon meaning this would be their last sortie for the day. The resentment grew knowing they would return to RAF Middle Wallop with their red gun port tapes still in place and unbroken. They had not fired a single round all day despite two very long sorties. The lack of action added to the frustration. None of them would be happy tonight.

The sun disappeared behind a low, thick layer of clouds to the west, as they landed at their Hampshire airfield. Brian taxied to his spot and shutdown his Merlin.

"No party today, aye, Mister Drummond," said Leading Aircraftman Bernard Gordon, as Brian unstrapped from his fighter.

"No, Bernie, and we are none too happy about it either."

"What happened?"

"We had to patrol over Brooklands and were forced, like some inhuman torture, to watch the damn Germans bomb the hell out of central London."

"Bloody hell."

"My thoughts exactly."

"Any squawks?"

"None. She's a dream."

"We shall give her a good drink of petrol and put her to bed for the night."

"Thanks, Bernie. Have a good night."

Brian joined some of the others at the tent. Corporal Warren was still at her post after a long day. The mood of the pilots took on an angry red tone.

"Damn fucking idiots," shouted Pilot Officer Frank 'Red' Burns, as he entered the tent and threw his helmet, mask and goggles in the direction of his equipment peg. "Why did they make us watch that and not let us do something about it?"

"I don't ever want to do that again," added Flying Sergeant Miles 'Fog' Johnson.

Others added their anger, frustration and resentment to the boiling stew before Squadron Leader Darling decided he had heard enough. "Listen up, gentlemen," he said, waiting for the huffing and puffing to diminish. "None of us likes to watch, but we had a mission to perform, and we fulfilled our mission completely. We protected Brooklands."

"Yeah, but Skipper. We didn't do anything," protested Burns.

"We did more than you think, 'Red,'" Darling said, as he looked to each of the pilots. "We were given a mission, and we accomplished it successfully. Many times in battle we must do things we cannot see the reasons for, but are important nonetheless. Now, this day is done."

"What about the debriefing, sir?" asked Flying Officer James Royster.

"Put in your report, we flew a second security patrol over Brooklands; we witnessed the criminal bombardment of London; and, we never fired a shot."

"But . . ."

"No, buts, Mister Royster. My pilots are tired and angry from a very frustrating and otherwise annoying day. There is nothing more to report."

The hurt expression of Royster's face told everyone he did not understand what Squadron Leader Darling had just done, but he was also not going to question the direction of a senior officer.

"Let's go have a bite and a few pints," said Darling. "We shall need a bit of the buzz."

"Hear, hear," several said with more anger than excitement.

Tuesday, 10.September.1940
Canning Town, London, England
10:30 hours

Winston Churchill knew the journey to the bomb-damaged East End would not be a pleasant trip. In a few cases, the London Fire Brigade was still working around the clock to extinguish major fires from the first day's bombing – three days ago. He could still see the smoke rising above parts of the city. The destruction from the second day's bombing mixed with the first. The Germans concentrated on the River Thames docks and harbor facilities, some dating back centuries, the primary supply line for London and a good portion of the country. They could manage without the port facilities of London, but the injury to the city's economy and the welfare of its people would be far greater. Families that had lived, worked and died in East End communities and neighborhoods for many generations would be displaced.

Normally, Winston thoroughly enjoyed getting out among the people, even when his audience was hostile. The experience gave him the touch, the feel, the mood of the people, and for a politician, public opinion was a crucial bellwether. He seemed to be riding a wave of support. People reacted to him as an inspiration, a guiding light, and the leader of the British people. He liked the feeling.

Although he had many things to tend to at this critical period in the nation's history, Winston knew and did not complain that he had to take the time for a motorcar excursion, so citizens could see him. As with virtually every field event since the Battle of Britain began, General Ismay joined him, this time in a modified convertible limousine with the top down – a black 1939 Rolls Royce Wraith touring automobile with a small Union Jack attached above the radiator grill.

Major General Sir Hastings Lionel 'Pug' Ismay, KCB, DSO, served as Principal Assistant to the Minister of Defence (Churchill), Secretary of the Imperial Defence Chiefs of Staff Committee, and Deputy Secretary of the War Cabinet. In essence, Ismay was the Prime Minister's military liaison officer.

They brought Winston's Assistant Private Secretary John Rupert 'Jock' Colville with them, to take notes on actions, observations and comments.

After so many bright, sunny days of summer, the dreariness and darkening clouds seemed odd and yet appropriate for the solemnity of recent events. The air smelled of rain, burnt wood and oil, and oddly enough, the earth itself, when they left No.10 Downing Street. The air turned heavy with brick dust, and then the stale smoky residue of urban fires.

Whitehall remained, as yet, untouched by German bombs. The Strand, the center of the London theater district, turned into Fleet Street, the newspaper capital of the world, also showed little damage other than some unusual debris like uncollected litter. However, as they passed St. Paul's Cathedral and joined Cheapside and Leadenhall Streets near White Chapel and the Tower of London, the unmistakable signs of bomb damage began to accumulate – walls collapsed into the street with small pathways cleared for emergency vehicles, the sickening black scars and smoldering residue of recent fire damage, and the soot smudged, nearly unconscious, expressions of exhaustion on the faces of people whose lives had been shattered. Commercial Road became East India Dock Road and the black smoke from the still burning docks dominated the gray sky.

The heavy, choking, acrid stain of smoke contaminated the air, making casual breathing difficult and uncomfortable. Winston kept his unlit, fresh, large, Havana cigar prominently displayed in his mouth. He tipped his characteristic black bowler hat as people recognized him and waved. Winston Churchill fought his emotions as he tried to smile a confident, encouraging and defiant smile, but the pain around him and the brutal injury to the beautiful old city made any smile agonizingly painful.

It was not until they crossed Bow Creek and entered Canning Town that the damage became nearly absolute. Entire buildings transformed into a massive pile of bricks and broken, charred timber. Smoke so thick, it choked off the oxygen, but people worked, moving one brick at a time to clear the road. Many searched through the debris of what had been their homes, their shops, and their lives. The Prime Minister ordered the car stopped. He started to get out to be closer to the people.

"Prime Minister," General Ismay said, grabbing Winston's elbow, "this may not be the best thing to do."

Winston glanced over his shoulder with a burst of anger in his eyes and heart, and then caught himself. His military aide was simply trying to protect him. "Nonsense," he answered with measured reserve. "This is my duty, my burden to bear for the people. If they are angry with me, then let them show it." Winston recalled their excursion to RAF Manston a fortnight earlier. "You know, as well as any, our destiny lies with the people and their confidence that we feel their pain." General Ismay released the Prime Minister's elbow and stood to join him.

At first, Winston felt awkward, out of place, like an intruder or interloper at the wrong party. He tipped his bowler, waved hesitantly, and tried to smile. The mounting sadness in his heart pulled down on his face. He did not try to interfere, as a politician might be tempted. He walked, observed and absorbed the minute scenes around him. Winston acknowledged a head nod, a slight wave or a feeble salute trying not to intrude in their tragedy. He swallowed hard several times as he fought his own tears of grief.

Churchill and Ismay came upon a silver haired man, at least Winston's 66 years or older, in the middle of an enormous pile of rubble. He was digging through the wood and bricks looking for something. He tried to straighten his bent and tired body. The pain in his nearly blackened face made Winston's bones ache. The old man glanced toward them, and then looked hard as he recognized the visitors. He pulled his shoulders back, straightened his back as best he could, came to a position of attention, and saluted in a crisp, palm out, traditional manner. Churchill and Ismay returned the salute, and Winston could no longer contain his grief. Tears streamed down his face, as he tried to choke back his emotions.

"We can take it, Mister Churchill," the old man said. "You just give it back to 'em."

"Hey, look, everyone," a middle-aged woman shouted, "it's Winnie himself come ta see us."

"We knew you'd come see us," said another.

Winston tried desperately to control his cascading emotions, as citizens began to gather around him. Their smiles, laughter and shouts of encouragement to him made his struggle even more arduous. He could not speak. He could only wave, nod his head and tip his hat to the small crowd around him.

"You're cryin,'" a woman said, as if he did not know what was happening.

"You see, everyone, he does care."

"Give it to 'em back in full measure, Mister Churchill."

"Never ya mind us. We can take whatever the Hun wants ta give us."

"He's cryin.' He feels our hurt."

"Don't give 'em an inch."

"Ya do care. He really does care 'bout what happens ta us."

"Let's take the fight to the friggin' bastards."

"We're here for ya."

The Prime Minister fought unsuccessfully against the convulsions of his sorrow and the overwhelming impact of their encouragement. He could not bear the weight of the injured city and accept the encouragement of the very people he was trying to bolster. They did not need to be supported. They buoyed him.

General Ismay, himself receiving pats on the back and handshakes, tried to move the Prime Minister back to the automobile. Winston was thankful for the assistance. He wanted to stay but knew he could do little to help. The driver had already turned the open, convertible limousine around pointing back toward Whitehall. Winston stepped in but remained standing. He faced to the rear and the now large gathering of cheering Londoners. He grasped the seat behind him. Winston waved and tipped his hat to the small gatherings of citizens. The cheers became a roar of defiance when Winston held his right hand high above his head, extended and spread his first and second fingers in the immortal V for Victory gesture. General Ismay tapped the driver's shoulder and told him to drive away slowly.

As the car passed behind the remains of a corner shop, Winston sank into the seat, placed his face in his hands and wept uncontrollably. General Ismay placed a firm hand on the shoulder of his Prime Minister. It took a few minutes for Winston Churchill to recover his composure and wipe away the tears.

"Dear God Almighty," he shouted, as if to an assembled audience, "how have I been so blessed," he said, as tears reappeared.

"It is rather remarkable."

"Remarkable! My God, man. It is nothing short of extraordinary . . . miraculous. I feel a duty to stiffen the resolve of our citizens, and it is they who have strengthened me. We are going to win this affair. We shall take back all that has been taken, even if we do it alone and it should take us a hundred years."

"Yes, sir. The resiliency of the people is truly extraordinary."

"I should like to see the Docklands district," Winston said.

The driver nodded his head and made the appropriate adjustments. They stopped several times to encourage weary fire fighters and constabulary, as others continued to labor against the flames around several oil storage facilities. The old wooden graving docks that used to hold mighty oak sailing ships had been severely damaged. Bomb craters and their ejecta made movement slow and circuitous. Bomb disposal crews sweated through the delicate task of disarming unexploded ordnance buried in buildings and warehouses. Winston Churchill knew that General Ismay's observation was precisely correct – the resiliency of the people was extraordinary.

A mist turned into a drizzle that became a steady rain. The convertible top was extended, stretched and fastened in place. The change in weather did not alter the activities of the people around them, but it was sufficient to persuade Winston to return to Whitehall.

———

Tuesday, 10.September.1940
No.10 Downing Street
Whitehall, London, England
11:45 hours

They reached No.10 Downing Street in the middle of a heavy downpour. Several servants reached for the car door with large umbrellas in an attempt to keep the two men relatively dry, as they stepped the short distance through the large, black, solid oak door with large polished brass numerals, 10, firmly affixed at the upper center of the door. Several people waited in the entry foyer. They all stood as the Prime Minister entered. He handed his hat and coat to the doorman.

Churchill's Principal Private Secretary John Miller Martin and Secretary to the War Cabinet Sir Edward Ettingdene Bridges, KCB, MC, FRS, stepped forward. Bridges nodded to Martin. "Would you like something to eat, Prime Minister?" asked Martin.

"Something light, John, if you please."

John Martin nodded his head and departed.

Sir Edward leaned forward to whisper to Winston. "Colonel Holderstam, from the War Office, has a very important personal for you from Mister Eden and Sir John," he said, referring to the War Minister and the Chief of the Imperial General Staff.

The Right Honorable and Gallant Robert Anthony Eden, MC, PC, Member of Parliament for Warwick and Leamington, Deputy Leader of the Conservative Party, had been appointed Secretary of State for War by Churchill in May, and served as the ministerial leader of the British Army.

General Sir John Greer 'Jack' Dill, KCB, CMG, DSO, had been elevated to Chief of Imperial General Staff (CIGS) in May as well – the ranking Army officer.

"In my office?" Winston asked. Sir Edward nodded his head. Churchill went to his private office. He smiled and nodded to each of the men waiting to see him.

The Army colonel in the uniform of the Royal Artillery waited until Sir Edward closed the door leaving the two men alone. "The Secretary of State for War and the Chief of the Imperial General Staff have asked me to personally convey the following message to the Prime Minister." Churchill nodded his acknowledgment. "We have confirmed reports of several known infiltrations by enemy forces. A small SS unit was captured after successfully parachuting into Essex, near Chelmsford. The team of eight specialists had orders to conduct surveillance of military and constabulary units. Mysteriously, the bodies of 38 soldiers and two officers in Army uniforms were found on the beach between Bognor Regis and Worthing. They appear to have drowned probably attempting a night landing. We also have three other men who were captured attempting to land by parachute in Oxfordshire, Cambridgeshire and Surrey. All three of these were in civilian clothes carrying false identification as British citizens. None has spoken a word, yet. They are probably Gestapo or SS spies, although we have no evidence as such."

As Winston Churchill listened to the report, his thoughts repeatedly returned to the disturbing ULTRA messages Stewart Menzies had shown him late last night. The thought of Heydrich and Himmler having their evil agents among them gave the Prime Minister a chill. Strangely, he recalled the ULTRA message last fall that had taken Bletchley Park three months to decypher. Canned Goods, he remembered. "Canned Goods," he muttered to himself as the image invoked by the details of Heydrich's orchestration of the catalyst for the Polish invasion and the sacrifice of those unfortunate Polish prisoners occupied his thoughts.

"Pardon me, sir?" asked the colonel.

"Nothing," answered Winston. "Another time, I should say."

"Do you wish me to take a response?"

"None required. I shall see them at War Cabinet this afternoon. However, please ensure MI5 and MI6 have this information, and full access to the prisoners."

"It was done first thing, sir."

"Thank you."

"Yes, sir. By your leave, then, sir."

"Certainly. Thank you for your patience."

As the colonel departed, Sir Edward reentered closing the door behind him. "The agenda for this evening's War Cabinet meeting." As the Prime Minister read the single page, Sir Edward continued, "The forecasters indicate we shall have thick cloud cover and rain most of the night. The Air Ministry believes London should have a quiet night, hopefully."

"The rain should help the Fire Brigade."

"Yes, indeed. How was your survey?"

"A very moving and emotional experience. The morale of those who have so recently suffered so much is almost beyond comprehension. I suspect I would not have believed it if I had not felt it myself. Most rewarding and encouraging."

"Good. Before I leave you alone for a few minutes, I do have one small item I thought you might enjoy after so much grievous news."

"Some good news would be most appreciated."

"We received confirmation this morning that Mister Joseph Kennedy, the American ambassador, has returned to America. Apparently, the bombing has frightened him off."

Joseph Patrick 'Joe' Kennedy, Sr., had been nominated by President Roosevelt and served as United States Ambassador to the Court of Saint James's since 8.March.1938, and had borne witness to the opening of the world war in Europe. "He does have children, Sir Edward."

"Sorry, sir, but I know he has been an enormous thorn in your side for quite some time."

"Indeed, he was practically licking his slobbering chops along with his erstwhile patron, *Herr* Hitler."

"Now, it is you who is being unfair."

Churchill chuckled. "Yes, I suppose you are correct, but it is good riddance as far as I am concerned. He was an obnoxious old sod, more enamored with his Irish heritage and German affiliation than he was with facts and reality."

"Hopefully, we shall see some good from this."

"We need more like this Murrow chap who seems to be having a positive effect upon U.S. public opinion. Kennedy always saw the negative, while Murrow has managed to capture so much that is good about our cause and our people."

Edward R. Murrow legally changed his name from Egbert Roscoe Murrow during his second year of college in 1928. He joined the Columbia Broadcasting System (CBS) in 1935, and was assigned to Europe, based in London, in 1938, just in time to cover the German annexation of Austria in March of that year. Murrow began broadcasting his *London After Dark* program, a joint venture radio program between CBS Radio and BBC Radio, in August 1940. One of his first broadcasts had been a captivating interview with a young American volunteer fighter pilot by the name of Drummond along with the war widow who had saved him from certain death. Murrow had attended the ceremony where King George VI awarded the George Cross to Mrs. Charlotte Palmer and the Distinguished Flying Cross to Pilot Officer Brian Drummond.

"Most impressive," added Bridges.

"Yes."

"I shall leave you. John has a tray of sandwiches and cheeses for you. I shall have him bring it in for you."

"Thank you, Ed. You have been most kind."

Less than a minute after Sir Edward left his office, his valet, Frank Sawyers, brought a small platter of food and a bottle of French Chardonnay. "Thank you, Sawyers."

"Yes, sir."

Churchill had not taken one bite of a sandwich triangle or even a sip of wine when Sir Edward returned. Sir Edward did not wait for the Prime Minister.

"We just received word from the Foreign Office, Prime Minister. President Roosevelt signed into law the Second Supplemental National Defense Appropriation Act yesterday afternoon. According to Lord Lothian, the new bill adds another five point five billion dollars of additional defense funding for their next fiscal year, which begins at the first of next month."

"Excellent. Good news, indeed. The President is making good on his promise to help us."

"Yes, it would appear so. Also, Lord Lothian also reported Sir Henry Tizard's team is now fully engaged with the Americans and our exchange items have been transferred to American custody."

"Brilliant . . . even more good news. There is sunshine among these dark clouds around us."

"Yes, sir."

"Now, Sir Edward, if you would be so kind, I would appreciate some quiet time to gather my thoughts from this morning's excursion and prepare for this afternoon's Cabinet meeting."

"Yes, sir, by all means. I will inform John and 'Jock' to guard your door."

"Thank you, Ed. See you in a few hours."

Sir Edward departed, closing the heavy door behind him. Winston took a chair facing his window and the garden beyond, as he took a sip of wine and a bite of his sandwich. The morning's events had a profound impact on the Prime Minister. He committed to himself that he must redouble his efforts to strengthen the resolve of his nation, the commonwealth and empire beyond their shores, and most importantly encourage and solidify the engagement of their American cousins across the Atlantic Ocean. He simply could not fail those dauntless citizens who defied the German brutality.

Chapter 3

From ghoulies and ghosties and long-leggety beasties
And things that go bump in the night, Good Lord deliver us!

-- Anonymous, Cornish prayer

Wednesday, 11.September.1940
Headquarters, No.11 Group
Uxbridge, Middlesex, England
13:45 hours

Week 10

Air Commodore John Spencer savored the strong, earthy flavor of the very rare, hot mug of South American coffee as well as the caffeine boost. The supply of tea remained quite good despite the defeats at sea at the hands of U-boats, merchant raiders and fast, hard hitting, capital ships of the German Navy. Coffee approached gold in its rarity in the British economy. Sleep always seemed to be in the same category as the other rare commodities, although he did manage a decent night's rest unfortunately without the comfort of Mary's presence and embrace. Maybe the long nights of winter and the prevention of an invasion would give him more time to spend with his wife. John had never thought of himself as a parent, but the prospect of being a father at 42 years of age with an infant daughter or son intrigued him. The importance of victory in the present battle became all the more important with Mary's announcement of her pregnancy. Their child, as all the other children evacuated from London, deserved freedom and safety. The lack of substantive enemy activity in the recent mornings combined with the images of Mary and child to bring thoughts of home. Perhaps he could accept Air Vice-Marshal Park's advice and go home for a night with Mary. He missed her. He missed her company, her wit, and her energy. Keith Park was right, as he usually was.

"Sir," said one of the controllers, a wing commander. John withdrew from his thoughts, looked at the map board to see if anything important might be happening, which it was not, and then looked up into the short, bald man's eyes. "We just received various bits of information from Fighter Command, and I took the liberty of calling a friend at Watnall," No.12 Group Headquarters in Nottinghamshire. "As best we can tell, we sent a large bomber raid to Germany last night. Weather was apparently perfect, and they hit Berlin and Bremen rather hard. The initial reports indicate the big boys hit the Focke-Wulf Works at Bremen and left it on fire."

"Good on Bomber Command."

"Quite. More importantly, as the raid was returning just prior to dawn, a half dozen Dornier One Sevens joined our lads, made it through Chain Home undetected apparently, and struck two Bomber Command aerodromes in the Midlands."

"What about Pipsqueak?"

"They don't know. Could be they managed to copy the appropriate signal, or maybe the operators were tired and missed the trailers among our chaps returning from Germany?"

"Radio procedures?" asked the Senior Controller, referring to specific daily code words used by Bomber Command pilots in addition to the electronic identification signal from the Pipsqueak equipment. The procedures had been firmly established after the dreadfully unfortunate incident at the start of the war the pilots called the Battle of Barking Creek, when friendly aircraft had been mistakenly shot down because of confusion and poor identification procedures.

"According to my friend at Watnall, the Germans made all the correct calls."

"Damn Germans, do not miss a trick, do they?"

"No, sir. Apparently not. One Two Group did not get fighters up until the attackers crossed the coast outbound. They got away Scot-free."

"This incident should be a interesting topic at HQ."

"Yes, sir." The wing commander nodded his head, as John Spencer turned his attention back to the still empty map board, and then returned to his post.

John's thoughts moved quickly to Northern France. What did the Germans have planned for the day's activities? There were still no indications on the map board, and it was not quite lunchtime. So far, things remained true to form based on the last few days. It would be another few hours before the Chain Home Radio Direction Finding stations on the cliffs of Southeast England would detect the first signs of the forming German bombing raids.

An attractive, young, Woman Auxiliary Air Force, corporal delivered a single piece of paper to Air Commodore John Spencer. She distributed Air Vice-Marshal Park's latest instruction. He generally knew what it would say. John Spencer knew Park felt the pressure from Air Vice-Marshal Leigh-Mallory and Squadron Leader Douglas Bader. He expected the instruction to be Park's first move to the use of fighter wings to engage the enemy bomber formations.

HEADQUARTERS, No. 11 GROUP

FIGHTER COMMAND

ROYAL AIR FORCE,

UXBRIDGE, MIDDLESEX

Telephone Nos. UXBRIDGE 2894 (4 lines)

UXBRIDGE 2896 (2 lines)

Telegraphic Address: "AIRGPLON UXBRIDGE."

Reference: -- FC11/P.11215

SECRET

11th September, 1940.

GROUP INSTRUCTION No. 16

This instruction is issued with immediate effect and shall remain so until further notice.

2. Squadrons are to attack in pairs wherever possible.

3. Readiness squadrons shall be positioned to attack 1st wave intruders. Spitfire squadrons from the Hornchurch and Biggin Hill Sectors shall engage the high fighter screen halfway between London and the coast, so as to enable Hurricanes from London Sector to engage the bomber formations and their close escort before they reach the line of fighter aerodromes east and south of London.

4. Available in 15 Minutes squadrons will be brought to Readiness to deal with the 2nd wave attack. Wherever possible, similar tactics should be utilised to allow maximum effectiveness against the bombers.

5. Available in 30 Minutes squadrons will be despatched singly to reinforce and protect aircraft factories.

6. If a 3rd wave should approach, remaining squadrons are to be paired from various stations, sectors, and group areas.

7. Fuel and ammunition permitting, maximum pressure should be maintained on enemy aircraft until return to base is warranted.

```
        8.   Controllers shall co-ordinate actions
with No.10 and No.12 Groups allowing for
distance to ensure continuous fighter engagement
excluding in range anti-aircraft areas.
Priority should be given to constant attack
efforts.
```

As ordered by,
Keith Park
Air Vice-Marshal,
Air Officer Commanding-in-Chief
No. 11 Group, Fighter Command, Royal Air Force

SECRET

John Spencer shook his head not in disagreement or disgust but in recognition of the power of peer pressure, politics and a survival situation. Air Vice-Marshal Park yielded to the cogent arguments of a vocal squadron leader at RAF Duxford, somewhat removed from the punishment of the No.11 Group airfields. As in all things human, situations were rarely clear. There were always versions of what could be reasons. How confident could anyone be in such an intense, confusing, survival situation as they were deeply immersed in at the moment?

The map board talkers placed the first h blocks over the French coast. John looked at the large wall clock.

14:10 hours

The Germans planned to start earlier this day. The Senior Controller stood, took another survey of the developing situation and moved to the controller's section of the gallery.

"Starting early," observed the wing commander, duty controller.

"Perhaps they have something else in mind," one of the junior controllers added.

"Perhaps. Let's get the lads up to Readiness," the Senior Controller directed.

The junior controllers responded with action, as they lifted the proper telephones to the Kenley, Biggin Hill and Hornchurch Sectors. The usual squadrons would be prepared to greet the developing raid. The deputy senior controller called both No.10 and No.12 Groups to alert them to the potential requirement for assistance, as they had done at the beginning of the day's enemy activity over the last month. The day's destruction and insult had become routine. Under a different, less personal, set of circumstances, the routine might even be called boring. The innovative tactics of August fell into normality. The sneaky trailer move in the early morning incident to the north offered the only ingenuity they had heard of in the last few weeks.

One of the young intelligence officers assigned to the Headquarters entered the controller's area. John nodded to him to state his business. "Sir, we just received a report from Y Section. The *Knickebein* beams are active and crossed over London."

"Any other cities?"

"None they have found or reported."

"So, it is all London this afternoon and tonight."

"It would appear so."

"Can they do anything?" John asked, referring to the efforts of the electronic countermeasures organization to jam or disrupt the radio beams.

"They will be doing what they can, however they do not believe it will be sufficient to completely disable the *Knickebein* system."

John nodded and leaned toward his assistant controller. "Notify Six Double Naught," No.600 Squadron, the sister night-fighter squadron of No.604 Squadron, "they will have business tonight. Also, contact One Oh Group; we shall need our specialists to coordinate the night-fighter operations tonight."

"Yes, sir."

The Senior Controller nodded his acknowledgment, and then nodded to the intelligence officer dismissing him. The young flying officer did not leave. John looked back at him again.

"One more item, if I may, sir." John nodded, again. "Just before I came down, the Y Section chaps intercepted a signal from the German control circuit. They told their crews to turn back if our defenses were too strong, or their fighter protection was too weak."

Air Commodore John Spencer smiled the broadest smile he had felt in months. "I trust you are not foolish enough to be kidding me?"

"No, sir, that is what the report stated."

"And, it was in open radio – not encrypted?"

"Yes, sir."

"I'll be damned," he said rather loudly. "Maybe we're winning this row, and we don't know it."

"Maybe," added the assistant controller.

"They are on the move, sir," said one of the junior controllers.

John turned his attention to the map board. The h713, h714, h717 and h721 blocks began moving across the Channel toward Southern England. He watched the movement for several minutes before giving the next command. "Let's bring the first line squadrons to Standby. Notify the boss. We have customers." He turned back to the intelligence officer standing at the edge of the booth as an observer. "Does Air Vice-Marshal Park know about the Y Section information?" The young man slowly returned his attention to John Spencer. A blank expression left an open question. "The Y Section report, lad," he barked, and then resented his abruptness and lack of patience.

"Sorry, sir. No, sir."

"Then, please inform him immediately," John said and waved his hand as if shooing a fly. He turned back to the map board. The German raiders passed the invisible line the controllers mentally etched on the board. "Scramble the first line, and let's get the second line to Readiness. Notify One Oh and One Two Groups to standby."

The action started before John finished his directions. The day's attacks had begun. Four raids totaling more than a thousand aircraft were inbound. Three more appeared, and then disappeared. Three others remained over Northern France and Belgium. Another big day and probably long night lay ahead.

Wednesday, 11.September.1940
RAF Middle Wallop
Middle Wallop, Hampshire, England
15:45 hours

Pilot Officer Brian Drummond and his squadron mates in No.609 Squadron held the noses of their Spitfire Mark IA fighters a good 15° above the hazy horizon with their Merlin engines boasting of the 1,030 horsepower they delivered to their masters. The still only nine aircraft were climbing as fast as they could to 22,000 feet to intercept the h720 raid of 100+ aircraft heading toward Southampton. They moved to the west to have the late afternoon sun at their backs when they turned to engage the enemy. Two Hurricane squadrons, No.238 and No.501 Squadrons, joined the engagement. The Hurricanes were to take on the bombers and close in fighters while No.609 Squadron engaged the high fighter screen.

Half of the approximately 50 or 60 Bf109 fighters turned toward them. There would be no surprise this day. *Nine Spitfires facing 30 Bf109s . . . not bad odds*, Brian said to himself. This was going to be a nasty fight with the other half of the high fighter screen looming unchallenged over the bomber formation and attacking Hurricanes. Darling did not have to command them as they spread out into their tactical formation, ready for the encounter.

The fighter engagement began head-on with no hits on either side as best Brian could determine. The high-speed initial pass with closure rates in excess of 700 miles an hour quickly deteriorated into the usual ball of twisting, turning, diving and climbing fighters mixed up in one big clump in the thin, clear sky above the Isle of Wight.

The calls of warning filled his headset earphones more than usual. The German fighters seemed more intense than usual, as if they had reached a new level of inspiration and motivation. Brian managed several quick hits on various targets, but could not achieve enough effect to claim a victory. The pace of the fight was instantaneous and relentless although fought to a draw until the German fighters disengaged *en masse*. The Spitfires of No.609 Squadron briefly tried to give chase without consequence.

Turning their attention to the retreating bombers, the 'PR' Spitfires joined the hunt with their brother Hurricanes. As Brian and the others dove for position, they scanned the sky for other fighters. None could be seen in the clear skies. Brian glanced toward Southampton.

Smoke rose from many sites. Winds from the west blew the thick black smoke from the docks across the Rivers Test and Itchen. He could not tell whether Woolston and the precious Supermarine Works had been hit. Smoke from an area northeast of the city made it look like Eastleigh Aerodrome, the Supermarine flight test center, might have been hit hard. Had they failed to protect Southampton?

The bombers grew rapidly in their windscreens. Brian returned his full attention to his targets. Hurricanes were everywhere all around the bombers. A squadron's worth of twin-engine, Me110 Destroyer fighters continued their desperate efforts to defend the bombers. No Bf109s could be seen. Brian felt the scent of a good hunt.

Brian searched the scene quickly, found an Me110 intent on his pursuit of a maneuvering Hurricane, and lined up his sight. He chose a slashing attack to arc his bullet stream across the planform of the German fighter turning into him. The large twin-engine, H tail, fighter filled his entire windscreen as he opened fire. No impacts could be observed. He adjusted his aim point slightly and flashes of molten metal dotted the wings, engines and cockpit of the Me110.

Brian passed above the German, as he pulled his nose up high to climb. He quickly scanned behind him to see if he had any attackers. None. He rolled onto his back. The now fully engulfed German fighter spiraled toward the waiting sea. At least the pilot was dead or incapacitated. He looked for a parachute from the rear seat gunner and observer. None came. Both Germans would not return home.

The 'PR-F' Spitfire continued its gentle arc, forming the top of half of a loop. Brian used his high perch position to find another target. A single Ju88 trailing smoke from its left engine and flying low on the water and without an executioner in pursuit presented a tempting target. However, he was high and the target was low. Before he chose another target, Sorbo Leader called for the squadron to break off their efforts, rejoin and return to Middle Wallop.

They gathered over the Western coast of the Isle of Wight, turned north and began a gradual descent. The skies quieted although the smoky remnants of the battle marred the scene. The Southampton docks and port facilities were on fire, again. As they passed to the west of the city, Brian looked for the Supermarine Works. The sawtooth roof of the largest building came into sight among the plumes of smoke. It took several minutes of intermittent observation to arrive at a conclusion of no identifiable damage. How could the Germans continue to miss the primary source and certainly the birthplace of the best British fighter? Brian checked the visible surfaces of his aircraft partly to see if he had any undetected damage, but also to remind him of the elegance of his machine and the importance of the Woolston factory.

The clear markings of heavy bomb damage covered the airfield of the Supermarine flight test center at Eastleigh. The landing area was unusable although crews were already at work filling in the many craters. All four large buildings including both of the Supermarine hangars were severely damaged, maybe even destroyed. Several Spitfires were still on fire and the flames had been extinguished on at least a half dozen more fighters. They had not been successful in preventing the damage and protecting the people. While the damage to Southampton and Eastleigh did not have quite the same level of impact as watching the bombing of London while on patrol over Brooklands, the heavy heart brought home to Brian once more the mortal struggle he was squarely in the middle of now. He remembered the words of his mentor and instructor, Malcolm Bainbridge – war was never a pleasant experience, only survival.

The reduced intensity of the ground crews told the pilots they did not have an immediate turn around to complete. The sun was lower on the horizon. They had an hour or two of daylight left – enough time for another mission in the South Central region but not enough time for a London trek.

"Good hunting, sir?" asked Leading Aircraftman Bernard Gordon.

"Another victory, I think."

"Smashing. Any problems?"

"No, Bernie. She ran like a charm. I don't think I took any hits."

"We shall give her a good look see."

Brian walked around his Spitfire just to look for himself. The sun-bleached and chipped paint, and eight streams of gun soot on the wings could not detract from the smooth lines of the beautiful machine.

Brian was the last pilot to complete his intelligence debriefing in the far corner of the dispersal tent. Corporal Jennifer Warren smiled at him as he approached her desk.

"What's our status, Jennifer?" he asked, using her given name as he usually did when the senior officers were not listening. After all, she was nearly ten years older than him.

"Available, sir."

"Do you think we're going again?"

"I shouldn't think so, sir, but you never know these days, now do we?"

"Oh so, right."

Brian joined the other pilots scattered about the grass in front of the tent. They were discussing the damage to Southampton and Eastleigh.

"How bad was Eastleigh hit?" asked Brian.

"Fifty to a hundred killed and double that number injured," answered Flight Lieutenant Robert Morrow.

"Out of commission for several days, at least," added Pilot Officer Roland Stockard.

"And, the bastards keep missing Woolston," shouted Flight Lieutenant Roger Beamish.

The telephone rang instantly stopping any conversation. They waited. Stockard stood and picked up his flight equipment.

"Skipper," called Warren. "It is for you. Sector wants a word."

No one spoke as they waited for Squadron Leader Darling to return with instructions, directions or information. As seconds became minutes, the pilots began to look at one another instead of the grass or the tent. Davies and Foxworth rose to peer into the tent. When Darling did exit the tent, all eyes were on him.

"Sector indicates London is taking another major dose from the Huns."

"Bloody criminals," mumbled Beamish.

"Another raid appears to be building over Cherbourg. They suspect it may be another attempt at Southampton. They are going to hold us back from supporting One One Group. Sector wanted us to be aware. The damage reports from Southampton confirm what we have heard. Last thing, the Prime Minister is making another evening radio address to the nation at eight, if you are so inclined."

"Then, we wait, Skipper?" asked Stockard.

"We remain at Available status. We wait until we are released."

Everyone settled back into the chairs or returned to their prone positions in the grass. Squadron Leader Darling walked toward the flight line and the ground crews finishing their methodical task of preparing the fighters for the next encounter.

'Boxer' Stockard mumbled loudly enough for Brian and several others to hear. "We sit here while London is flattened."

Brian looked at the older pugilist who had been with No.609 Squadron since December 1938, when it was converted from a bomber to a fighter squadron in the Royal Auxiliary Air Force Reserve. He had two victories. While not the most successful fighter pilot in terms of aerial victories, he was a survivor and a skillful fighter pilot. Brian recognized the experience of 'Boxer' Stockard in many arenas, but the implication of his mumbling did not sit well. Roland Stockard sensed Brian's thoughts.

"What?" he nearly shouted at Brian. "You would rather sit here than defend London."

"Roland," barked Jonathan Kensington.

"The hell you say," Roger Beamish added.

"Poor form," said Robert Morrow like a referee.

"He is the best damn fighter pilot we have, and you accuse him of falling from the fight," continued Beamish.

"Oh sod off," Stockard said, finally having enough and recognizing his mistake. "My apologies, Brian. No offense intended."

"It's all right, Roland. We're all a little tight."

"Maybe an understatement," Frank Burns decided to contribute.

The telephone rang one more time. With the sun nearly touching the western horizon, it was probably their release call. They did not have to wait long. Corporal Warren confirmed the call, as she jogged off to the flight line to tell Squadron Leader Darling.

18:30 hours

The evening meal was unusually quiet, even with many of the other squadron pilots mixed in among them. Brian thought several times, it did not matter what they did, they could never do it right. The successes of the great fighter pilots around them, 'Sailor' Malan, 'Bobby' Stanford-Tuck, 'Tin-Legs' Bader, 'Sawn-Off' Lock, 'Nine-Lives' Deere, 'Ginger' Lacey, grew like mighty trees. The names kept coming to him, and yet the number of Germans they faced each day seemed to grow as well. The bomber and fighter formations were larger, more sophisticated and certainly more intense. To Brian and the other pilots he talked to, the situation they faced each day continued to get worse, and now they were bombing London, the most beautiful city the young American had ever seen. What else could go wrong or get worse?

A half dozen of the squadron pilots celebrated Brian's 16th aerial victory. Several believed he would soon get his second Distinguished Flying Cross. Brian did not like the talk surrounding him, but he did not protest the celebration. It was as much a part of the process as anything else they did around flying.

20:00 hours

"Quiet please, gentlemen," shouted the sergeant tending the bar. "The Prime Minister's about to speak," he said, as he turned the radio volume up.

"Why doesn't he come here if he wants to talk to us?" said one of the other squadron pilots.

"This is the BBC World Service. This evening we broadcast a message from the Prime Minister. Ladies and gentlemen, Mister Churchill." The announcement was simple and direct.

The characteristic solemn but strong, deep, somewhat mumbled voice of the Prime Minister, the Right Honorable Winston Churchill filled the airways. "When I said in the House of Commons the other day that I thought it improbable that the enemy's air attack in September could be more than three times as great as it was in August, I was not, of course, referring to barbarous attacks upon the civil population, but to the great air battle which is being fought out between our fighters and the German Air Force.

"You will understand that whenever the weather is favorable, waves of German bombers, protected by fighters, often three or four hundred at a time, surge over this island, especially the promontory of Kent, in the hope of attacking military and other objectives by daylight. However, they are met by our fighter squadrons and nearly always broken up; and their losses average three to one in machines and six to one in pilots.

"This effort of the Germans to secure daylight mastery of the air over England is, of course, the crux of the whole war. So far it has failed conspicuously. It has cost them very dearly, and we have felt stronger, and actually are relatively a good deal stronger, than when the hard fighting began in July. There is no doubt that *Herr* Hitler is using up his fighter force at a very high rate, and that if he goes on for many more weeks he will wear down and ruin this vital part of his air force. That will give us a very great advantage.

"On the other hand, for him to try to invade this country without having secured mastery in the air would be a very dangerous undertaking. Nevertheless, all his preparations for invasion on a great scale are steadily going forward. Several hundreds of self-propelled barges are moving down the coasts of Europe, from the German and Dutch harbors to the ports of Northern

France; from Dunkirk to Brest; and beyond Brest to the French harbors in the Bay of Biscay."

"Bloody heathens are gathering at the gates."

". . . for a clout in the puss."

". . . and a kick in the pants."

"Shhhh," several listeners tried to quiet the commentators.

"Behind these clusters of ships or barges, there stand very large numbers of German troops, awaiting the order to go on board and set out on their very dangerous and uncertain voyage across the seas. We cannot tell when they will try to come; we cannot be sure that in fact they will try at all; but no one should blind himself to the fact that a heavy, full-scale invasion of this island is being prepared with all the usual German thoroughness and method, and that it may be launched now – upon England, upon Scotland, or upon Ireland, or upon all three.

"If this invasion is going to be tried at all, it does not seem that it can be long delayed. The weather may break at any time. Beside this, it is difficult for the enemy to keep these gatherings of ships waiting about indefinitely, while they are bombed every night by our bombers, and very often shelled by our warships, which are waiting for them outside.

"Therefore, we must regard the next week or so as a very important period in our history. It ranks with the days when the Spanish Armada was approaching the Channel, and Drake was finishing his game of bowls; or when Nelson stood between us and Napoleon's Grand Army at Boulogne. We have read all about this in the history books; but what is happening now is on a far greater scale and of far more consequence to the life and future of the world and its civilization than these brave old days of the past.

"Every man and woman will therefore prepare himself to do his duty, whatever it may be, with special pride and care. Our fleets and flotillas are very powerful and numerous; our Air Force is at the highest strength it has ever reached, and it is conscious of its proved superiority, not indeed in numbers, but in men and machines. Our shores are well fortified and strongly manned, and behind them, ready to attack the invaders, we have a far larger and better-equipped mobile army than we have ever had before.

"Besides this, we have more than a million and a half men of the Home Guard, who are just as much soldiers of the Regular Army as the Grenadier Guards, and who are determined to fight for every inch of the ground in every village and in every street."

"Armed with wooden rifles," someone said, evoking some nervous laughter.

"Quiet," someone else demanded.

"It is with devout but sure confidence that I say: Let God defend the right.

"These cruel, wanton, indiscriminate bombings of London are, of course, a part of Hitler's invasion plans. He hopes, by killing large numbers of civilians, and women and children, that he will terrorize and cow the people of this mighty imperial city, and make them a burden and anxiety to the Government and thus distract our attention unduly from the ferocious onslaught he is preparing. Little does he know the spirit of the British nation, or the tough fiber of the Londoners, whose forebears played a leading part in the establishment of parliamentary institutions and who have been bred to value freedom far above their lives. This wicked man, the repository and embodiment of many forms of soul-destroying hatred, this monstrous product of former wrongs and shame, has now resolved to try to break our famous island race by a process of indiscriminate slaughter and destruction. What he has done is to kindle a fire in British hearts, here and all over the world, which will glow long after all traces of the conflagration he has caused in London have been removed. He has lighted a fire which will burn with a steady and consuming flame until the last vestiges of Nawzee tyranny have been burnt out of Europe, and until the Old World – and the New – can join hands to rebuild the temples of man's freedom and man's honor, upon foundations which will not soon or easily be overthrown.

"This is a time for everyone to stand together, and hold firm, as they are doing. I express my admiration for the exemplary manner in which all the Air Raid Precautions services of London are being discharged, especially the Fire Brigade, whose work has been so heavy and also dangerous. All the world that is still free marvels at the composure and fortitude with which the citizens of London are facing and surmounting the great ordeal to which they are subjected, the end of which or the severity to which cannot yet be foreseen.

"It is a message of good cheer to our fighting forces on the seas, in the air, and in our waiting armies in all their posts and stations, that we send them from this capital city. They know that they have behind them a people who will not flinch or weary of the struggle – hard and protracted though it will be; but that we shall rather draw from the heart of suffering itself the means of inspiration and survival, and of a victory won not only for ourselves but for all – a victory won not only for our own time, but for the long and better days that are to come."

"At least, Mister Churchill thinks we are going to win this."

"So he says."

"We are."

"Yeah, but what is he thinking?"

"What do you mean?"

"Look at us. Look at all of us. Do we look like we are stronger?"

"Weaker."

"We are struggling just to get up in the morning."

Brian listened to the banter with a rather distant ear. Invasion remained the primary topic of discussion in virtually every conversation in any location. Talk of rumors, preparations, observations and hypotheses surpassed the experience of flight even for the pilots. Everyone, from small children to feeble retirees, sensed the tension around them. Brian's thoughts turned to Mary Spencer carrying his child and presenting it as a product of her marriage. He thought of Charlotte Palmer who held his heart. He could not escape the pleasant memories of Anne Booth, and certainly not the bold treatment of Rosemary Kensington when he lay in the Winchester Hospital bed. He wanted to call Mary to make sure she had not been injured in all the London bombing. Most of the bombing was focused on Central London, so she was most likely still safe. He also needed to hear Charlotte's voice, to feel her warmth and tenderness as well as her strength even if only a reproduced representation.

As the conversations of invasion and consumption of alcohol continued, Brian excused himself. The urge for sleep overpowered camaraderie. He went to sleep that night with visions of Charlotte Palmer in his mind.

———

Thursday, 12.September.1940
Dungeness Rail Station
Dungeness, Kent, England
14:30 hours

Churchill and his senior entourage boarded the Prime Minister's special train exactly where they had left it three hours ago – the siding at the small, very rural, rail station. The Army squad of armed sentries assigned to the train kept curious citizens away from any car of the train, from the locomotive and tender to the three special cars. Word had apparently passed swiftly among the locals that Prime Minister Churchill was among them. A small gathering of citizens waited patiently for a glimpse of the King's first minister.

The inspection tour had begun in the early morning. With the Prime Minister on the tour were:

-- First Lord of the Admiralty Albert Victor 'A.V.' Alexander, PC, Member of
     Parliament for Sheffield, Hillsborough, a Labour Party member and
     ministerial leader of the Royal Navy;

-- First Sea Lord Admiral of the Fleet Sir Alfred Dudley Pickman Rogers Pound, GCB, GCVO;
-- Chief of the Imperial General Staff General Sir John Dill;
-- Commander-in-Chief Home Forces Lieutenant General Sir Alan Francis Brooke, KCB, DSO; and,
-- Major General Sir Hastings 'Pug' Ismay.

The men had just completed the inspection tour that began at North Foreland on the south side of the Thames Estuary. They stopped at Dover to visit the headquarters of Vice Admiral Sir Bertram Home Ramsay, KCB, MVO, the commander of Dover area operations and responsible for protecting the Straight. Ramsay led the extraordinary naval effort to evacuate the British Expeditionary Force from the beaches and surrounded enclave at Dunkirk, and had been immediately knighted by King George VI for the successful planning and execution of Operation DYNAMO – the Miracle of Dunkirk. At each stop, they surveyed the tank traps, minefields, trenching, fortifications and deployments of precious artillery, infantry and what armored forces remained in England. Churchill remained deeply engaged, offered a flood of comments and suggestions for improvement, and serious criticism that made each of the general-grade officers quite uncomfortable. They all felt the urgency of what was before them. The signs of impending invasion remained extraordinarily dark, and worse, the Home Forces were in a very diminished state after the collapse of France, but could have been verging on tragic if it had not been for the successful evacuation of the Dunkirk enclave – a third of a million precious soldiers saved from death or capture.

"What are we waiting for?" Churchill asked with impatience.

'Jock' Colville stood, "Perhaps lunch would be in order while I sort out this delay."

"Indeed. We are late to lunch," Churchill responded and stood. "Shall we gentlemen," he said, gesturing toward the combination conference and dining car.

The chef and stewards were ready for their important principals. Churchill was uncharacteristically withdrawn and quiet, which in turn allowed them all to enjoy their lamb chops, mushy peas and roasted potatoes. They finished their meal. The stewards cleared the service. They were sipping a snifter of brandy when Colville returned. Churchill gestured for him to report to the group.

"The engineer is ready to depart. The control messages have been confirmed. However, we have been advised to wait here for a special courier on his way from London."

"How much longer do we have to wait?"

"Based on the latest information," Colville began and paused to look at his wristwatch, "received 15 minutes ago, we have another hour wait."

"No!" Churchill exclaimed. "We are done here. Tell the engineer to move back to London and work out a mutual intermediate stop to meet the courier."

"Yes, sir." Colville departed.

Churchill stood and began to pace. "What could be that important to freeze us on the southeast coast?" The other senior officers and minister took the Prime Minister's question as rhetorical. A few minutes later, the train began to move. "Shall we retire to the salon car to be more comfortable?" Churchill did not wait for any response. The other men follow the Prime Minister to the next car.

Again, Colville returned. "We will meet the special courier at Ashford."

"Very well. Thank you, 'Jock.'"

Ashford Rail Station
Ashford, Kent, England
15:55 hours

As the train approached the station, an army sergeant with a courier satchel and two additional armed guards waited for the Prime Minister's train to stop and boarded immediately. Colville guided the men. The sergeant saluted crisply as he entered the salon car. Churchill returned the man's salute. The sergeant moved to the Prime Minister, unlocked his satchel, retrieved a single sheet message, and handed it to Churchill.

MOST SECRET - ULTRA

```
SECRET
DATE 10TH SEPTEMBER 1940
TO ARMY GROUP A
FROM HEADQUARTERS ARMY
OPERATION SEALION
BEGIN
CONFIRM CHIEF TO ARRIVE AM THURSDAY FOR PLANNED
COMBAT ENGINEER EXERCISES WITH 8 ENG GRP BREAK
EXPECT CONFERENCE ON SITE AFTER EXERCISE BREAK
HAIL HITLER
END
SECRET
```

MOST SECRET - ULTRA

Churchill passed the message to Alexander. Each of the officers read the message as well, and then returned it to Churchill, who returned it to the courier.

"Is the chief referred to here von Brauchitsch?" asked Churchill.

Sir John responded. "Based on the addresses, yes . . . my German counterpart Field Marshal Walther von Brauchitsch, Commander-in-Chief of the German Army. That would be my assessment."

"Excuse me, sir. I also have a companion message passed to me by 'C' before I left Broadway House."

Churchill extended his right hand. The courier sergeant retrieved the second message and handed it to the Prime Minister. Churchill unfolded the single piece of paper.

MOST SECRET – PRIME MINISTER EYES ONLY

```
DATE 12 SEPTEMBER 1940
TO PRIME MINISTER
FROM WAR MINISTER AIR MINISTER
BEGIN
SIS LOCATES 8 ENGINEER GROUP AT ESCALLES BREAK
JOINTLY REQUEST AUTHORITY TO ATTACK ESCALLES
IMMEDIATELY BREAK WAR CABINET CONCURS BREAK
BOMBERS AND FIGHTERS STANDING BY FOR YOUR
AUTHORISATION
END
```

MOST SECRET – PRIME MINISTER EYES ONLY

The Prime Minister chose to share the contents of the second message as well. Once the others had read the second message, Churchill said, "As I recall, Escalles is on the coast between Callais and Cap Gris Nez."

"Correct, sir," answered 'Pug' Ismay.

"Sergeant, if you would excuse us for a moment. 'Jock,' please secure this compartment." The Prime Minister directed the rail car compartment to be cleared of everyone except the principals and the doors secured. "Given the hour, we may have missed a very narrow window. Should we take this shot? If we did, would that expose 'Boniface' – ULTRA?"

The men looked to each other, as if they were searching for their answer. General Dill was the first to respond.

"To be frank, I have mixed feelings about such an attack . . . not least of which is, he is me. Are we to conduct warfare by assassination? My opinion at the moment is no, we should not. While the target is tempting, the risks are too great."

"I agree with Sir John," offered Alexander. "We would need numerous companion targets along that portion of the coastline to mask any foreknowledge of who was there and what we were attempting to do."

"Admiral Pound, General Brooke, your opinions?"

General Brooke waited for the senior officer to reply.

"We have faced this question before. My opinion remains the same. While there are palpable risks associated with such missions, the temporary disruption to the German high command at this critical juncture, with Sealion looming before us, might be just enough to cause them hesitation."

"So, your answer is yes?" asked Churchill, for clear confirmation.

"Correct."

"I concur with Sir Dudley," General Brooke added without hesitation. "As we have seen and you have so precisely noted on today's field inspection, we are extraordinarily thin and quite uncomfortably vulnerable. We need any edge, however so slight, to prevent any landing of German combat forces on the Home Islands."

Churchill nodded his head. "General Ismay?"

"Without further details regarding the masking of our specificity of the target, we must err on the side of caution to protect ULTRA. I suspect, as you noted at the outset, our window to catch him and the Group 'A' commanders may have already closed. I believe the risk exceeds the potential reward."

"Very well." Churchill took the fountain pen from the table and scrawled 'Not Approved,' allowed the ink to dry and refolded the paper. "'Pug,' if you would be so kind to retrieve the courier and Colville. Thank you."

Churchill handed the folded paper back to the courier. "Please remain, just for a moment longer, if you would sergeant. I would like you to hear my response, so that you can convey it to 'C' and the War Cabinet. 'Jock,' please send an immediate priority, open source, reply message to the War Minister and Air Minister, 'Request of 12th September message not approved, repeat not approved.'" Colville wrote down the words. "Any questions?"

"No, sir."

"Send that immediately via the railway network, and then it's get back to London."

Colville followed the courier sergeant.

Churchill watched the courier and two guards board their single car train. They did not depart, presumably to wait for the Prime Minister to go first. A few minutes later, the Prime Minister's train began to move out of Ashford Station, before Colville returned to the compartment. Churchill's assistant private secretary confirmed the completion of his task and the remainder of their itinerary. Barring any unforeseen obstacles, they should arrive at Victoria Station at 17:30, and be at the Cabinet War Rooms by 18:00.

—

Thursday, 12.September.1940
Headquarters, No.11 Group
Uxbridge, Middlesex, England
16:10 hours

John Spencer saw the rain outside the converted mansion that served as the group headquarters. London would hopefully experience at least a significant reduction in the horrific bombardment. The headquarters building seemed unusually quiet even in the early morning hour. John looked both directions down the long, high hallway. No one could be seen. John stretched backwards, and to each side as he loosened his wire taut muscles and sore joints from the night on the small bed in the officer's quarters. He walked through several doors to the corner reception and informal conference room. The tall windows had been uncovered, with the heavy blackout curtains pulled back. The grounds did not have the floral colors he remembered from the spring. The rain made the brownish tints of summer appear less brown. As he looked out the windows across the gardens, for the first time in many months, John felt optimism about their prospects in the present battle. The wave of optimism came over him like the warm air from a kiln.

John removed his uniform tunic and necktie. He opened the tall, glass door and walked out into the garden and the rain. The crunch of gravel under his feet soon gave way to the gurgle and rush of falling water from the multilayer fountain. He stood near the fountain, stretched his arms out with his palms up, closed his eyes and raised his face to the heavens. The rain tickled his face and soaked his clothes. "Dear God, it is great to be alive," he said aloud, although no one could hear him. The vision of Mary and his unborn child brought added enjoyment to John. There would be a peaceful tomorrow, and they would be free. John now knew inside they were going to win this fight. It might take many years, but they were going to prevail in this fight.

John closed his eyes to feel the rain soaked chill, and listen to the gurgling water of the fountain and pitter-patter of the rain falling around him. John Spencer absorbed hope and the first threads of the future. The rain felt like life for him, for his family, and for his nation.

"John, have you gone daft?"

He turned to see his commander, Air Vice-Marshal Park standing in the doorway with a puzzled expression on his lean face. John did not respond other than to face his commander and lower his hands. There could be no acceptable explanation for a senior officer to be standing in the early morning rain. He walked calmly, as though nothing whatsoever was out of place, toward the open door. Park stood back to let him in the room.

"Are you all right, John?"

"Yes, quite, sir. Thank you for asking." John looked over his shoulder at the fountain and the rain. "I felt this urge with the exquisite rain and the respite it brings to feel the weather. We are going to win this row."

"I am not quite so sure," Park said, talking to his senior controller as though everything was perfectly normal and ignoring the growing puddle on the wood floor. "We continue to lose more pilots than we gain. You know that as well as I do, John. It shan't do us much good for Lord Beaverbrook to fill the flight lines with aeroplanes, if we have no pilots to fly them."

Minister of Aircraft Production Lord Beaverbrook was actually William Maxwell 'Max' Aitken, Bart, Kt, PC, ONB, 1st Baron Beaverbrook of Beaverbrook in the Province of New Brunswick in the Dominion of Canada and of Cherkley in the County of Surrey. He moved from his native Canada to England in 1910, and became a Member of Parliament for the Conservative Party in the same year. Aitken was knighted in 1911, and elevated to his peerage and moved to the House of Lords in 1916.

"You and I should have a go at Gerry."

Park smiled, looked John in the eyes and laughed modestly. "You know, I would probably feel better and quite a bit more helpful if we did." His expression returned to his stone cold determination. "It might just come down to that. Better to die in battle aloft than in some dank Nazi prison."

"But, we are winning."

"How so?"

"Did you receive the Y Section intercept yesterday, instructing the German bombers to turn around if our defenses were too strong?"

"Yes."

"Doesn't that seem to indicate they may feel we are stronger than we are?"

Park considered the words. He moved a few steps to John's left and toward the tall glass door. John Spencer began to feel a deeper chill. He needed to get out of his wet clothes and into the fresh uniform he kept in his office, but he also wanted the private chat with the accomplished New Zealander. Park turned to face John, and then smiled again.

"Maybe I should doff my tunic and stand out in the rain," he said, and then added a slight chuckle.

"It did feel quite invigorating, if I do say so."

Park considered the notion for a moment. He shook his head, apparently to dissipate the image. "We received notice from HQ an hour ago," Park said. "Gerry bombed Buckingham Palace yesterday . . . substantial damage but no injuries reported among the royal family or their staff. The King and Queen were in residence, and now according to my discussion with Sir Hugh, the King refuses to leave the palace."

"I imagine the prime minister . . ."

"Your uncle?"

John chuckled. "Yes, sir, my uncle, I imagine he feels the same, like some foolish act of defiance."

"Nonetheless, it is an inspirational gesture to the people who must endure this bombardment."

"Quite so."

"You are beginning to shiver," Park observed. "You had better dry off before you catch your death."

"Yes, sir. Thank you, sir," John said, nodded his head and walked across the room with squishy footfalls and a trail of drippings.

"Oh, John." The No.11 Group Senior Controller turned to face his commander. "I now know why Sir Hugh was so reluctant to let you go."

John smiled at the indirect compliment from the respected commander of fighter pilots. "Thank you, sir."

"No, thank you, John. I do believe you may be correct. We shall endeavor to make it so."

"As you say, sir."

Air Vice-Marshal Park nodded his head and turned his gaze toward the same fountain John had just left. He left Park in the corner room closing the door quietly behind him. He managed to make it down the hall and up the stairs to his above ground office, a converted bedroom, with only the duty WAAF receptionist in the foyer seeing him. She watched, mouth agape, in dumbstruck curiosity she could not satisfy.

As he changed his uniform in the privacy of his office, Keith Park's words returned to him as well as his thoughts of Mary. He wanted to be right. He felt right, but did Park know more than he did. By definition, he must.

He was the commander. John wanted to be right. He had to be right. There would be no acceptable future if he was not correct.

———

Friday, 13.September.1940
RAF Middle Wallop
Middle Wallop, Hampshire, England

The rain from the previous day had turned to a light drizzle as the day had begun for the pilots of No.609 Squadron. By the time, Pilot Officers Brian Drummond and Jonathan Kensington arrived at the dispersal tent in the southwest corner of RAF Middle Wallop the rain stopped and the clouds began to break up. Rays of sunlight pierced the cloud layers slanted to the west.

They would not get a break today with the clearing clouds. Yesterday's single sortie and the strain of maintaining formation with eight other fighters did not provide much rest despite the lack of combat. They hunted among the clouds for a small band of raiders they never found. Fortunately, there were no indications the bombers could find their targets either.

They sat quietly in the poor light but dry interior of the tent at Available status. A resolute patience kept them calm, inward and poised. Like their brothers at 40 other Fighter Command airfields, there was no doubt they would fill their small cockpits, place their hands on the sticks and throttles, and direct their fighters to the skies above and the waiting enemy.

"Good morning, gentlemen," said Flying Officer James Royster, the squadron's assigned intelligence officer. Several grumbles and grunts rose from the frozen men, but no one threw anything. "I have come with information for your consideration and use."

"Yes, yes, Mister Royster," said Squadron Leader Darling with some impatience, "get on with it."

"We have two reports of a Blenheim Mark IV strafing the port facilities at Dover and dropping several bombs."

"The hell you say."

"While few believe the reports to be true . . . ."

"Some bloke must have lost his head."

"No. The prevailing opinion is, if true, the Germans may be operating a captured aircraft or possibly a modified German or French machine painted in RAF colors."

"And, we thought we had seen it all," said Roger Beamish.

"The Air Ministry cautions you to confirm hostile intent before engagement."

"Bloody hell."

"Fuckin' Krauts can't face us fair and square," reacted 'Red' Burns.

"Easy, lads," interjected Darling.

Burns caught himself and nodded to Corporal Warren. "My apologies, Miss Warren."

"No offense. I've certainly heard much worse on the docks of Liverpool."

"Anything else, Mister Royster?" asked Darling.

"No, sir, that was it."

"Thank you, then," the commander said, indirectly dismissing the intelligence officer. Royster nodded his head to acknowledge Darling's words, and then left the tent.

"As always, gentlemen, we simply do our duty, and all shall be right," said Darling.

"Sure, Skipper, but how many more obstacles must we face?" said 'Boxer' Stockard, who could usually be counted on for a caustic or sarcastic comment.

"As many as are placed before us, Mister Stockard." The more formal reference communicated to everyone, other than maybe Stockard, Squadron Leader Darling's irritation with the triviality. "Let us not forget, we did not ask for this fight. It is the enemy that decides the extent of battle at the moment. Our sole mission in life is to defend the freedom of this nation. We shall not bow against the wind."

The ring of the field telephone stopped the conversation. "Squadron to Standby," announced Corporal Warren.

Without additional words, the pilots retrieved their flight equipment and walked to their aircraft. The ground crews were already among the aircraft. They launched ten minutes later for what became two long sorties to engage the Germans flying to and from London on the day's insult. They landed at RAF Middle Wallop an hour before sunset. By the time they reached the Officer's Mess, the dinner hour was nearly complete. They did not have time to clean up or change their sweat soaked uniforms.

The pilots gathered in the bar after the meal. The smoke of cigarettes and cigars began to fill the rectangular room. The dominant topic of discussion remained the invasion. It was on everyone's mind. The scenes of old men in villages across the country standing in lines with wooden, pretend rifles and pitchforks or other farm implements as weapons could not be ignored by any of them. Many of the pilots felt Fighter Command was the only barrier between the enemy and what appeared to be certain defeat. The growing number of barricades, lines of tank traps, barbed wire obstacles and other impediments to any invaders gave them some assurance the fighter pilots were actually not the only defenses. The burden remained heavy but largely laughed at through the many intoxicated conversations around the land.

Brian looked among the crowd for Jonathan. He found his friend embroiled in some debate. The American volunteer did not move closer, but waited until he caught Jonathan's eye, and then motioned with his head that he was going upstairs to his room. Jonathan nodded his acknowledgment.

Brian had his uniform tunic and tie off before he reached his small room. He sat on his bed, tempted to just lay back. He would probably be asleep before his head hit the pillow. As he unbuttoned his shirt, Brian noticed a letter on his small desk. It was from the U.S. Embassy in London.

"What in God's name are they going to harass me about now?" he asked himself aloud.

He opened the envelope. It contained another envelope, however the second envelope was from the White House. His heart quickened as he considered the possibilities. He tore open the second envelope nearly tearing the letter it contained.

The White House
Washington, DC

September 3, 1940

To: Mr. Brian A. Drummond
No.609 Squadron
Fighter Command, Royal Air Force
c/o: The United States Embassy, London
Dear Mr. Drummond:

Yesterday, many of your fellow citizens
listened to a broadcast report by the CBS
London correspondent, Mr. Edward R. Murrow,
about your heroism in the face of daunting odds
and your miraculous rescue. Your recognition
and decoration by King George VI are public
acknowledgment of your skill and courage. You
have made us all very proud to be Americans.
Your sacrifice on behalf of freedom shall be
recorded in history as a most noble act. I
have taken the liberty to write to your parents
as well. I trust you will not object. You
may never realize or appreciate the profound
contribution you are making to return the world

to peace and safety. The First Lady and I wish
you the best of luck for your safe return home.
By the hand of the
President of the United States of America
Franklin D. Roosevelt

Tears filled his eyes as his mind gained a second wind. Someone out there actually cared about what they were doing. The President of the United States of America knew what he was doing. He wanted to show Jonathan. He wanted to tell his friends, colleagues, lovers and acquaintances. The single piece of paper meant as much to Brian as the violet and white, diagonal striped, ribbon underneath his RAF pilot's wings. Instead, he decided to write his parents. He would tell them, even though the President indicated he had written to them as well.

RAF Middle Wallop
Hampshire
13 September 1940
Dear Mom and Dad,

First, let me tell you I am fine. I am tired from all the flying, but I am in good health. I know you worry about me, and for that I am thankful. These days I must always apologize for not writing more often. We have been very busy.

I am very excited. I just received a letter from the President, Mr. Roosevelt. He said there was a radio broadcast in America about me. I never heard it, so I don't know what was said. I am thankful that the President knows. He should know. Everyone in America should know what is happening here. The Germans have started to bomb London. They can't beat us, so they are going to pick on innocent people like big bullies. This is a very important thing we do, Mom. I hope you understand how important this is to all of us, to you as well.

My best friend, Jonathan, I told you about him before, feels the same way, not about America, but about the importance of this battle and war. Mr. Churchill made a very good radio speech earlier this week. His words give us so much strength. I don't know if you get to hear his speeches, but they are very good, let me tell you. Everyone is worried about invasion, but we are going to stop them.

I'm very tired. I'm going to bed. I hope you both are well. Take care.

Your loving son,
Brian

———

Brian sealed and addressed the envelope to his parents. It would be mailed the next day. He re-read the letter from the President, and then placed it in the only drawer the desk had along with other important letters. He readied himself for bed, and quickly drifted off to sleep with a smile on his face and in his heart.

———

Saturday, 14.September.1940
Cabinet War Rooms
New Public Offices
Westminster, London, England

The small room that served as bedroom, private quarters and personal office allocated for the prime minister still possessed a cold, isolated distance for Winston Churchill. Clementine and the children alternated between Blenheim Palace, his birthplace and the home of the Duke of Marlborough, and her family home in the Midlands. He, as the leader of the British nation, was relegated to the reinforced underground bunker and office complex, newly placed in operation. The prospect of spending the war cornered in these small rooms did not appeal to Winston in any possible way. He would much rather take his chances at No.10 Downing Street, or even Chequers or Chartwell. The War Cabinet and Defense Committee along with the associated intelligence, operations and logistics staffs now occupied the protected seat of His Majesty's Government.

Winston looked across the room to the cardboard sign recently added to his desk. The sign occupied a prominent place on the desk, so anyone entering the room could not escape its message. In clear, bold, block letters, it said:

'Please understand there is no pessimism in this house, and we are not interested in the possibilities of defeat; they do not exist.'

The famous words of Queen Victoria continued to give Winston strength.

His radio broadcast of three nights ago had gone reasonably well, based on the comments of friends and colleagues. The newspapers universally reported a very positive reception of his radio address and the message he communicated.

He hoped the people and the armed forces would gain strength to stand against the storm rolling in upon them.

The knock at the door ended his thoughts. "Yes," he answered.

Sir Edward Bridges entered the room and closed the door behind him. "'C' is here with some important information for you, however I wanted to catch you first. This afternoon's War Cabinet meeting is to be held in the conference room. I have taken the liberty to reduce the attendance due to the limited space."

"By all means, Ed. In the next few days, you might consider other arrangements to allow maximum utilization of that limited space. We do need the Defense Committee along with the War Cabinet in these meetings."

"We shall try several seating arrangements to see if we can accommodate everyone."

"Thank you."

"The King and Queen began a three day excursion through the Midlands to bolster morale."

"We must be thankful for their Majesties. They are the perfect examples for our people."

"Indeed."

"Anything else, Ed?"

"No, sir."

"Would you be so kind to show Stew in?"

Bridges nodded his head in acknowledgment and stood at the door until Colonel Menzies entered the Prime Minister's small underground office. Bridges closed the door behind them.

"Good afternoon, Stew."

"Good afternoon to you, Prime Minister."

"How is the weather above ground?" asked Winston, since it had been more than a day since he had been outside.

"Intermittent, isolated showers, but mostly quite pleasant and occasionally sunny."

"Then, we shall have our afternoon visitors."

"I suspect so, Prime Minister."

"What do you have for me today?" asked Winston, showing some impatience.

"I thought I would bring the ULTRA to you myself." Menzies opened the manacled case, extracted a single sheet of paper, and handed it to the Prime Minister.

MOST SECRET - ULTRA

```
SECRET
DATE 11TH SEPTEMBER 1940
TO OKH OKM OKL
FROM OKW
BREAK
OPERATION SEALION
THE LEADER HAS POSTPONED THE DECISION TO
EXECUTE OPERATION SEALION UNTIL 17TH SEPTEMBER
BREAK ALL KEY DATES ARE SHIFTED SIX DAYS BREAK
ALL PREPARATIONS ARE TO BE CONTINUED BREAK THE
AIR ATTACKS AGAINST LONDON ARE TO BE CONTINUED
AND THE TARGET AREA EXPANDED AGAINST MILITARY
AND OTHER VITAL INSTALLATIONS BREAK HAIL HITLER
END
SECRET
```

MOST SECRET - ULTRA

Churchill looked into the dark eyes of Menzies. "So, they are ready, and they have not made their decision on the 11th as they stipulated."

"As it would appear."

"What do you think?"

"That is why I decided to come myself. Although ULTRA has not yielded any definitive issue for the Germans, we seem to have many otherwise unimportant communications, many not coded, that lead us to believe we may be having more impact on them than we may have otherwise suspected. They seem to be profoundly concerned about our fighters. We suspect they may have seriously underestimated our strength, and either overestimated or missed our weaknesses. This," he said holding up the ULTRA message, "is the first formal postponement of an invasion decision, but it is also at least the fourth informal delay. They originally wanted an early September landing. As I recall, there was even some mention of an anniversary commemoration."

". . . of the war?"

"Yes."

"How quaint?"

"Indeed."

"If we are fortunate, the rough weather of autumn and winter could be with us any day now."

"The records say by the third or fourth week of September."

"None too soon for any of us, I should say."

"How is the Air Force doing?" asked Menzies as he sought his own information.

"Hanging by a thread, I'm afraid. One One Group, the southeast fighter unit, has reached the end of the rope. Sir Hugh Dowding appears to be doing everything humanly possible to keep pilots in the air. Fighter pilots, at least the number of them, remain our greatest vulnerability."

Menzies considered his words. "Pardon my bluntness, but your public pronouncements certainly lead the citizenry to a different conclusion."

Winston smiled. "It is a risk, Stew, but I am trying to find the single thread of hope in this mire of disappointment. Furthermore, my radio words are heard by that foul bastard, as well. I certainly do not want anyone to know how close we are to the edge of the cliff."

"Are we going to make it to the bad weather?"

"Dear God above, I hope so. We must all pray for those young pilots rising each day to fight the enemy in the skies above us."

Menzies shifted his weight several times. "We also have several bits of information that the raid on Berlin three days ago was probably more successful than we thought."

"How so?"

"We have the first public acknowledgment of the damage as being serious, which says quite a lot for the Germans who usually try to understate or ignore such news. We also have evidence they have diverted several anti-aircraft batteries from France to Berlin."

"Good. It is about time they tasted the sour sting of their own medicine."

"Indeed," responded Menzies. "Has the War Office informed you of events in Egypt?"

"Not as yet, I'm afraid. I imagine they will at the Cabinet meeting this afternoon. What happened?"

"General Berti and his 10th Army initiated a broad attack eastward toward Egypt. Early reports suggest General Wavell was surprised and is conducting a fighting withdrawal."

"Then, it will definitely be a topic at this afternoon's meeting. This appears to be the major offensive we have suspected for months."

"And Wavell was not ready?"

"Such a conclusion may be rather premature, Winston. It is simply too early to tell. I imagine the War Office is trying to sort things out as we speak."

"We do not need this now, with this damnable invasion looming, but these are the cards that are played and we shall play ours."

"Do you need anything else from SIS for this afternoon's War Cabinet meeting?" asked Menzies.

"A miracle," shot Winston.

Colonel Menzies laughed, as did the Prime Minister. "I shall see what we can do."

"Thank you, Stew. Keep on the ULTRA traffic. Spare no resources. That little box may hold the only key."

"As you say, sir. With that, I bid you good day."

Alone again, Winston stood squarely in front of a large map of Europe to the Ural Mountains and Africa. Admiral Raeder, among all the political and military leaders of Germany, maintained a healthy respect for the Royal Navy and the narrow body of water known as the English Channel. Winston returned to history as he did in his speech of the previous evening. Raeder knew, maybe alone among the German leadership, any crossing of the Channel would present the first real opportunity for failure. They had seen his reservations in various ULTRA communications outside the Navy. They had to reinforce the German admiral's apprehension.

Chapter 4

Aide-toi et Dieu t'aidera.
(God helps those that help themselves)
-- Proverb

Sunday, 15.September.1940
RAF Middle Wallop
Middle Wallop, Hampshire, England
05:30 hours

Week 11

The morning's wake-up call came much earlier than Brian's body and mind were ready for, after weeks of early mornings, long days and intense aerial combat. His muscles and joints ached, and his brain retained the fog and dull pain of an abused gourd. He sat on the edge of his bed, as he did most mornings these days, fighting the magnetic attraction of his pillow and still warm sheets. The dim light of dawn outside his window was sufficient to tell him the weather would probably be a good day for flying.

The shower and shave helped him shake some of the residual fatigue effects. A good morning meal would provide the last of any gain he would have for the day. He looked at his desk drawer to consider whether he should take the letter from the President to show Jonathan. He shook his head and closed the door behind him.

06:30 hours

Most of the pilots sat around the long table at various stages of consumption. He found an open place next to 'Boxer' Stockard. No words came other than his order for scrambled eggs, sausage, fried potatoes and toast.

Brian managed only several bites before the duty corporal entered the dining room with a concerned expression on his face and in his eyes. He struck a hand-held bell as a traditional attention-getter.

"The Station Commander has ordered all day-fighter pilots to their respective dispersal points," the corporal said.

The unusual announcement did not surprise the pilots, although several grumbled at the interruption to their morning routine. Brian stuffed his mouth and chewed quickly as he stood. It was going to be a very long day – good weather and 12 hours of daylight. The Germans would not likely pass up the opportunity.

Trucks picked up pilots for each of the squadrons. They arrived to hear Corporal Warren tell them No.609 Squadron had been ordered to Standby status. Darling told her to report the squadron as responding. They grabbed their flight equipment and headed toward their aircraft. The sun was above the horizon but behind distant clouds to the East. The ground crews had no time to complete the morning engine run-ups and pre-flight checks. They had been through the routine previously, although it was more the exception than the norm. Brian dropped into the cockpit and readied himself for flight. He waited for Leading Aircraftman Bernard Gordon's signal. The engine started promptly. Brian completed the ignition and propeller checks quickly to let the four men draped across his horizontal tail move to the next fighter. He finished the systems checks, ensured his engine temperatures were within normal limits and shutdown the systems and engine. Brian reset his switches to ready the fighter for launch. He waited in the cockpit for the command to go.

"Everything OK, sir?"

"Perfect, as always, Bernie."

"You going to stay with her?"

"We're at Standby, so we sit."

Bernie Gordon nodded his head and jumped off the wing. Brian sat with his thoughts, as each of the No.609 Squadron pilots did the same. The scent of the hunt eliminated the last vestiges of Brian's aches and pains, or at least banished them from his consciousness. Brian was ready for battle.

The controllers were usually careful about bringing crews to high alert levels too early. The mental processes began to dull the longer they waited in the cockpit for what seemed to the pilots as eternity. Brian tried not to think of anything other than his aircraft and the hunt. He could not tell how much time had passed when he noticed Corporal Warren jog out to the 'PR-A' Spitfire. The conversation with Squadron Leader Darling lasted only several sentences, as Brian watched across several fighters. Warren moved off the wing. Darling stood up and stepped out of his cockpit. He looked across the other aircraft, moved his hand across his throat, the universal cut or stop signal, before he returned to the ground. The pilots joined their leader as they walked back to the dispersal tent.

"We are back to Available, lads," he told them.

No one reacted to the change. The ebb and flow of battle as well as the common changes of direction had become the norm. Several months ago the grousing at such changes reflected the taut nerves of the unseasoned fighter pilots. They knew the character of air battle all too well these days. The pilots settled into their chairs or other relaxing activities without words of complaint or protest. Squadron Leader Darling went directly to the telephone, as he often did to learn more about the tactical situation.

Their leader returned. "Headquarters saw the build-up begin at dawn. They thought the Germans shifted their large raids to the morning hours, however they did not continue. Several low-level attacks are in progress in Kent and Sussex. Weather is fine all over the South, nearly perfect actually. They expect this to be a long day."

No one reacted to the words. The pilots settled into their chairs. Brian dozed for an indeterminate amount of time and woke with an odd uneasiness.
08:45 hours

Brian wanted to think of something else. He leaned over to Jonathan Kensington. "How is Rosemary?" For some odd reason, Brian's memory filled with his experiences with Jonathan's sister.

Jonathan looked into his friend's eyes, as if to ask him why he was asking. "She is back at Oxford, although I tried to tell her to skip a semester or two until this situation has passed."

"Is she OK?"

"Sure."

"I miss her."

"Along with all your other women," Jonathan said, partially joking and partially serious.

"That's not fair, Jon. I didn't ask for any of it."

"Nonetheless," he stopped to scan for listeners, and then continued in a whisper, "you have a married woman pregnant, your once-lover executed for spying, you are sweet on a much older woman who saved your young skin, and now you say, you are still interested in my sister."

Presented in Jonathan's words, Brian's relationships sounded like some sordid mess from a risqué romance novel. Was his love life really that strange, he asked himself? "I was just thinking of her," Brian said feebly.

"Look, Brian, what you and my sister do is between the two of you. She is an adult woman, able to make her own decisions. I just don't want you to use her, and then throw her aside."

"Jesus, Jon! Is that what you think I do – use women and throw them away?"

"No. I just worry about my sister."

"That's good, but she started it, and she is an exciting person to be around."

"So you say. I have never thought of her that way. She has always been my bratty little sister."

"She is not that, now."

"As you say, then."

"Neither one of us has seen her . . . ."

The telephone rang freezing all conversations and thoughts.

"Scramble the squadron," Corporal Warren shouted from inside the tent.

The pilots sprang from their positions toward the fighters. "We'll finish this later," Brian shouted at his friend, as they joined the others running toward their airplanes. Jonathan only waved his hand.

Engines began firing off. No.238 Squadron, ten Hurricanes, raced across the airfield taking to the air ahead of Brian and his mates. As the squadron reached the top of the slight hill and turned for their takeoff, the Spitfires of No.152 Squadron were starting up as well. Yes, as Darling told them, this was going to be a big day. Brian could not remember all the Middle Wallop squadrons starting up at the same time. As the landing gear locked into the wings and Brian adjusted his position off the left wing of 'Jackstay' Beamish, he wondered what they would see before lunchtime. Darling turned the squadron to the east. It was probably London, again. Would they be placed on patrol over Brooklands and have to watch the bombing of London again, or would they be in the middle of another large raid on the capital city.

"Sorbo Leader, this is Horse calling."

"Sorbo Leader, go ahead."

"Climb angels two eight, vector now one oh five. The raid is . . ." They saw the mass of German aircraft stacked up across the sky ahead of them before the controller finished his transmission. ". . . very large, perhaps 1,000 plus." *Yeah, like 5,000 plus*, Brian said to himself. He instinctively tightened his mask on his face to ensure he got all the oxygen the aircraft delivered. There were other risks at high altitude beyond the bullets.

"Tally ho."

"Sorbo, you are joining eight other squadrons, so far."

"So far," Brian repeated aloud to himself. His mind considered what lay ahead. "Three Spits high and five Hurris low."

"Roger, Horse. We have the party."

"Short Jack will pick you up. Good hunting."

The other squadrons began the engagement. The orderly formations of German bombers and fighters progressively deteriorated as the RAF fighters plunged into the pool of targets. No.609 Squadron watched the developing battle as they raced at full power to join their brothers. Brian continued to scan the sky all around them expecting to see German fighters appearing from the ether.

Darling maneuvered the squadron to join the developing ball of fighters. They were going to dive into the fight from the south, behind the attackers.

"Let's get to it, lads," Darling radioed. He rolled his Spitfire left and led the other No.609 Squadron Spitfires into battle.

———

Sunday, 15.September.1940
Headquarters, No.11 Group
Uxbridge, Middlesex, England
09:15 hours

John Spencer watched the mass of h and F blocks moving into Kent and Sussex. "We had better get the boss down here," John said to one of the junior controllers, who did not wait to respond. The large raid came from four different assembly areas in Northern France and Belgium, joined over the mid-Channel north of Boulogne and headed directly toward London as a large mass.

John glanced up to the status boards on the opposite wall. Three quarters of the group's twenty-three squadrons were indicated as airborne, plus three from No.10 Group's ten squadrons and two requested but not yet reported as airborne from No.12 Group's fifteen squadrons. Twenty-one fighter squadrons, twelve Hurricanes and the rest Spitfires, closed with the approaching mass of nearly 2,000 enemy intruders. They had 205 fighters with 1,640 machine guns and 492,000 rounds of ammunition against so many enemy aircraft. Ten to one odds, it just did not seem like a fair fight. Even if there were no bombers, there were probably more German fighters in the raiding party than in all of Fighter Command.

Leaning over the controller table, John said, "Let's bring the remaining five to Readiness. Inform Watnall we will most likely need at least five more from them as well. We will wait until the leading enemy elements are over Redhill, to launch the others."

"That would not leave us much margin," one of the controllers stated the obvious.

"Quite right. Let's make sure the auxiliary aerodromes are ready. We will probably have to use alternate landing areas. We have the squadrons spaced out in time rather well. Good Lord willing, we should be all right."

John noticed Air Vice-Marshal Park enter the observation gallery. He joined the commander in the middle of the balcony.

"Rather large commitment," said Park.

"Yes, sir. We have less than half a dozen waiting to cover the withdrawal. We are still finding some resistance from Watnall."

Park looked into the eyes of his senior controller. "Sir Hugh has emphasized the importance of their support. I trust Leigh-Mallory will meet the expectations."

"I hope so. We could use all of Ten Group and better than half of One Two Group, and still not have enough."

"We shall persevere."

"How long?" John asked almost absentmindedly.

"We started this fight with nearly a thousand pilots. We are down to just over 500. We have tapped every source any of us can think of, and we are still bleeding." Park looked again at John Spencer, and then lowered his voice to a whisper. "I would say, if you want my candid opinion, we are within days of being rendered ineffective. We are teetering on the brink."

"Do Sir Hugh and Sir Cyril share your opinion?"

Park smiled, and John regretted the impertinent question. "Yes, John. I do believe they do. We discussed our resources just last night. Sir Cyril is working with the other chiefs of staff as well as the ministers to act upon every recommendation I have made. The only one we seem to be having any difficulty with is the support from One Two Group."

"I suppose Air Vice-Marshal Leigh-Mallory must feel as we did at Dunkirk. Maybe he does not want to expend his resources to a battle already lost."

Park turned to face the smaller John Spencer squarely. He leaned forward slightly to minimize the distance his words had to travel. "If we lose the Southeast, his bloody fifteen squadrons are not going to stop them."

"We are not going to lose the Southeast," John answered, more instinctively than logically.

Park stood up straight. "I truly hope you are correct."

The muffled sounds of the talkers communicating with Fighter Command and moving the h and F blocks across the map board changed. John turned his attention to the situation. The lead h block reached Redhill on its course toward London. He looked to the controller's booth. The controllers all looked to him. John nodded his head.

Telephones were raised and the commands given to launch the remaining squadrons as well as the additional requested squadrons from the adjacent groups. John knew they would soon have more fighters aloft than any time in the history of the RAF. Maybe Park was right, John said to himself. Maybe this was the day . . . their salvation or their damnation. Along with many others, John Spencer watched the status lights progress down the boards. There were no lights lit in the top half of any of the tote boards. The battle was joined, and it was not yet noon. The imagination of what lay ahead had no bounds that morning.

———

Sunday, 15.September.1940
RAF Northolt
Northolt, London, England
11:05 hours

The nine Spitfires of No.609 Squadron landed for the first time this Sunday morning at a familiar airfield. The 'YO' Hurricanes of the No.1 Squadron, Royal Canadian Air Force, retracted their wheels for the second time this day. The first sortie for Pilot Officer Drummond and his squadron mates had been successful. He thought at least two of the pilots added victories to their tally. Brian achieved hits on at least four German Bf109 fighters but could not claim a victory. He felt good about their performance as he sat in the cockpit of his Spitfire. The ground crews raced to reload and refuel his aircraft. Several of the other pilots took the opportunity to stretch their muscles and flex their joints while Brian closed his eyes, ignored the metallic and gushing fluid sounds generated by the ground crews, and took advantage of the moment for a quick nap.

The sounds of Merlin engines firing off close by brought Brian back to consciousness. He leaned out both sides of the open canopy to check the area forward of the wing and around the propeller arc. "Clear," Brian shouted to warn anyone close by that he was starting his engine. He pushed the starter button, watched three blades pass the vertical position and switched the ignition switch to BOTH. The Merlin III fired, choked a few times, and then roared to life in a billow of blue-gray smoke.

Once in the air and the wheels fully locked in the wings, Brian searched the skies. The contrails to the south of the city looked like a splattering of spaghetti across the roof of the sky. As they rose out of the haze, the mixture of incoming and withdrawing German bombers along with their antagonists came into view. They were vectored to the south-southwest, as though they were en route to RAF Tangmere, but the controllers were simply giving them space to gain precious altitude. Behind his left wing, Brian could see the smoke rising from London. The smoke indicated that the fires did not appear to be quite as bad as on the 7th, or the several days in between, but it was confirmation the enemy had been at least partially successful.

Brian watched the battle above them as they climbed. He tried to pick out targets, wanted to call out to several RAF pilots being chased, and watched as pilots on both sides died. The scene, while new and unique, was also quite similar to so many before it. The life struggle in the air had become the expected, maybe not the usual, but no longer held the awe of a few months ago. This was simply business, now.

No.609 Squadron joined the battle as the German fighters were trying to fight their way south. Darling decided to engage from sections. Brian followed 'Jackstay' Beamish along with 'Crazy' Kradilcek into the fight. It was like jumping into a muddy river or pond, everything collapsed to a sky filled with fighters, green and brown among the gray and black.

Brian had just finished a good pass on a turning German. As he arced over the top inverted and looked for his next target, he saw a nightmare unfolding – the distinctive tail letters, 'PR-D,' with one, red nose, Bf109 on his tail and another closing just behind him.

"'Jackstay,'" shouted Brian over the radio. "On your tail."

Brian pulled the stick back to his lap. The nose of the Spitfire responded, but shook and shuddered near maneuvering stall. He pulled the nose to line up a long, diving shot in the hopes of distracting Beamish's attackers. He fired. The bullet arced off well behind his targets. He dove with his engine at emergency power. Speed built up rapidly. The first German shot off several pieces of Roger's Spitfire. Brian fired again as he closed – still behind. He was now closing very rapidly. Brian considered trying to thread the needle by diving through the attackers – too risky – a mid-air collision at these speeds would be fatal for everyone. The 'PR-F' Spitfire slashed through the two Germans. He saw several flashes on the lead German, but the pass was too fast. Brian immediately pulled hard with both arms on the spade circle of the stick, and yelled hard straining against the heavy 'g' forces of the high-speed pull-up. He felt enormous stress on his neck muscles and spine, as he tried to look through the overhead canopy to reacquire his targets.

As Brian grunted and strained, he also noticed he had his own tail, not close enough for an accurate shot, but definitely trying to close. It took several long seconds for Brian to find the now smoking 'PR-D' Spitfire. Only one attacker remained. Roger Beamish continued to maneuver although the motions were less crisp. Brian lined his nose up well above the horizon trying to get a climbing shot. He quickly glanced over both shoulders for his own attacker. The German did not make as tight of a turn as he did. He was out of position. The rapidly slowing airspeed would help the German trying to shoot him. Brian pushed harder on the throttle already against the emergency stop. The telltale sounds of rushing air dissipated leaving only the roar of the engine. He was not going to make it. Brian quickly aligned his sights with the passing target trying to make sure he did not hit Beamish's Spitfire. As airspeed rapidly dropped toward zero, Brian squeezed off a long burst. Several hits into the belly of the German flashed before him as his Spitfire stalled and fell.

Brian instantly pulled this throttle back to idle and neutralized his controls trying to avoid entering a spin. The nose dropped sharply slamming him into the metal sides of the cockpit and against the canopy. The nose pointed toward the brownish-green earth below. The reassuring rush of air told him he was gaining airspeed. Brian smoothly pushed the throttle forward to emergency power. The engine groaned but responded. Brian took what energy the airplane could give him to pull the nose up holding just above 100 mph straining to get back in the fight.

He checked his tail – no fighters. His head and eyes swiveled rapidly. Fighters still filled the sky. Now, white and yellow nose Germans were among them. Green on Gray. Gray on Green. Brian tried to find Beamish. The growing and blackening smoke trail led his eyes to the 'PR-D' Spitfire. Beamish was in a moderate descent, wide left hand turn. At least he had no assailants. Brian maneuvered aggressively and precisely to protect his leader. He watched Beamish, as he scanned the sky around them. No change in his flight path was not a good sign.

"'Jackstay,'" he radioed with a strong, confident voice, "level your wings."

The sickening sensation in his gut added to the mortality of the progressive descent toward the ground below. They passed through several scattered layers of German bombers and British Hurricanes. The smoke coming from all the cowling edges and openings meant a slow smoldering engine fire. The propeller continued to turn. The engine was still running although at fairly low power.

"Roger, you've got to pull out."

Still nothing. They were passing 10,000 feet. No one was around them. Brian maneuvered his Spitfire close in, off Beamish's right wing. The Scotsman's head was slumped forward onto his chest. Brian could see holes in the canopy and blood or oil splattered on the transparency.

"Roger, wake up," Brian shouted, as if he were yelling in Beamish's ear.

The agonizingly slow, descending spiral continued to eat precious altitude. Brian watched for several more turns trying to see any response. The smoke pouring from the engine compartment remained unchanged and did not appear in the cockpit – at least one good sign. Something had to change or Flight Lieutenant Roger 'Jackstay' Beamish, DFC, would soon impact the approaching earth.

Brian maneuvered his aircraft to place his left wing tip over Beamish's right wing tip. Several thuds and jolts convinced Brian to move more cautiously. Every muscle in Brian's body was tight and aching. He backed off slightly, took several deep breaths of oxygen from his mask, and remembered the words of Flight Lieutenant Lord Jeremy 'Mud' Morrison, regarding close formation flight – he moved back into position.

Gently, Brian used his wing to push down Beamish's right wing. Although the effort bumped and joggled both aircraft, the wings did come level. Brian checked his heading – northwest. The nose of the 'PR-D' Spitfire was still below the horizon – 4,000 feet and descending at 700 feet per minute. The hills of Northern Kent and Sussex lay ahead. He only had several more minutes to keep Beamish in the air hopefully long enough for him to regain consciousness, if he ever would.

Brian released his position on the right wing of the 'PR-D' Spitfire, which rose again starting into the left turn. Somehow Beamish was trimmed into the left hand turn. Again, Brian leveled the wings. They were headed directly for the escarpment south of RAF Croydon.

"Roger, wake up, damn it," radioed Brian again. "Jesus, Roger. You've got to wake up. You're going to hit the ground." There was no reaction from the adjacent aircraft.

The nose of the stricken fighter had to come up to give him more time to regain consciousness, if he could. Brian needed to bump his tail down to get his nose up enough to stay airborne, but every time he reduced the pressure on Beamish's right wing, it would rise. Brian tried to nudge the backside of the wing, but hit the aileron every time causing several potentially serious impacts to both aircraft. The nose came up slightly. They would at least miss the face of the Kent escarpment.

Beamish was running out of time. Brian fought with both aircraft taking more risks as the ground approached. He was making some progress but not enough. They passed several hundred feet over the trees at the top of the rise. The ground sloped away. He could see aircraft landing at RAF Croydon.

Brian held Roger Beamish's wings level until the last instant, and then he pulled back and pushed his throttle forward. The 'PR-D' Spitfire disappeared from view. "Oh God," Brian unknowingly transmitted. Brian banked as hard as he dared at slow speed. The Spitfire responded perfectly. He looked out the top of his canopy as he turned to see the cloud of dirt and trees. As the debris began to settle, the wreckage burst into fire. "No!" screamed Brian.

He circled the spot of the crash. Brian considered trying to land his fighter in the sloping and somewhat rough field to help his friend. Amazingly, a fire engine raced across the field within minutes of the crash. The Fire Brigade vehicle must have been passing by and saw the impact. Brian continued to circle as he watched the firefighters perform their risky duty. They quickly extinguished the flames. Two of the men reached the upright cockpit. They tried to open the canopy. It was jammed. They called for an ax that another man brought to the cockpit. Several good whacks opened a hole large enough for them to pull Beamish free. They carried him clear of the wreckage, lay him on the ground and checked him for life signs. All three men stood up and looked toward Brian circling overhead. One of the men swept his hand level and shook his head. Roger Beamish was dead.

Brian continued to circle, not wanting to leave his friend and leader. The men on the ground completed their tasks and carried Beamish's body to their vehicle. Brian checked his fuel. He was nearly empty. "Sorbo Leader, this is Sorbo Green Three calling." The radio chatter of combat came to him as he turned his attention to this headset. No answer. "Sorbo Leader, Sorbo Green Three." Still, no answer. They must have disengaged to land somewhere. He had no idea where. He needed to find a place to land soon.

Croydon and Kenley were close. Brian climbed just enough to relocate RAF Croydon. He did not have the frequency for the control tower. He used smoke and dust to find the wind direction, checked for traffic, spaced himself behind a flight of Hurricanes and landed. The amount of visible damage to the airfield surprised Brian.

———

Sunday, 15.September.1940
RAF Croydon
Croydon, London, England
13:20 hours

Brian landed without incident, taxied around craters and debris to the Operations tower, found a reasonable spot and shutdown. Several enlisted men ran to his aircraft. He was on the ground when a leading aircraftman and two aircraftmen arrived.

"Do you need any help, sir?"

"I need petrol and ammunition."

"We shall take care of it, straight away."

Brian walked around his aircraft to inspect the damage he undoubtedly picked up during the flight. *Several new bullet holes. I don't remember getting hit.* His left wing tip showed several wrinkles and indentations from his contact with Beamish's aircraft. *All of it looks service-able.* He nodded to the leading aircraftman, and then walked into the Operations building, told the clerk what had happened and asked him to contact RAF Middle Wallop. The corporal lifted the handset, asked for the appropriate connections and handed the handset to Brian.

"Six Oh Nine, please?" he asked.

The connection was quickly made. "Six Oh Nine Squadron. May I help you, sir?" said the familiar voice of Corporal Jennifer Warren

"Jenny, this is Brian Drummond."

"Where are you, sir?" she asked, but did not wait for an answer. "Both you and Mister Beamish were listed as missing."

"I was with Roger. He crashed just south of Croydon, where I am now."

"Did he make it?"

"No," Brian said solemnly. "I don't think so."

"Are you all right?"

"Yes. My bird is damaged, but it is being fueled and loaded. What does the Skipper want me to do?"

"Let me give you to him."

"Mister Darling," she called away from the telephone. "It is Mister Drummond."

Squadron Leader Horatio 'Spike' Darling came on the line. Brian relayed the sequence of events and his present location. Darling indicated there was a break in the raids. The squadron was at Available status although they were trying to find something to eat and drink. He suggested Brian should try to return to RAF Middle Wallop as soon as possible.

A squadron of Hurricanes was landing, as he stepped outside the building. He returned to his fighter. The leading aircraftman waited for him. "The underside of your left wing tip is rather smashed up, sir."

"Can I fly?"

"Well, yes, sir, I believe you can, however you really should get the damage repaired." The man anticipated Brian's question. "It will most likely require a wing change."

Oh great! Bernie is not going to be happy with me, and the maintenance staff is not going to appreciate another wing change on my fighter. "Thank you, then I'll be off."

"As you wish, sir. I would try to avoid a fight, if I could suggest."

"Most definitely."

Brian started and took off. He found 'Jackstay's crash site, circled several times looking at the wreckage of the 'PR-D' Spitfire like a tombstone, and then turned west flying low level toward his home base.

Sunday, 15.September.1940
RAF Middle Wallop
Middle Wallop, Hampshire, England
14:35 hours

Brian landed at RAF Middle Wallop and noticed most of the assigned aircraft were on the ground in their proper places. He taxied across the field to his parking spot and could see a small welcoming committee – Kensington, Burns, Johnson, Davies, and even 'Curly' Mansek, who must have just returned. His ground crew greeted him first.

"You should not scare us like that, sir," said Gordon.

"Sorry, Bernie. I stayed with Mister Beamish. He crashed and didn't make it."

"I am so sorry. A great loss to us all."

"Yep," Brian answered, but wanted to change the subject. "We need to check my left wing tip. It got banged up pretty good trying to save 'Jackstay.'"

They all took a look, kneeling down under the left wing tip. The dented skin and scraped paint looked bad, but it was the several tears in the aluminum skin and the bent rib that Gordon touched like a sore wound. He whistled as he finished his inspection.

"Bashed it in rather well, did you now."

Brian felt the need to explain to his crew chief. "I was trying to tip his wing to keep him from hitting the ground."

"You did what?" Gordon asked with surprise maybe verging on horror.

"I tried to level his wings."

"Jesus, son of Mary, you could have crashed as well. You might not have survived loosing your wing tip. You must refrain from such heroics."

Brian felt a surge of resentment. It was his life, and just a little damage to the metal skin of an airplane. He had tried to save his section leader's life, or at least that's what he thought at the time. Roger Beamish had probably been mortally wounded, if not dead, at altitude, but Brian did not know that at the time. He just remembered the sickening feeling of watching his friend descend slowly toward the ground. "Is it flyable?" snapped Brian, as he stood up.

"I shall need the engineer to make a determination."

"Then, let's hop to it. We probably have more damn Germans to kill this afternoon," he snapped again, as he walked away toward the tent. Brian's fatigue allowed his anger to bubble to the surface. As he walked, he resented his loss of control. Brian stopped, kicked himself, and then turned around toward Bernie. "I'm sorry. I didn't mean to be so sharp."

"You have had a rough day, sir."

"Thanks, Bernie," said Brian with what warmth as he could offer.

Brian retold the story several times to mixed reviews. Some thought he was insane, risking so much when all the signs appeared to indicate Beamish had been killed high in the sky. Others praised him as heroic. He did not think of himself as heroic, no more than he did the incident several weeks previous when he launched by himself to attack the enemy bombing their airfield. He had not considered the consequences. He simply reacted as best he could. There was no premeditation about his actions, simply instinct.

The maintenance chief, a warrant officer, arrived at the dispersal tent. The wing tip might tear rapidly under maneuvering loads. They were going to attempt to replace the tip if the underlying structure was not seriously damaged. They had a new Spitfire, unmarked and still in the factory paint, assigned to Brian until his aircraft was repaired.

Squadron Leader Darling gathered the pilots. "The loss of a friend is always tragic, and especially someone as loved as Roger. However, we have a war to fight, and we owe our friends, loved ones, citizens and families the best we can do. We shall mourn our friend, but in its time. We will undoubtedly have more sorties today, and if not, many more tomorrow. Let us keep our heads straight to the task." Darling scanned the faces and eyes. Satisfied the message got through, he continued. "Jonathan, you are the next senior pilot. You will take Green Section." He turned directly to Pilot Officer Kormer Mansek. "Are you ready to fly, 'Curly?'"

"Yes. I am."

"Good. I am afraid we do not have time for an evaluation or warm-up flight."

Mansek nodded his head. Jonathan Kensington found his friend's eyes, and gave him a wink and nod. For Brian, it would not be appreciably different flying Jonathan's wing than it was flying Roger's left wing position.

They did not have long to consider the loss of their comrade or all the changes occurring around them. The telephone rang. "Scramble the squadron," shouted Corporal Warren.

Sunday, 15.September.1940
Cabinet War Rooms
New Public Offices
Westminster, London, England
15:20 hours

Winston Churchill had watched the morning's air battle from the roof of the New Public Offices building. The drama played out in the skies above and south of the city reminded him of an elaborate stage play carefully choreographed to heighten the play's adventure and conflict. He sat with growing irritation through the afternoon's War Cabinet meeting, listening to intelligence reports that carried new information but the same message, as well as endless invasion preparation reports. The dull thuds of exploding bombs propagated through the ground itself to the underground bunker stimulated Churchill's impatience. It was the status report from the Chief of the Air Staff Air Chief Marshal Sir Cyril Newall that took the Prime Minister right over the top. The afternoon's battle was even larger and more dramatic than the enormous morning raid.

"Gentlemen," he interjected, as his tolerance broke, "I am terribly sorry. History is being created this afternoon. I can no longer sit here in this hole as our young men create the legends that shall be told for hundreds of years. I must adjourn this meeting." Winston stood with void, shocked or confused faces watching him. "I beg your forgiveness. Sir Edward, please reschedule this evening's meeting and notify the participants."

He was out of the conference room before anyone could say a word. He nearly ran to his office, dropped his papers on his desk and grabbed his bowler hat. Only a few of the ministers and officers had left the conference room. "General Ismay," he said, as he walked swiftly past the still stunned leaders of His Majesty's Government and Armed Forces. "Would you be so kind to join me?"

Ismay took a few quick steps to catch his prime minister before they started up the stairs to the entrance of the bunker. As they reached the entrance lobby with its armed guards and entry maze, Churchill ordered his car. It would take a minute or two. Churchill squinted his eyes and pulled his hat down as he stepped out into the bright sunlight of the late summer day. He paced impatiently and did not talk. General Ismay stood by the curb near the entrance to the bunker. When Churchill's eyes adjusted, he used his hat to shade his eyes, as he looked skyward. The bombers and weaving fighters were clearly visible above the city through the array of barrage balloons. He could see the bombs falling, and a short time later hear and feel the explosions mainly south and east of Whitehall. He debated whether to see the full scope of the battle from the Fighter Command Operations Room at Stanmore, or the No.11 Group Operations Room at Uxbridge. He remembered his nephew, John Spencer, had been transferred to Uxbridge, plus most of the action was in the Southeast area of operations under the control of Park's Group. Detective-Inspector Walter Henry Thompson was already in the automobile when it arrived.

Thompson had been assigned out of retirement from the Metropolitan Police, Special Branch, to be Winston Churchill's personal bodyguard and physical protector, since August of 1939, before his charge had returned to His Majesty's Government from the ostracism of his Wilderness Years.

"Where to, sir?" the driver asked, as the two men settled into the back seat and Ismay closed the door.

"One One Group Headquarters at Uxbridge, and make it as fast as you can," Winston said.

The driver smiled before he turned completely around in his seat. The sergeant had just been given the license that he rarely received, to use his siren. He sped off as the siren blared their approach. Traffic, what there was of it, parted allowing them to make the 30 to 45 minute journey in less than 20 minutes. The Prime Minister walked swiftly through the Headquarters building startling officers and enlisted personnel alike. Churchill and Ismay joined Air Vice-Marshal Keith Park in the commander's observation gallery.

Park casually briefed the Prime Minister on the situation displayed before them on the map board. The group commander summarized the morning's action, as well as recreated the genesis of the enormous afternoon raid. Nearly 5,000 enemy aircraft assembled from all points on the European continent across the English Channel now streamed toward London. They had set a record in the morning for the most RAF fighter aircraft airborne at one time. The record would soon be broken.

Winston looked across the balcony. His nephew, Air Commodore John Spencer, sat perched near the railing with his chin resting in his hands. He remained intensely focused on the map board, oblivious to anything and everything around him. Several times he turned to give instructions to his staff. Maybe later Winston might have a word with him, if there was a break in the action.

Air Vice-Marshal Park excused himself to join his senior controller. The hum of muffled conversations filled the room, but Winston could hear the bombs going off in the buildings and streets of London. He felt the pain of the city like needles were being stuck into his flesh.

The complex mass of blocks, some marked with small h's and others designated with capital F's, moved across the map board between Europe and London. Several of the WAAF talkers kept an impromptu chalkboard score as they listened to radio calls. The board showed nearly a four to one victory difference in favor of No.11 Group despite being out-numbered.

Winston absorbed the scene in intimate detail. The large clock with its red and blue triangles marked every five-minute segment. The row of squadron status boards with all the lights near the bottom of the stack. The constant activity of everyone in the room. The concentration of every person, each focused on their part of a complex task.

Air Vice-Marshal Park returned to the commander's gallery. The tall, thin, New Zealander grasped the railing in front of him. Winston noticed the whiteness around his knuckles. Park maintained his concentration. Winston waited for the first sign of a break.

"How many reserves have we?" asked the Prime Minister.

Park looked Churchill in the eyes for several seconds, and then turned back to the map board and lowered his head. "There are none," he said simply and softly.

Winston's heart skipped a beat. He remembered instantly standing in *le Quai d'Orsay*, the headquarters of *le Armée de Terre* in the middle of May, less than a week after the invasion of France, and asking General Gamelin a similar question. Winston received the same answer – *aucune* – none. Had the British arrived at the same point the French had four months earlier? *It was not the same*, Winston told himself. The British had the English Channel, and the Germans still had to contend with the Royal Navy, if the Air Force collapsed. The young warriors – The Few the press now called them – had borne the entire brunt of the German assault. Now, the citizens of London shared their pain. They now had an adequate quantity of fighter airplanes, thanks to Lord Beaverbrook, but they were dreadfully short of pilots.

"May we listen to the radio?" asked Churchill.

Park nodded his head and leaned over to a small box beside him. The decorated fighter pilot and commander of fighters turned the volume knob. The words came in a flood. Fear mixed with calm. Rejoicing overlapped warning and loss. Victory commingled with defeat. Hunter and hunted became one. Nearly every fighter squadron south of Manchester now flew in this enormous air battle. The commands of battle directed the fighters against elements of the enemy invaders. The young voices of men hardened by months of battle kept a calm, skillful hand on the reins as the scared cries of new, inadequately trained pilots grappled with the intensity, energy and danger all around them. Winston felt tears return to his eyes, as he listened to the human drama played out in the sky. Everything Newall, Dowding, Park and the others had predicted moved before them. He now realized why Park did not listen to the radio – joy and death, opposite hands, gripped your heart right through your ribs.

The battle continued into the evening. Then, one by one, the blocks were pulled off the map board table. They all watched the far side of the map – Europe. Were any other attackers going to come as the RAF refueled and

rearmed? There was still enough daylight for another raid. If the Germans were going to catch Fighter Command on the ground, they would have to come soon. They watched and waited. The British fighters fought the retreating enemy until the last of the fighters had to disengage to avoid running out of fuel.
15:50 hours

They did not come. Winston felt exhausted, and he had not fired a shot. It was as if he felt the fatigue of the young pilots. His head ached. Winston placed his left hand on Park's shoulder. The commander turned away from the board.

"Your pilots have won a historic victory," said the Prime Minister.

"I hope so," Park answered.

"Most of the bombers did not reach their targets. The tally board shows the enemy was punished today. If they were hoping to gain supremacy in the air or cow Londoners, they have failed. They will postpone their invasion. Winter will soon be upon us. We will gain strength from here. We shall consume them from here."

"I hope you are correct, Prime Minister. I truly hope you are right."

"I know I am."

Winston Churchill left the group headquarters and returned to London. He took a nap and slept like a rock. This had been a very long day. They had the night raiders to look forward to for sometime. The day's damage to London, so far, was less than at the beginning of the Blitz, as citizens were now calling the city bombing, on the 7th of September. The German losses had been reported to be considerable, but they had a large air force that could tolerate the losses. To the RAF and especially Fighter Command, every loss added to the precarious situation. They needed time to train more pilots and build more airplanes. They needed time to convince America to wake-up and join the fight against tyranny.

Sunday, 15.September.1940
RAF Middle Wallop
Middle Wallop, Hampshire, England
16:30 hours

The general ache inside his skull and the weariness of all his muscles substantiated the residual product of three, long, combat sorties for Pilot Officer Brian Drummond. They landed 25 minutes earlier. This time they absorbed the luxury of laying in the grass with a cool breeze cooling their sweat soaked uniforms. None of the pilots wore their neckties any more. A royal blue silk scarf, purchased by Mrs. Emily Darling in Andover, filled the space in their open collars to reduce the chafing of the uniform shirts. Brian wanted to groan as he laid down but kept it to himself.

He reflew the last sortie with Pilot Officer Jonathan Kensington leading Green Section. They chased the last of the Bf109 fighters across the Channel, and even managed to pound some bullets into several withdrawing bombers. The most amazing sight came just before they turned for home. Brian followed Jonathan into a sweeping turn to line up on a Do17. Jonathan had a near perfect angle with a slight dive from the left rear quadrant of the bomber. As Brian lined up behind Jonathan for a pass at the bomber, one of the gunners decided he had had enough punishment. He jumped out, fell several hundred feet and popped his parachute. Brian could only imagine what it must look like to a bomber's gunner watching two sleek Spitfires closing rapidly on your aircraft, and feeling the attacker's gunsight directly on your chest. Brian smiled. Now, if they could just get the pilots to jump as well, they might win this thing without firing another shot.

"Rather amazing watching that poor bloke jump out before we fired," Jonathan said with his eyes closed.

"I was just thinking the same thing. We musta scared the piss out of him."

"Or something else." They both chuckled. "Maybe the poor bastard wanted to change sides, and he felt it was his last chance."

"Hell of a thought. Maybe we're winning this fight after all."

"Indeed. Maybe."

The sound of Merlin's starting across the airfield took them to a different awareness. A squadron's worth of engines came to life. Brian thought, as probably most of the pilots did, their call would follow shortly. When the telephone rang, Brian and several of the other pilots headed toward their aircraft. Brian's temporary machine looked rather odd among the squadron aircraft, but it performed well on their first outing. The announcement followed them as well as the other pilots. The 'UM' Spitfires of No.152 Squadron took off first.

The surprise came when they received vectors to intercept a low-level raid due south, and they were one of six Spitfire squadrons scrambled for this task. Brian instinctively knew what they were flying to protect – the Supermarine Spitfire Works at Woolston across the River Itchen from Southampton. Although no clues were radioed, Brian suspected they were about to engage a renowned German Air Force unit. They had all heard the stories and received the intelligence briefing about the special *Luftwaffe* unit, *Erprobungsgruppe* 210 or Erpro210 as the intelligence folks referred to the special operations squadron led by the well-known *Hauptmann* Walter Rubensdörffer. The unit had been seen many times since early August. The former experimental test unit, pressed into operational duty, flew distinctive, twin-engine, Me110C-4/B Destroyer fighter-bombers painted with red propeller spinners and a black nose, and armed with two 20mm cannons and four 7.9mm machine-guns plus a 7.9mm, rear-facing, swivel machine-gun. They specialized in high-speed, low-level, bombing missions against specific hard targets. They made one unopposed and thankfully unsuccessful attempt on the Woolston Works a week earlier, on the 7th. If this raid was Erpro210 and their target was the Supermarine factory, they would undoubtedly get much closer to their target.

Squadron Leader Darling must have sensed the same thing. They were running at 500 feet with emergency power. Darling considered the risk of overheating the engines to be appropriate, which probably meant he thought they faced a very serious threat. As they raced toward Southampton, the little, ugly balls of anti-aircraft shell bursts pocked the scene behind the barrage balloons. Two Spitfire squadrons approached above 5,000 feet in case there were other fighters. Brian only found one other low squadron.

The markings on the twin-engine Me110 attackers confirmed Brian's suspicions. At least the German aircraft held their track flying at near wave-top level up the estuary from the Solent between the Isle of Wight and Portsmouth.

"Horse, this is Sorbo Leader calling. Tally-ho. Sorbo, take sections in trail." Darling maneuvered to place the squadron between the intruders and their target. "Maida Leader, this is Sorbo Leader, we are coming in head-on."

Jonathan repositioned Green Section behind Blue Section and slightly to the left. They picked out their targets. The other sections fell back taking their positions.

"Sorbo, Maida. We have you. We will pull off as you engage."

Sweat gushed from every pore as they dropped down the River Itchen into the estuary. The barrage balloons scared every pilot, especially pilots flying with other aircraft close-by and those invisible cables. They turned the corner at the Supermarine plant. The Germans were clearly illuminated in the afternoon sun. The two groups were closing at full speed. The Spitfires of No.152 Squadron pulled up from behind the attackers. This was absolutely crazy. There was so little margin for error.

Both groups opened fire early. Small geysers popped up ahead of them as bullets hit the water. The tactic worked. The Germans began to break their formation and maneuver to find an opening. Several Me110s pulled up sharply making themselves quick targets for the No.152 Squadron pilots. The sky instantly filled with aircraft.

Brian banked hard, but did not climb much, just enough to allow his two mates to pass under him. He followed 'Harness' and 'Crazy' through the turn looking for their targets. Spitfires swarmed around the German squadron. Two of the attackers had already gone down. The first Me110 strafed the shoreline portions of the Supermarine factory and dropped a 250 kg bomb that shot a huge geyser from the seaplane ramp as it past over it. The second Me110, with a Spitfire blasting away at it, dropped its bomb short and nearly took out the pursuer.

There were so many aircraft in the sliver of sky just above Woolston and Southampton. Brian had to maneuver sharply numerous times to avoid colliding with other aircraft both friendly and hostile. His frustration mounted with his fatigue and the streams of sweat in his mask and goggles. Dodging fast aircraft with guns, barrage balloon retention cables, the ground and other obstacles became more fearsome than bullets. Among all the confusion, the Germans managed several more passes at the Supermarine factory. There was no other target for Erpro210. The Germans paid a heavy price for the damage they caused. Small wafts of black smoke rose from two parts of the factory.

Just as quickly as the engagement began, the Me110s of Erpro210 dropped onto the water heading back out the way they came. The Spitfires gave chase as the Germans flew into the Solent and passed the Isle of Wight and into the English Channel fighting their way home. Brian chose to take a higher line waiting for an opportunity. The enemy aircraft had plenty of company. The British protectors kept up the chase until near mid-channel. No.609 Squadron broke off the engagement along with the other squadrons. As they rejoined for the return flight, the sun touched the western horizon. The overwhelming odds in favor of the defenders felt good, actually refreshing.

As they passed over Southampton, Brian evaluated the Woolston plant. There did not appear to be any serious damage. They would un-doubtedly hear the results when they landed.

The condition of the Woolston Works was paramount on everyone's mind. The debriefings were unusually short. They wanted to know if they still had a supply source for Spitfires. It took half an hour and several telephone calls before they received definitive information. The plant had sustained some shrapnel and bullet damage but nothing serious.

The worst of the news . . . the Germans now clearly recognized the importance of the Supermarine Works. There was no doubt about the target. The use of the elite Erpro210 unit on such a risky approach added to the importance. The fight to defend the factory would be much more difficult from now on.

Chapter 5

Those who profess to favor freedom, and
yet deprecate agitation, are men who
want crops without plowing up the ground,
they want rain without thunder and lightning.

-- Frederick Douglass

Monday, 16.September.1940
RAF Middle Wallop
Middle Wallop, Hampshire, England
07:15 hours

Week 11

Rain beat on the window like little beads thrown by the wind. The wake-up call kept Brian confused through several sets of thoughts. While there was light outside, it was heavy gray. The wind and rain meant they would not fly until afternoon at the earliest. The weather swirled around his still cloudy mind, throbbing from the abuse of the last few days. His body told him to lie back down for as many more hours of sleep as he could steal. The routine told him each successive call would be more dramatic. All the pilots experienced their personal form of physiological reaction to protracted fatigue, both mental and physical. It was the mental stress of combat, especially that insane low-level mission the previous afternoon that took the greatest toll on their concentration and resiliency.

Of all the losses in his young life, Brian returned vividly to yesterday. Roger Beamish had always been a quiet, methodical, successful, Scottish fighter pilot. *He protected me*, Brian said to himself, *when I needed protection. He supported me when I needed to be supported. God, he was a good friend, and he did not need to die like that – the agonizingly slow descent – the red smears along with black smoke – the sliding impact into the trees.* Brian forced his legs to the floor and sat up, and then slumped over placing his head in his hands. Roger's limp and battered body pulled from the wreckage by the emergency services personnel. Brian felt the suction of the first true waves of grief. The loss of such good men – Billy Fiske, Ceasar Hull, and now Roger Beamish – so many good men. Brian felt his eyes water.

The door opened. "Better get to it, lad," came the voice of Flight Lieutenant Robert 'Sparky' Morrow.

Brian swallowed hard to choke back his inner grief. "Why?"

"Oh, I don't know. Someone said something about an invasion or some such. Maybe it was a war on."

"It's raining for God's sake," Brian nearly shouted at the senior officer. The welling moisture in his eyes must have caught the flight leader's attention.

Morrow stepped into Brian's small room and closed the door. "We all miss him, Brian." He bent over toward Brian. "His loss does not change a thing. We have lost others – 'Mongo,' 'Organ,' 'Junior,' 'Angle,' even your countrymen, 'Hank' and 'Slim.'" He gritted his teeth as the tendons and veins of his neck jumped out. "We owe it to them to win this frigging fight." Morrow regained his control, stood up straight, pulled down on his tunic to flatten it against his chest and cleared his throat. "Now, I would suggest you get dressed quickly. You have missed breakfast, and the lorry to dispersal will depart shortly."

Brian nodded his head and stood up with just his underwear covering him. "How much time do I have?"

"Maybe a few minutes."

"I'll be right there."

Morrow left the room and closed the door behind him. Brian knew he did not have time for a shower or shave. He dressed as fast as he could, and ran downstairs and outside. The pilots were seated in the covered rear section of the large truck. The rain felt good on his face. He looked up, closed his eyes and let the rain beat on his face.

"In the lorry, mate," someone shouted.

"Oh, leave the kid alone," came the only other American accent of 'Red' Burns.

"Who the hell are you," shot back another voice, "his keeper."

"Knock it off, all of you," growled Morrow.

The only seat was forward on the right, next to Pilot Officer Janus 'Crazy' Kradilcek. The truck jerked forward before Brian could sit. He stumbled, stepping on several of the pilots, producing grunts and groans of protestation. Brian eventually made it to his seat.

As they turned south on the A343 toward their dispersal point, 'Crazy' leaned over to whisper in Brian's ear. The peculiar combination of strong garlic and the residue of still dissipating alcohol turned Brian's empty stomach. "You no mind the bastards. Losing friend never easy," he said, as he tried to smile and nod his head for emphasis. "We to . . . ge . . . ther for to kill fucking Germans." He pulled back and smiled broadly.

Brian returned the smile although it was the furthest response from his heart. "Thanks, 'Crazy.' We'll kill us some Germans for 'Jackstay.'"

"For 'Jackstay,'" he responded, leaning left into his compatriot 'Curly' Mansek, so he could pat Brian on the back of the shoulder.

When the truck stopped, the pilots jumped from the bed, ran through the opening in the bordering hedgerow and into the dispersal tent that served as their operations or ready room. Brian trailed the group, and again waited just outside the tent entrance to let the rain beat upon his face. The cool wetness soaking his uniform felt refreshing, rejuvenating for the young fighter pilot. Those raindrops contained a life force that dulled the grief and washed away the dirty surface of fatigue.

"My God, boy," shouted Flight Lieutenant Davies, "are you turning into a duck?"

Brian looked into his eyes. He was not smiling. Brian could not and did not respond. He simply walked into the tent with its dim light. Less than half the faces watched Brian drip on the wooden boards that made the floor and kept the mud out.

"Have you lost your friggin' mind," barked Davies. "If we launch, you will surely freeze to the controls or simply freeze solid, full stop."

"Will you guys leave the fucking kid alone," Burns said, returning to Brian's defense.

"Stop!" commanded Darling.

All heads and eyes turned to their commander, and everyone and everything froze in place. Darling engaged each set of eyes before he spoke. "First, we will keep our civility despite the madness around us. 'Red,' you will apologize to Corporal Warren."

"She's heard worse," responded Burns without thinking and still wanting to defend Brian. Jennifer Warren nodded her head in agreement although discreetly.

"Pilot Officer Burns, this is not a debate," Darling said with a noticeable deepening to his voice.

Burns received the message this time. He looked directly to Corporal Jennifer Warren. "My humblest apologies for my improper word choice," he said, and then bowed to her. She only smiled slightly and nodded her head once.

"Now," Darling continued, "I know we are all tired, and I know we feel the loss of our brother in arms, but we cannot allow ourselves to be distracted. Simply put, I will not tolerate any bickering amongst ourselves or any of our brethren. We are engaged in mortal combat in defense of this country and freedom itself. The enemy is out there," he said, pointing to the sky outside, "not in here. Do I make myself clear?"

Brian looked around the tent as everyone nodded their heads. No one spoke. Brian could not escape the possibility all the other pilots felt Roger Beamish's loss as he did and more acutely than the many losses before him.

"I would like each of you to read this message. We must focus on our accomplishments and dedicate ourselves to the task at hand. We have several more weeks, a month perhaps, to keep Gerry at bay, and then the weather should keep them off." Darling passed the single piece of paper to Flight Lieutenant Morrow first. "No one told us this task would be easy. We must hold our course and stand the line. Our families, our loved ones, our nation and empire are counting on us to succeed. We will not be deflected from this task."

Brian waited for his turn to read the message. Most of the pilots showed no reaction to the content. Some smiled as they passed the paper.

```
ZZZZ/5139GGE7639/AM-AC/5600832/ZZZZ
FROM:  CHIEF OF THE AIR STAFF
TO:  FIGHTER COMMAND, ALL UNITS
DATE:  16.09.40, 0730 HOURS
SUBJECT:  CONGRATULATIONS
BEGIN
BY ALL REPORTS, ACTION OF 15TH SEPTEMBER WAS
AN EXTRAORDINARY SUCCESS.  WHILE ENEMY RAIDS
WERE ENORMOUS, MOST BOMBERS DID NOT REACH THEIR
TARGETS CHOOSING TO DISCARD THEIR PAYLOADS IN
OPEN COUNTRY AND RUN FOR SAFETY.  LOSS OF ENEMY
AIRCRAFT FAR EXCEEDED OUR OWN.  THE NATION WILL
BE PROUD OF YOUR ACCOMPLISHMENT AND WILL LONG
REMEMBER THIS DAY.
THE PRIME MINISTER, WAR CABINET AND AIR
MINISTER SEND THEIR HEARTFELT GRATITUDE AS
WELL.
GOD BLESS YOU ALL AND KEEP YOU SAFE.  NEWALL
END
ZZZZ/5139GGE7639/AM-AC/5600832/ZZZZ
```

The tent remained devoid of words and the rain continued to drum on the canvas around them. Each of the pilots withdrew into his thoughts. Brian fought against his memory, but could not hold off the vivid images of yesterday's events and his failed struggle to save his section leader. He wanted to cry, to wash away his grief, but knew those emotions were not an option available to him.

The rain subsided near mid-morning. Patches of drizzle intermittently enveloped the fighter base in fog. The Meteorological Office held to their projections that the poor weather over nearly the whole of England and Wales would remain through the night and probably through tomorrow.

The squadron was released from duty just prior to noon. Brian knew exactly what he wanted and needed. Charlotte sounded relieved, pleased and anxious for him to join her at Standing Oak Farm. He ate lunch with Jonathan who intended to make his own visit, to Linda Mason in London. He took a long, hot shower, cleaned himself and donned a fresh uniform.

Brian thought about asking Squadron Leader Darling if he could borrow Roger Beamish's Morgan. The thought passed quickly as his stomach twisted itself into knots and turned sour. He ordered a taxi from Salisbury. The journey would cost him nearly three pounds, a large sum for him.

———

Monday, 16.September.1940
Standing Oak Farm
Winchester, Hampshire, England
13:30 hours

Charlotte Palmer must have waited for his arrival. She opened the front door and stood just back from the splash of the rain from the porch as the taxi drove up. Brian paid the driver and stepped quickly to the porch. Charlotte took a couple of short steps backward to let Brian into the house. She smiled and reached with both her warm, soft hands to gently grasp his face. Leaning forward slightly, she kissed him in a delicate, almost feathery manner that ignited his heart. Several times Brian started to speak, only to find her finger over his lips. They allowed the passion of their separation as well as life and death around them to fill the void they both felt since their last meeting. The caresses and intimacies pushed out the pain, the anxiety, the concerns, each of them brought to this point.

In time, words began to return as they lay in bed entwined into one body. They shared his pain. Brian exorcised his demons through the re-creation of Roger Beamish's last minutes and the struggle to save him. Charlotte listened as she stroked his hair and kissed away the tears. Her soft, warm caring hands accomplished what nothing else could for Brian. As she gradually returned him to the pleasures of the flesh, Brian reacquired the will and purpose he needed to face the danger. He also recognized Charlotte's carefully hidden but not invisible troubles.

"Here I've been babbling on like a baby. Something's on your mind, Charlotte. What is it?"

Charlotte looked deeply into his eyes, leaned toward him, pressed her body against his and kissed his chest gently. Brian's sense of foreboding brought him more awareness. He sensed a dissonance within her.

"Charlotte, what's wrong?"

She drew back just enough to see his full face. Her eyes danced back and forth as she searched for some clue, a sign. "It's nothing."

"Yes, there is," he said, as he sat up and separated their skin. "Something is bothering you." He waited for a response. "Is it the farm? Are you having money problems?"

She smiled and even chuckled inside. "It is nothing, Brian. Certainly nothing like you must face each day, like watching your friend's crash."

"That's crap. Now, are you going to tell me or not."

Charlotte thought about his question and decided against any disclosure or discussion. She rose, stood before him for a brief moment, and then wrapped a thick, heavy robe around her. "How about some tea and biscuits?" she said, as she walked toward the bedroom door.

Brian wanted to stop her, to coax her into sharing her concerns, as she had done for him. He waited for her return. The unattended whistle of the kettle boiling forced him to put his pants on and go to the kitchen.

He saw the jet of steam rising from the kettle before he saw her leaning over the sink with her arms straight. She had not heard him join her, or she chose to ignore his presence. Brian knew there was something not right within her. He also knew she really did not want to talk about it. Her reticence made his earlier confession and exorcism seem selfish and less mature. He needed to support her as she supported him. Brian walked to the stove, turned off the gas flame and removed the teapot. The removal of the whistle turned her around sharply.

Brian checked the teapot with the strainer ball filled with leaf tea and poured the hot water into it. Placing the kettle back on the stove, Brian turned to her. They faced each other, looking deeply into each other without words or movement. The urge to hold her, to caress her as she had caressed him, pushed him, but he did not move.

"Brian," she said, opening the words. He nodded his head. "I will say this, but for many reasons I would rather not talk about what I do not know." She stopped to search his eyes for any reaction. Brian remained stoic. "I want you to survive this dreadful war. I think I know the key to your survival is your concentration on the task at hand. Distractions can only add to your danger." Charlotte hesitated, this time to gather her thoughts. "I am worried about becoming a distraction."

Brian laughed loudly as though some huge pressure relief valve let go excessive steam pressure. "That's ridiculous. You're not a distraction."

Charlotte remained serious and focused. "Maybe not, but I could be," she paused to search his eyes. "Brian, I love you. I realize I love you more than I ever thought I could love anyone again, maybe more than I have ever loved."

"And, I love you."

"I have lost too many loved ones in my life. I am not sure I could bear the loss of another. I am so afraid you might not be on the peak of your game, if you worry about me."

Brian smiled and contained his urge to laugh again. "I was taught at an early age . . . ," he stopped with realization the words sounded funny. He was still quite young. "I was taught by my instructor, I think I told you about him, Malcolm Bainbridge." She nodded. "He taught me to eliminate everything from my mind when I'm flying, to concentrate on only the flying. I've done pretty well. You shouldn't worry about me."

"Now, it is my turn to laugh. Maybe you have never really loved anyone. I cannot stop thinking of you."

Brian felt a stab of ridicule. "It's not the same. I have to put things in their proper place, or I could make a mistake or miss an opponent."

"My point precisely."

He went to her, held her shoulders and pulled her into his arms. They held each other. Brian whispered into her sweet smelling hair. "Nothing is going to happen to me."

She opened her robe. Their skin fused and relit the fires of their passion. The remainder of the day was spent in bed except for the afternoon milking of the cows, a light meal with some red wine and a joint playful bath. Neither one of them returned to the seriousness of their earlier words. The rest of the world and all its troubles disappeared from their lives.

Tuesday, 17.September.1940
Cabinet War Rooms
New Public Offices
Westminster, London, England

The gray skies and intermittent showers greeted the day for Londoners. Winston Churchill was just another Londoner as he walked alone through the glistening trees and wet grass. After the relative success of Sunday's aerial combat, the rain took on new meaning. Yesterday's foul weather kept the huge formations of bombers away. Bombs continued to fall through the clouds, but they amounted to mere fractions of the good weather, daylight raids. The rain meant relief and with each additional day the protection of winter. So, for Winston, the rain felt exceptionally good.

He passed a few citizens, older men and a few women. They recognized the leader of His Majesty's Government, and even offered greetings of encouragement and support. The rejuvenating force of the morning walk carried mystical powers for the Prime Minister. The deep, rich, Big Ben bell in the Westminster Palace tower announced the approach of the morning War Cabinet meeting. He returned to the entrance of the underground bunker at the northeast corner of the New Public Offices building.

The two Home Guard soldiers snapped to attention. The Prime Minister saluted his soldiers as he moved through the entrance and the security anteroom. The stairs down were poorly lit but sufficient to keep from stumbling. Yet, another set of guards kept watch over the start of the L shaped grouping of rooms and hallway. The first door on his right was the War Cabinet conference room. He greeted members of the staff with a smile and a nod of his round head.

Secret Intelligence Service Director-General Colonel Stewart Menzies, otherwise known as simply 'C' in the world of spies, stood as Winston approached his combination office-bedroom. Seeing the intelligence chief with the security case so early in the day meant he brought either good news or really bad news.

"Good morning, Stewart."

"And a wonderful good morning to you, Prime Minister."

"Then, it must be good news."

"You shall be the judge," he said, motioning to the small office. "Shall we?" The two men entered the room. Menzies closed the door behind them and immediately opened his case. When the Prime Minister finished hanging his wet overcoat and hat, Menzies handed him a single piece of paper. "I think you will find this interesting."

MOST SECRET - ULTRA

```
SECRET
DATE 17TH SEPTEMBER 1940
TO OKH OKM OKL
FROM OKW
BREAK
OPERATION SEALION
THE LEADER HAS DECIDED TO POSTPONE OPERATION
SEALION INDEFINITELY BREAK AIR FORCE OPERATIONS
```

```
AGAINST THE ENEMY MUST CONTINUE UNABATED BREAK
AIR CAMPAIGN OBJECTIVES REMAIN UNCHANGED BREAK
THE NAVY MUST DISPERSE THE INVASION FLEET
TO REDUCE VULNERABILITY TO ENEMY AIR RAIDS
BREAK THE ARMY MUST REDEPLOY FORCES TO REDUCE
CONCENTRATIONS AND CONTINUE TRAINING FOR
INVASION OPERATIONS IN THE SPRING
END
SECRET
```

MOST SECRET - ULTRA

The smile on Winston's face strained his facial muscles. "Please tell me this is not a hoax," Winston said into the smiling eyes of Menzies.

"This is one of those cases where we have substantial corroborating evidence. This particular message was intercepted just after midnight and decyphered four hours later. This is the source dispatch. We have numerous other wireless transmissions this morning that disseminate this directive. We have not been able to fly any photo recce missions due to cloud cover over the French and Belgium coasts, but we expect aerial photography to confirm the order. This is obviously very fresh information, and I would suggest we refrain from any celebration. Enemy agents of any kind could sense a dramatic change and link it closely to the postponement. The compromise of Enigma is a consequence."

"Yes, yes, quite right, Stewart, but dear God Almighty in heaven above, we have stopped the foul smelling villain."

"It would appear so."

"How tempting it would be to rejoice at our victory from such dismal beginnings. After the embarrassment of Dunkirk, begging for war supplies from the Americans, and the horrific losses of our youth in Fighter Command, it is so exquisitely delightful to have kept the Hun at bay. Do we have any indications of his next move?"

Menzies returned the Enigma intercept to his locked case. "It is too early to tell, actually. However, from this message," he said holding up the message case, "continued bombing of our cities is a virtual certainty, potentially with even more vigor now that the invasion attempt has apparently failed. The prevailing opinion amongst our analysts is, one of two likely alternatives. The most obvious is a renewed invasion attempt in the late spring when the weather clears and the land dries. Hitler could never feel safe with a free England on a significant flank."

"Agreed, and we shall become a more annoying thorn in his side with each day as we gather strength. The second?"

"While the Germans will continue to consolidate in the Balkans, most probably Greece and Northern Africa, they cannot entertain their real objective, Arabian Gulf oil, without subduing Soviet Russia. We suspect Hitler may turn east, if he or his generals feel they cannot beat us across the Channel. Of course, it could be Raeder who feels he could not fend off the Royal Navy."

"Russia, you say?" Winston uttered, as he contemplated the possibilities.

"Yes, sir, that is one of the two predominate scenarios." Winston rubbed his chin and began a slow methodical pacing in the small room. "My opinion, for what it is worth . . . he would be foolish in the extreme, if he took the second course."

"How so?" asked Churchill.

"He must recognize we will use every day of respite to rearm and strengthen our forces. As long as we remain militarily capable and so close to the Fatherland, he would not likely achieve his professed desire for a united Europe. Furthermore, although the Soviet Union remains pacified and unalert from everything we have seen, the expanse of the Russian countryside just to the Urals and the long, severe winters must be considered major obstacles. He may surprise the Russians, but I doubt he will be anymore successful at subduing Russia than Napoleon was a hundred and a quarter years ago. I cannot imagine Stalin sitting idle as the Germans progress further along their southern flank."

Winston moved slowly back and forth along the length of the small room as he contemplated the possibilities. The image of Hitler turning his attention away from Great Britain and toward the Soviet Union had many attractions, in fact, too many attractions. The thought of substantially reduced pressure and attacks on England meant relief for Britons and a real opportunity to recover from the debacle of the Battle of France. It would be a possibility almost too good to be true.

"Do you seriously think he would be so foolish to attack Russia?"

"I know it is implausible, but it is a possibility, and maybe even a requirement, if he hopes to secure Arabian oil. The Germans will undoubtedly press to control the Mediterranean coastline, but not likely press on into sub-Saharan Africa. Yes, I think we are serious. Russia and oil make sense from his perspective."

"What a godsend that would be."

"Quite right."

"And, thank God for the Royal Navy, if it is Raeder advising against crossing the Channel."

"Indeed."

"I should say you brought good news, Stew. The first *bona fide* good news since the sinking of the *Graf Spee* last winter." Winston nearly danced with an almost boyish excitement. "This is his first real military defeat . . . his first, Stew."

"Yes, sir."

"Anything else?" asked Winston, feeling an urge to share the information with a few of the closest appropriate confidants.

"One last item, sir." Churchill nodded his consent. "We received confirmation from Intrepid last night that President Roosevelt signed the Selective Training and Service Act of 1940, which to me suggests the Americans are taking another step toward joining us."

"I have not read the language, as yet, or heard from the Foreign Office. The bill may just authorize the President to draft men at which time he deems it appropriate."

"Yes, sir, but even so, authorization is better than doing nothing."

"True. We should have official public notification from Lords Halifax and Lothian by Cabinet time this evening. Nonetheless, I suspect your information to be accurate and your assessment to be appropriate."

Edward Frederick Lindley Wood, KG, GCSI, GCMG, GCIE, TD, PC; 3rd Viscount Halifax of Monk Bretton served in the ministerial position of Secretary of State for Foreign Affairs. He was also a member of the Conservative Party and a Peer in the House of Lords, and popularly known as Lord Halifax.

"Thank you, Prime Minister."

"Thank you, Stewart. As I know you will, we must look for corroborating evidence from other sources, excluding Enigma, regarding the enemy's invasion preparations, so we can withdraw the Cromwell invasion alert. Also, we must refocus whatever assets you have to determine his next move. If you can obtain such corroboration prior to Cabinet time, please contact Sir Edward."

"As you wish, sir."

"Smashing, absolutely smashing, but as you say, we cannot display any hint of foreknowledge."

"We shall keep a vigilant eye."

Winston Churchill thanked the chief of the Secret Intelligence Service and shook his hand vigorously. He remained in his office and almost absent-mindedly went through the morning Cabinet meeting agenda with Sir Edward Bridges. He wanted to tell everyone what he saw, or to talk about the incredibly good news he received that morning, but in the end, he wisely chose to contain his elation and renewed optimism. Winston knew the battle in the air would continue to rage, but the removal of the invasion threat changed the situation the British nation faced.

———

Wednesday, 18.September.1940
RAF Middle Wallop
Middle Wallop, Hampshire, England

The second day of being held at Available status, while passing showers kept the airfield just wet enough to make takeoff and landing a serious risk, wore down the patience of the pilots. Pilot Officer Brian Drummond could not forget the afternoon and night spent with Charlotte Palmer at Standing Oak Farm, west of Winchester. The pleasure, comfort, warmth and caring of two days previous made the tension of the wait compounded by the marginal weather more difficult to accept. *Why don't they just release us*, he asked himself many times, *so I can go to her, to feel the same ecstasy and separation from the war.*

Pilot Officer Jonathan Kensington, now Brian's section leader as well as his best friend, joined Brian outside during one of the morning pauses between showers. "Does not look like we shall have a break in the weather," said Jonathan to start a conversation.

"Nope."

"Perhaps the foul weather of autumn comes early this year after such a long, hot summer."

"Perhaps."

"Talk seems to suggest a squadron raid on London tonight, should we be released at a reasonable hour." Brian just nodded his head. "What is on your mind?"

"Nothing."

"That is what you always say, Brian. Now, what is it?"

"I'm not really interested in going to London."

"As you say, then. It seems like you have been off to see your Mrs. Palmer every night this week. Fortunately for all of us, we have not faced multiple sorties, or I'm afraid you would be totally exhausted."

Brian looked at his best friend wondering what point he was trying to make. He had not been with her since Monday. Was he concerned about his health, his flying performance, his squadron socialization, or his love life? He did not particularly want to discuss Charlotte with Jonathan. His new section leader would probably give him some type of lecture about their age difference, or his susceptibility to female charms. Brian just wanted the sanctuary of Charlotte's embrace and caress. "I'm doing OK."

"Certainly, and we have only had one sortie since Sunday and no combat."

"Meaning?"

"Meaning, as a friend, I am concerned about the hours you spend on the road and with her when you should be sleeping."

"It's only been since the rain."

"What happens when the weather breaks and the major raids return?"

"I'll deal with it."

"Will you?"

Brian felt irritation with Jonathan he had not felt before, like an older brother's annoying ridicule of his newly acquired dating. He looked over his shoulder toward the tent to ensure no one else was within earshot. "Look, Jonathan, I love her."

Kensington smiled and appeared to laugh although no sound came out. "That is what you have said about all your women including my sister."

"No need to get personal, here. Charlotte is different."

"May be, however, maybe it is you who is different."

Brian considered the possibilities. In the last year of his life, he transitioned from biplanes to the fastest fighters in the world. He experienced aerial combat beyond any description. He had been shot down three times, nearly died and been rescued by his current lover. He had shot down at least 16 documented enemy aircraft and been awarded the Distinguished Flying Cross by the King of England for his aerial heroism. Only the second woman he had ever known intimacy with was arrested for spying and hanged. The wife of his RAF benefactor and mentor was pregnant with his child. He watched many men die in the skies, including the close impact of Roger Beamish. Yes, he was different, quite different from when they first met in training at RAF Hawarden in North Wales. How much more life could he experience? Would he survive this conflict to tell his grandchildren about the great war and the massive air battle in the summer of 1940, and of so few against so many? He felt safe in Charlotte's arms. There was no crime in that.

The first drops of another shower forced them both to the tent as the morning mail delivery arrived. As Corporal Warren sorted through the mail, handing several official correspondence items to Squadron Leader Darling. Brian knew Jonathan was correct. He was not getting enough sleep. When he was away from the airfield, he wanted to spend all his spare time with Charlotte. Maybe they would not be faced with three or four combat sorties in a single day until the long days of next summer. Maybe he could spend more time with her and not diminish his alertness and focus for success in the air.

Before Corporal Warren could complete the distribution of the personal mail, Squadron Leader Darling stood and knocked on the operations desk. "I have two announcements, if I may have your attention for a moment. First, in addition to the congratulations of two days ago from the Chief of the Air Staff, we have received another congratulatory message in the form of a personal letter to the Air Officer Commanding-in-Chief of Fighter Command from Queen Wilhelmina of Holland. It seems she bore personal witness to Sunday's air battle overhead her cottage."

"Cottage, my arse. Probably a sodding mansion," mumbled Roland Stockard, producing rolling laughter. Even Darling smiled.

"Nonetheless, the Queen wishes to convey her pride, appreciation and respect for our flying to protect this country and her family. Also, Air Vice-Marshal Brand asks that we read the latest epistle from the One One Group commander, Air Vice-Marshal Park. Since we have been involved in some, if not most, of this, you should find it interesting."

HEADQUARTERS, No. 11 GROUP

FIGHTER COMMAND

ROYAL AIR FORCE,

UXBRIDGE, MIDDLESEX

Telephone Nos. UXBRIDGE 2894 (4 lines)

UXBRIDGE 2896 (2 lines)

Telegraphic Address: "AIRGPLON UXBRIDGE."

Reference: -- FC11/P.11233

CONFIDENTIAL

16th September, 1940.

GROUP DIRECTIVE No. 18

Recent operations have illuminated numerous shortcomings.

2. Single squadrons are failing to rendezvous.

3. Single squadrons are being detailed to large raids.

4. Paired squadrons rendezvous too far forward and too low.

5. High flying massed formations of German fighters attract Group fighters while enemy bombers get through to their targets.

```
        6.  Delays in vectoring paired squadrons
to raids prevent effective engagement.
        7.  Errors in sector reports on pilot
and aircraft strengths lead controllers to
inappropriate assignments for interception.
        8.  Efforts must be redoubled to carry
out the tactics set forth in Group Directive
No.16.
```

As ordered by,

Keith Park

Air Vice-Marshal,

Air Officer Commanding-in-Chief

No. 11 Group, Fighter Command, Royal Air Force.

CONFIDENTIAL

"Rather demanding old coot, isn't he?" said Stockard.

"Critical I think is the word," Burns added.

"Looks like 'Tin Legs' Bader is winning the big wing tactics debate," commented Flight Lieutenant Morrow.

"May I remind you malcontents," interjected Darling with some levity to his words, "that Air Vice-Marshal Park, in addition to being a senior officer in the Royal Air Force, is a holder of a Military Cross and a Distinguished Flying Cross as well as a current Hurricane pilot."

"Maybe so, Skipper, but does he know what it's like up there?" asked Stockard.

"He was in a Hurricane above London the night the London bombing began."

The image of a senior RAF officer and commander of the vanguard fighter group flying around in a single seat, monoplane fighter brought a smile to his face. For Brian, it meant he could fly as long as he was able.

Corporal Warren took the pause as an opportunity to complete the distribution of the morning's mail. The letter from his parents, more precisely his mother, surprised him. How was he going to tell his parents that he was in love with a woman eight years older than their son? Maybe he did not have to tell them. This crazy war kept everything uncertain. He had not told his parents about Anne, or Rosemary, and especially not Mary Spencer. There was a difference. This time he wanted to tell his parents. He wanted them to know how happy he was with Charlotte. But, what if something happened? Brian opened the letter.

September 4, 1940

Dear Son,

It is hard to contain our excitement and pride, Brian. Yesterday, we received several telephone calls from neighbors and friends as well as Gertrude Bainbridge about a CBS Radio News broadcast by correspondent Edward R. Murrow. Then, today, we received a telegram from President Roosevelt himself congratulating us on the success and heroism of our son. We just returned from the radio station in Wichita. They played a recording of the News broadcast.

It was like a dream, Brian. We couldn't believe Mr. Murrow was actually talking about our son. He said your name several times. He said you had been awarded something called a distinguished flying cross, or something like that, for bravery in aerial combat, and by the King of England, no less. He also said you were shot down and rescued by a farm woman who was awarded the second highest medal for courage. Mr. Murrow said you and the other six American volunteers represented all that is good in America. But, probably more importantly, he said he talked to you, and you looked dashing and healthy in your uniform. It was in the newspaper as well. We saved all of them for your return.

Anyway, I must hurry. Dad has stopped Mr. Jordan, the mailman, so we can mail this letter today. I hope this letter reaches you quickly and finds you safe. We are proud of you. The whole town is proud of you. We love you very much, Son. Please be careful and come home soon.

Love,

Mom & Dad

"That must be good," Jonathan said nodding to the letter, "the way you are smiling."

"I guess so. It's from my parents." He handed the letter to his best friend who quickly read it.

"Dashing, aye?"

Brian stood up straight, pulled back his shoulder and pushed out his chest. "Why not?"

"So, you brave Americans are saving our beloved England," Jonathan said with a slight edge.

Brian looked into Jonathan's eyes to ascertain the seriousness of his biting remark. "It's not like that and you know it. My mother is just proud of her son. Your mother would feel the same way, if it was you."

A smile washed over Jonathan's face. "I was just kidding with you, you silly old sod."

"I don't think it is so funny. This is serious business, and we're all in this together."

"My, aren't we testy. Of course we are, you twit. Are you going to tell your parents about Charlotte?"

"They all ready know, according to the letter."

"It doesn't say anything about your involvement with that woman."

"I've thought about it," Brian said, as he inspected his boot tops.

"Surely you jest."

"Why? I love her."

"So you said earlier, but it's lust, Brian. It is lust you feel, not love. There is no love. There can be no love with this bloody war going on. A pressing of the flesh keeps us alive inside with all this death around us, don't you know."

"I do."

"I do, what?"

"Love her."

Jonathan shook his head in humorous disgust. Brian wanted to tell him, but he quickly realized their conversation had already drawn too much attention. Any further discussion would only instigate unwanted comments from the others.

The squadron remained on alert through the afternoon showers. They never passed the Readiness level. The pilots were finally released at mid-afternoon. The group decided they did not have enough time for the journey into London, just to drink beer with other fighter pilots. They would go to their favorite local establishment, the Black Swan, in Andover. Jonathan Kensington and Robert Morrow coerced Brian, as the only hold out, to go with them and forgo his evening dose of earthly delights. Brian called Charlotte, who did not hide her disappointment with the change. They made a date to meet in Salisbury tomorrow night at the White Horse Inn. She resisted, but then agreed when she decided to ask her elderly, occasional helpers to cover the evening and morning milking chores. With the arrangements made, the pilots of No.609 Squadron moved up the A343 toward Andover making several brief stops along the way to sample the fare at their usual way-station pubs. Brian joined in, although without the usual enthusiasm as his mind spent most of his conscious time at a country farm near Winchester.

———

Chapter 6

Veni, vidi, vici
(I came, I saw, I conquered)
-- Julius Caesar

Thursday, 19.September.1940
RAF Middle Wallop
Middle Wallop, Hampshire, England

Week 11

The day seemed much longer than it was. The squadron had remained at Available status from 07:00 to 15:15 with only the one step down to Available in 30 Minutes status to allow the pilots some lunch at the Officer's Mess.

For the first time since the end of the Phony War, the pilots of No.609 Squadron began to think there might be an end to the intensity of aerial combat that characterized the extended daylight of summer. The rain that began Monday morning remained as the answer to the collective prayers of the pilots, military leaders, politicians and the Prime Minister. The enormous conflagration of Sunday took on the vestiges of the dying flailings of a mortally wounded and beaten carnivore. They also prayed their impressions were correct and not part of some deceptive faint of an injured but still viable and now angry beast.

Although the pilots did not get to enjoy true rest and relaxation away from duty, they had only flown one sortie with no contact in the last four days. It was the longest period of non-combat since the beginning of the Battle of Britain, excluding Brian's stint in the hospital after being shot down at the end of July. They were all grateful for the break. A new form of euphoria showed through like the relief one must feel when a beating stops. Words of success and even victory crept into conversations, although none of the pilots thought the fight was over. The shortened days along with the cold rain and thick layers of clouds brought a scent of change, a shift in the fortunes of war.

Pilot Officer Brian Drummond still felt the tentacle's grip from the dirty fog of last night's binge with his brethren. The telephone call with Charlotte helped to shake the grip, but only her embrace could vanquish the demons.

Thursday, 19.September.1940
White Horse Inn
Salisbury, Wiltshire, England
16:45 hours

The taxi ride to Salisbury had not taken long, actually comparatively short by wartime standards that Brian knew. The White Horse Inn, a centuries old way-station house now nestled among the shops of the city, represented a brief holiday amid the chaos.

Brian was the first to arrive. He chose a street side room on the top floor of the narrow, four-story hotel. The manager fussed about, like Brian was a war hero honoring his establishment with his patronage. Several citizens, male and female, shook his hand and offered their words of appreciation. Two of the men recognized the violet and white striped ribbon beneath his RAF pilot's wings. Brian accepted the attention although he yearned for anonymity, so he could fade into the background when Charlotte arrived. The rich aroma of freshly baked scones and bread filled the lobby and could not mask the smell of a beef roast nearly ripe for the evening meal.

The RAF fighter pilot spent more time standing to acknowledge the well wishes of people than he did sitting in the comfortable chair near the stairs. Brian smiled and nodded his head, as he talked to an elderly couple when Charlotte walked into the lobby. The solid, medium blue suit brought out the blue tints of her eyes like beacons in the night. The grays in her hair and her eyes disappeared into the blues. Her face radiated life and energy. The red lipstick on her full lips accentuated the brilliant white of her teeth and her smile.

Brian excused himself, walked toward her and swallowed her into his arms. They kissed, instigating the muffled cheers of several witnesses as well as a couple of 'good show' comments. Brian nodded his head and bowed slightly as if accepting the accolades of his audience at the curtain call. The rain stopped, and Brian suggested a walk to allow the attention to dissipate. He wanted her, to feel her, but discretion seemed appropriate.

They walked down the narrow street arm-in-arm. He could feel the fullness of her breast against his arm. The temptation of a boy in a candy shop became unbearable. He wanted the heavenly bliss of carnal union and satisfaction. They meandered, talking only of sights and smells from the shops along the narrow street. The fresh air felt good to both of them.

The two lovers returned to the hotel and were the first to be seated, as the evening meal was served. Neither of them showed much interest in the food although it was well prepared, tasted excellent and was pleasantly served. Without saying as much, they wanted the last of their necessary contacts with the outside world to be completed as quickly as possible. They had other needs to be fulfilled without interruption or interference no matter how well intentioned.

Brian used the last of their public time to show Charlotte the letter from his parents. When she finished and handed the letter back to him, he looked deeply into her eyes. "I also told them about you, about us."

"Brian, you didn't."

"Charlotte, I love you very much, more than I have loved any person in my life. We were made for each other."

"I have grown to love you, as well."

Brian looked around the small, hotel restaurant. Two other couples sat several tables away. He thought about the images in his head when he had written his parents last night before starting the acts of camaraderie. Brian felt the tremors of indecision, sensations quite foreign to the accomplished fighter pilot. He took a deep breath and told himself the risk was worth the reward.

"I don't exactly know what to say. Malcolm always told me to shoot straight and don't try to get fancy. That advice has kept me alive in the air. Maybe it will keep me alive on the ground." He paused to see if she might react. The puzzled expression on her face and the searching of her eyes gave him the extra push he needed. "Charlotte, will you marry me?"

Charlotte Palmer gasped and covered her mouth. Shock filled her eyes, as if she had just witnessed a terrible accident. Brian struggled to determine whether it was the reaction of excited surprise or repulsed horror. He opened his mouth to speak, but no words would sound. He wished he could take back the last sentence. The remembrance of her slap when he kissed her for the first time in the barn at Standing Oak Farm rushed back. *Have I made a dreadful mistake? Have I misread the signs of love?* His inexperience in the ways of the world erupted into self-doubt and regret. Words of apology welled up within him as he fought for control of his emotions.

"I'm sorry, Charlotte," he said finally. "I made a mistake, again, it seems. Please forgive me."

"I will not," she said, behind her hand still covering her mouth.

"What do you mean?" *Dear God*, no, he said to himself. *Please do not take her from me.*

"I won't forgive you," she said as a broad smile bloomed across her face. "I must apologize to you. I did not expect a marriage proposal. It caught me by surprise like your first kiss."

"I shouldn't have."

"Brian, stop," she protested, "or, you will make me feel quite grotesque." It was her turn to look around the restaurant. Brian kept his eyes on her. "I don't know what to say." Her face drained of life. "It has only been three months since Ian died," she whispered, not to him, but more to herself.

"I'm sorry. I shouldn't have said that."

"Will you stop!" she commanded, tensing every muscle in her body. She leaned forward, so she could speak as softly as possible. "Now, I want you to listen carefully. I do not want to hear anymore of these . . . these . . . these infernal apologies. I am simply not accustomed to such directness. It is actually one of the attributes I find so invigorating in you. It is me who should apologize. You have never kept anything hidden. In many ways, I have come to see in you the embodiment of life itself. I love you very much, Brian. I have grown to love you more than I have any man. It is this damnable war that causes me to hesitate." She searched his eyes as he waited for her. "Could I ask you for some days to think things through?"

Brian breathed a sigh of what he thought was relief. "You can have all the time you want. Just don't tell me, no."

She sat back in her chair although a serious expression remained on her face. "I think I know how important your concentration is to your survival. I must do everything to not cause you any distraction."

"You've already said that, so tell me, yes."

"You are a gem. Now, I have a more serious and immediate need that demands your undivided attention."

Brian grinned, winked to her and paid the bill. He followed her up the stairs. She accentuated the movement of her hips to focus him. Brian lost his connection to the rest of the world. They wasted no time reaching for the peaks of pleasure.

———

Friday, 20.September.1940
RAF Middle Wallop
Middle Wallop, Hampshire, England
05:00 hours

Now, Brian felt the lack of sleep Jonathan had talked about on Wednesday. Charlotte and he spent the majority of the night doing what they spent most of their time doing. She understood the necessity for him to leave so early. She knew it was his duty. Charlotte had given him a memorable departure gift before he dressed. Brian had kissed her gently and tucked her back into bed for a few more hours sleep.

Brian had to ring the desk bell several times to rouse the proprietor. After explaining that his lady friend would depart later in the morning, Brian paid the bill and thanked the man for the hospitality of the inn.

Trying to find a taxi so early in the morning proved more difficult than he had expected. He considered several times going back up stairs to wake Charlotte to ask her to drive him back to Middle Wallop. The sleepy, barely coherent, night manager eventually found a driver. The time to make the short journey from Salisbury required him to be dropped off at the dispersal tent.

The other pilots were already at the dispersal tent when he arrived to a scowl from Squadron Leader Darling, as he called in his unit's readiness for duty. No one made any comments.

No.609 Squadron bounced from Available to Readiness status several times during the morning. The showers disappeared although the cloud cover remained. The grass field that served as their take off and landing area dried quickly in the light, steady breeze. The fighter base was ready for operations. Brian used every available minute to sleep and recover his alertness. With the slowly improving weather, combat could not be far off. It was not until after lunch that Brian finally talked to someone.

"Where were you last night?" asked his friend. "With her?"

"Yes."

"Brian, you simply must get a grip on this. You know as well as anyone what distractions do to fighter pilots. Slow this down a little."

"Jonathan, we haven't been flying. The weather has been crappy. She makes me feel good – alive."

"I am certain she does, but she can also wear you out and give you something other than killing Germans to think about."

"Don't worry. I'll be OK."

"You've been asleep most of the morning. What if we had to launch and had two or more sorties by now? Would you have been ready?"

Brian knew his friend was precisely correct. If he had flown in the morning, he would not have been at his best. The feelings for Charlotte grew stronger by the day, by the moment. He wanted to spend all his time with her. She was a physical person, who loved the subtlety, splendor and pleasure of touching. She was gentle and caring. Even the depth of his experience with Anne Booth or the energy of Rosemary Kensington could not cast a shadow upon Charlotte Palmer. They were good for each other. They needed each other.

"All right. I'll slow it down. It looks like the weather is breaking, so we should be back in business soon."

"As you say."

14:17 hours

They stepped to Readiness status.

14:33 hours

Standby.

Brian greeted each of his ground crew, Leading Aircraftman Bernard Gordon, his crew chief; Aircraftman Jordan Toldson, his rigger; and Aircraftman Colin Jenkins, his armorer. The respite of the last week brought freshness to their faces, and energy to their step and attitude. Brian had been strapped into his cockpit for several minutes when the launch command came. They took to the air in short order.

"Bandy, this is Sorbo Leader calling," radioed 'Spike' Darling, as the nine 'PR' Spitfires climbed at full power through 1,000 feet.

"Sorbo Leader, this is Bandy. Vector two oh five, angels two six. We have a fifty plus raid inbound."

"Roger, Bandy."

And so it begins, Brian said to himself. He rechecked his oxygen mask connections as well as the arming of the fighter. A raid of 50 intruders was substantially less than the thousands they had seen earlier in September. This actually might be fun. The odds were certainly better.

The squadron leveled off at 26,000 feet. Most of the clouds lay below them although a thin layer of wispy, Cirrus clouds dulled the sun. As they closed with their intercept point, the cold of high altitude flight refreshed Brian's memory of the hazards in the thin, cold air.

"Sorbo Leader, this is Bandy calling."

"Go ahead."

"We have reports from the east of large, fighter only, sweeps. You are requested to identify your bandits prior to engagement. If there are no bombers, you are directed to remain clear and not, repeat not, engage enemy fighters."

"Bloody hell," someone radioed.

"I understand we are only to engage if bombers are present," responded Darling.

"Correct, Sorbo Leader."

"Understood."

The same thoughts had to be running through the minds of all of the pilots, as they were with Pilot Officer Drummond. What the hell were they doing flying around German fighters and not trying to down as many as possible? Were they not in the business of shooting down enemy aircraft regardless of type? Why were they being prohibited from engaging if no bombers were present? Brian wondered what they would find.

In a few miles, they had their answer. "Tally ho," broadcast Darling. "We have 50 to 60 One Oh Nines. No bombers."

"Roger, Sorbo Leader. Stand clear. If they make any move to attack ground targets, you are clear to engage. If not, I repeat, stand clear."

"Roger."

Four squadrons of Bf109E-4 fighters spread out across the sky ready for a fight. Brian scanned the whole sky around them. No bombers. Squadron Leader Darling altered course several times to avoid an intercept as well as take up a position several miles away on the left, rear quarter flank of the German formation. They could see another Spitfire squadron off the right flank. The Germans tried several times to engage the British aircraft. They upheld their orders although the temptation was great. Seeing German fighters so close and yet not be able to mix it up with them produced a whole new combination of anxiety, frustration and curiosity.

The Germans turned east for ten miles or more before turning south to return to their bases in Northern France. No.609 Squadron tailed the Germans until they were about halfway across the English Channel. The urge to follow them to their base and attack them as they tried to land added some tension to Brian's thoughts. It would sure be nice to take the fight to the enemy instead of waiting for them to come all the time.

As soon as they landed at RAF Middle Wallop, Squadron Leader Darling walked smartly to the dispersal tent. He was on the telephone listening. He had spoken his words before Brian walked into the tent. Several additional partial questions interspersed the listening, and then signed off.

"So, what's the deal, Skipper?" asked 'Red' Burns, on behalf of the others.

"It appears Gerry has adopted a new tactic to lure our fighters up. The thinking is, they want to deplete our fighter resources without risking their bombers. One One Group had two waves of larger fighter sweeps. As tempting as those sodding bastards are, we must not give them the satisfaction. As happened this afternoon, they did no harm. We are to preserve our fighter resources."

"Precious few," said Davies.

"Yes, quite right, precious few, which is why we are to save our bullets for the bombers or the invasion fleet."

"Helluva deal, that's all I can say," commented Burns.

"Yes, well, we have our orders."

"Mister Darling," interrupted Corporal Warren. "A warrant was delivered this afternoon."

Darling retrieved it from their operations clerk. Warrants usually meant promotions although the squadron had not seen one in many months. They sometimes meant awards. Brian wondered who it might be.

The squadron leader knocked on the wooden desk, as if it was a door. "It seems we have a promotion. Mister Kensington, you have been promoted to Flying Officer and officially assigned as the Green Section Leader."

Darling shook his hand first. Brian was second. They would have a good reason to celebrate tonight.

"One other bit of relevant news, if I may." Darling waited for the attention of everyone. "The Air Ministry informs us of the formation of a new fighter squadron, yesterday, designated Seven One Squadron. It will be the first of what they expect to be several squadrons comprised of American volunteer pilots. They will join the Czech and Polish squadrons that are just now becoming operational. Since we have both Czech and American volunteer pilots, we are notified to expect orders soon, transferring said pilots to their respective new units."

"Well, I'll be," pronounced 'Red' Burns.

"That will leave us with half a squadron," observed 'Waggle' Davies.

"Yes, well, I would presume the Air Ministry has thought this through, and they are aware of replacement pilots to fill the seats.

"How soon, Skipper?" asked 'Red.'

"The message does not say . . . only soon."

The news gave them all, more to think about. Brian did not want to leave the only operational squadron he had ever known. He recognized that things change, but he did want it to happen. His potential reassignment would probably take him farther away from Charlotte and his brothers-in-arms. His mind ground through so many possible obstacles rising up between them. Brian did not want to tell Charlotte about all this and certainly did not want to tell her he would not be able to see her tonight, but he also knew he had to tell her. They had made no specific plans, other than to take every moment together they could find. She would understand . . . he hoped.

Friday, 20.September.1940
Cabinet War Rooms
New Public Offices
Whitehall, London, England
14:45 hours

The dull, muffled thuds of bombs exploding not far away kept Prime Minister Churchill underground in his small but functional office. John Martin and 'Jock' Colville kept a constant stream of paper moving across his desk. The paper of governance had to be served, and Churchill's sense of history kept his curiosity active and well fed. Winston relied on his private secretaries, who were quite proficient at reading documents, sorting, cataloguing, organizing and presenting the paper life-blood of His Majesty's Government, as well as the personal communications between Winston Churchill and friends, world leaders, concerned citizens and acquaintances. A relay of cleared, government stenographers kept pace with the Prime Minister's frenetic dictation regarding the daily communications. Many of the decisions were simple statements: "Action this day," "Study & recommend," "Solutions?" and "Work with so & so." Others were more considered responses, especially to friends and acquaintances. Martin and Colville were relentless with their follow-up on designated messages. They had completed the daily dose shortly after lunch.

The Prime Minister planned to leave early for the peace of the country at Chequers and a more comfortable environment for the social side of his calendar. 'Jock' Colville and two of the stenographers would accompany him to keep up with communications for the short weekend in the country. His personal staff would precede him to Chequers. As usual, Inspector Thompson and General Ismay would ride with him for the hour or two, automobile journey to the country estate.

Sir Edward knocked on his door-frame and announced, "The War Cabinet has assembled, Prime Minister."

"Excellent. Thank you, Ed. Let's make quick work of today's agenda."

"Yes, sir. We should have you out of here by four."

Churchill nodded his agreement, stood and followed Sir Edward down the left passageway, around the corner and to the simple doorway, not far from the security lobby, and into the Cabinet Conference Room. Only the War Cabinet and the Secretariat staff were present.

"Let us get started," Churchill said, as he moved to his chair at the center-head of the U-shaped conference table.

Sir Edward jumped in. "We are missing the Home Secretary this afternoon. However, I believe the Foreign Secretary can update the War Cabinet."

"Yes, certainly," responded Lord Halifax. "Sir John intended to be here, however, he is attending to several important matters within his domain. He informed me of his status an hour or so ago, so that I might inform the War Cabinet on his behalf. A German spy parachuted into a farmer's field last night on the outskirts of Willingham, Cambridgeshire. Local farmers happened to observe the man's descent and detained the man until the constabulary officers arrived. The spy was immediately transfered under guard to MI5, who in turn moved him to Wormwood Scrubs Prison for interrogation. As of an hour ago, the Security Service believes they are making progress with the man. Both Sir John and Brigadier Harker agree with the interrogation team and the chief of Department Twenty. This fellow may be a candidate for the Double Cross program."

Brigadier Oswald Allen 'Jasper' Harker, CBE, served the King as Director-General of the Security Service otherwise known as MI5. The MI5 director-general was often referred to by the simple initial 'K' in similar fashion as the director-general of MI6 was referred to as 'C.' Only a few of the senior leadership of His Majesty's Government knew that Harker's position was temporary until David Petrie completed his in-depth study of the Security Service and accepted the assignment, at which time Harker would return to his post as deputy. MI5's Department 20 was also known as Department XX, using Roman numerals, and thus provided the objective of the group – turning enemy agents for Allied intelligence purposes and often colloquially referred as the Double Cross Committee. As war appeared inevitable last year, MI5 moved their operations out of Whitehall to new buildings on the grounds of Wormwood Scrubs Prison on the outskirts of London, to keep operations of the Security Service functional in the event of aerial bombardment of the capital city.

"If he was only captured last night, turning him seems a bit premature," observed the Right Honorable Arthur Greenwood, Member of Parliament for Wakefield – Minister without Portfolio and Deputy Leader of the Labour Party.

"At face value," interjected Churchill, "you would be correct. However, I suspect they have heard responses early in the game that lead them to believe he will be cooperative and malleable. These are professional men who have been doing this sort of thing for quite some time now. I suggest we let them do their job. They are by nature a suspicious lot and they will undoubtedly test this enemy agent more than a few times to establish his reliability for such work."

"Double cross, indeed," Greenwood mumbled.

"Yes, quite the program, I must say. The program may pay off handsomely as we march to victory."

"We are not out of this fight, yet, Winston," cautioned Attlee.

"You must admit, Clement, we appear to have turned the corner in the current fracas."

"Perhaps, but we must be cautious not to ascribe our deliverance, just yet."

"I have another tragic bit of news, I'm afraid," Lord Halifax said, bringing them all back to the agenda. He waited for their attention. "The Foreign Office received notification from the Admiralty that the *City of Benares* was sunk three days ago."

"We lose ships every day," Churchill growled.

"Yes, this one demands special attention. The ship was making poor headway in a gale when a U-boat managed to put two torpedoes in her side near midnight. The weather and darkness made the abandon ship process enormously more difficult. Another ship – the *Marina* – was also hit and sunk at about the same time. HMS *Hurricane* remained on scene for 24 hours searching for survivors. The convoy – OB-213 – had to continue to avoid losing more ships. There are still too many missing, I am afraid. The First Lord and I confirmed the details by separate means, and we are very sad to report that the *Benares* was bound for Canada and transporting 90 evacuation children under Operation PIED PIPER."

A deathly silence fell upon the men. Only the distant hum of the air handling system could be heard.

Attlee was the first to speak after several minutes of personal contemplation. "That program was supposed to keep our children safe."

"You say three days ago?" asked Churchill.

"Yes. Coastal Command has sent several aircraft to search the area without success," Halifax answered and added.

"Bloody Germans," mumbled Greenwood.

"Such loss is not tolerable," Churchill proclaimed. "Unless there is any objection, we must terminate any further ship-borne transport of our evacuation children – too risky for the children. We must find another way to protect them from the ravages of war."

"With the Germans expanding their bombardment of our cities, they surely cannot remain in any city of more than village size," Attlee said.

"And, the country homes have so generously welcomed our children, but they are at capacity," added Greenwood. "If we were to build camps for the children, they would become targets."

"Arthur," Churchill said, "would you be so kind to work with the Home Office to find some other means to evacuate our children from the cities without sending them aboard ship to Commonwealth countries. The *Benares* must be the last tragedy of its kind. I know the Admiralty is hard pressed to escort these vital convoys, so I am afraid we must challenge Coastal Command to redouble their search efforts. We simply must hope there are still survivors to be rescued, and they do not have much time in the North Atlantic."

The Prime Minister nodded to Sir Edward, who noted the action to be taken on behalf of the War Cabinet. There was not much enthusiasm for the remainder of the agenda. The War Cabinet concluded their afternoon meeting. After a few short matters with Sir Edward, Churchill had Colville call for his car. Inspector Thompson met Winston and 'Jock' at the limousine. They were off to Chequers.

———

Friday, 20.September.1940
RAF Middle Wallop
Middle Wallop, Hampshire, England
16:15 hours

The squadron launched one more time that afternoon. The reported medium altitude of the raid meant their targets would be in and out of the clouds. They spent most of the flight zigzagging around the sky trying to find their adversaries. More than an hour passed until they spotted what looked like a squadron of Do17s with a very light fighter escort heading south and several miles away. The 'PR' fighters throttled up to full power in pursuit. The German fighters either did not see the approaching British defenders, or sensed there was not much they could do. The closure rate was very slow. The German bombers and fighters were light and much faster than usual. With less than a mile to intercept from the rear, the German fighters began to deploy. Brian checked his instruments. He had only 20 minutes of fuel remaining – not a healthy condition to start a fight. Squadron Leader Darling recognized the situation and broke off the attack.

Two sorties with the enemy within reach and not one bullet had been fired, spending all that time and effort only to return with the gun-port tapes still intact did not feel right. The frustration passed quickly as they landed and taxied to their parking spaces.

The squadron pilots ate evening meal together. Brian called Charlotte Palmer. As expected, she was disappointed but understood Brian's absence. They drank their beer in the Officer's Mess bar, congratulated Flying Officer Jonathan Kensington and traded observations on the day's frustration. The pilots did not last long as fatigue began to claim their consciousness.

———

Friday, 20.September.1940
The White House
Washington, District of Columbia
United States of America
20:30 hours

The President of the United States of America asked for the small, private, evening meeting with Brigadier General George Veazey Strong, Assistant Chief of the War Plans Division and Head of the U.S. Military Mission to the United Kingdom. Roosevelt asked for only four other people to attend the meeting. Beside the host and principal guest, the attendees included:

-- Secretary of War Henry Lewis Stimson, a Republican politician from New
    York, confirmed by the Senate two months earlier;

-- Harry Lloyd Hopkins, special adviser and long-term confidant of the President;

-- Colonel William Joseph 'Bill' Donovan, a Medal of Honor recipient, New York lawyer, and also known as 'Big Bill' to the British and 'Wild Bill' to his detractors; and,

-- William Samuel Stephenson, MC, DFC, the round faced, but trim former Canadian newspaper publisher, and now special envoy from Prime Minister Winston Churchill. Stephenson came to be an important link between Churchill and Roosevelt. The Prime Minister referred to him as 'Intrepid,' while many others called him 'Little Bill' in counterpoint to Donovan's size.

After the introductions and coffee service, the President began the informal meeting once the door closed. "I appreciate the extra time from each of you this evening to permit this little chat. I invited General Strong to give us a briefing on his assessment of the situation in England. Now, General," Roosevelt said, looking directly to Strong, "I understand you know both Bill Donovan and Bill Stephenson."

"Yes, sir. I met them during Colonel Donovan's visit to England in July."

"Excellent. What you may not know is, I sent Donovan, and Stephenson accompanied him, to learn about British counter-intelligence operations and to absorb what he could about the viability of the British situation. I say this because I need your frank perspective, and I want you to have my personal assurance that everyone in this room is fully aware and involved in this question. I also asked Bill," he said, nodding to Stephenson, "to join us so he may pass first hand information back to Prime Minister Churchill. I want Mister Churchill to know what we are thinking. This is not an inquisition, George," everyone laughed, "but rather, an informal situation assessment, which is also why I did not invite Lord Lothian, I might add. I would encourage each of us to be frank and forthright, and to ask questions as they come."

"Yes, sir. I understand."

Roosevelt looked into the eyes of each man. "Any questions?" Each of them shook their heads in the negative. "Good, then, General Strong, if you would begin."

"Thank you for the opportunity, Mister President. As you know, I have been assigned as an observer in England since just after the invasion of France. Several senior officers including Brigadier Emmons of the Air Corps and Rear Admiral Ghormley as well as a dozen or so field grade officers, off and on, have been warmly received, with every door opened to us by His Majesty's Government. The acceptance of our presence especially in light of our acknowledged assignment has been most impressive. I would like to personally convey my appreciation to Prime Minister Churchill, Air Minister Sir Archibald Sinclair, and Chief of the Air Staff Air Chief Marshal Sir Cyril Newall. Without their support, we would have a vastly incomplete view of the military and political situation."

The Right Honorable Sir Archibald Henry Macdonald 'Archie' Sinclair, Bart, PC, CMG, Member of Parliament for Caithness and Sutherland and 4th Baronet of Ulbster had been Secretary of State for Air from the outset of Churchill's premiership, the ministerial leader of the Royal Air Force, and he remained as Leader of the Liberal Party.

"Please pass my sincerest appreciation as well," Roosevelt said to Stephenson.

"I will, Mister President."

Roosevelt nodded, smiled, and then nodded to Strong to continue.

"I will start at the top, and we can work our way into the details, as you wish," Strong said, and received a nod from Roosevelt. "From all the information we have, it appears they have won the Battle of Britain. Recent aerial photographic reconnaissance indicates Hitler has postponed the invasion attempt. Forces in Northern France, Belgium and Holland are being dispersed. Although Fighter Command was nearly decimated and hanging on by a thread, they are now gaining strength. The turning point had to be the decision of Hitler or Göring to shift their targeting from Fighter Command resources to London and the other major cities. It gave the British just enough room to conserve their drastically dwindling numbers of fighters and pilots, and in fact, rebuild their strength. As of the moment of my departure, yesterday, Fighter Command recovered about 10% of its strength. The low point was near the beginning of what the newspapers are now calling the Blitz, the bombing of London and the cities. We see considerable evidence that they won."

"Considering the state of His Majesty's armed forces three months ago, this is nothing short of miraculous – a genuine tribute to the courage of the pilots," said Stimson.

"Some of your citizens as well," added Stephenson.

"Yes, indeed. Does anyone know how the American volunteers are doing?"

Strong cleared his throat. "Seven participated in the battle proper. Four of them joined from the remains of *le Armée de l'Air*. Four have been killed in action."

"How about the young Kansan recognized by Ed Murrow?"

"What was his name?" asked Strong.

"Drummond, I believe," Stimson answered.

"I believe he is doing OK with his British squadron. I will check on all of them when I return, Mister President."

"Mister Churchill asked me to meet him," added Donovan. "Both Bill and I received a briefing from him on his Spitfire fighter airplane – a most impressive young man, I must say."

"Excellent. Someday, I would like to meet that young man. Ed Murrow was sure impressed by him as well, and Ed is not easily impressed. I also sent him a personal note on his accomplishment."

"We'll see if we can make the arrangements when it is appropriate, Mister President," answered Stimson.

"Thank you, Henry." The President turned back to General Strong. "Are we learning as much as we can about successful fighter tactics against the Germans?"

"Yes, thanks to the generosity of the British."

"Have we made sure this knowledge, this experience, is being incorporated in our pilot training courses?"

"Yes, sir."

"I would amplify General Strong's response," said Stimson. "We have active communications channels at various levels in all the services. We are learning as much as there is to learn. We are far stronger today than we would have been without British cooperation."

"The Germans continue to bomb London and other major cities. The daylight raids are diminishing, while the night attacks are increasing. The British newspapers have begun calling the bombardment 'The Blitz.'"

"This is the face of modern warfare," Roosevelt observed.

"Unfortunately, so it appears . . . total war, I believe they are calling it. The destruction is truly tragic, so indiscriminate. According to the Air Ministry, the Germans have perfected an electronic means of conducting night and foul weather bombing with fairly good accuracy . . . if we call hitting a city the size of London, hitting the target. They apparently use intersecting radio beams to establish their aiming point. They call it the Broken Leg System."

"*Knickebein* is the German word," interjected Stephenson.

"That's it. The British are working feverishly on counter-measures and night-fighter intercept equipment and procedures, but they have not found the trick, as yet."

"So, they will continue to suffer this outrage?"

"I am afraid so, Mister President. There is no sign the Germans will alter their current operations. However, I think most of us who were there on the

7th would say we were shocked when the Germans changed their focus. They were within reach of air superiority by their concentrated attacks on fighter resources . . . so close, yet so far. At any rate, we did not see their change of tactics coming on the 7th, so I doubt we will see any change coming on their current bombing of the cities."

"How long?" asked the President.

"There is no way to know, sir. Based on their nightly commitments and their available resources, these night raids could go on indefinitely. Weather is already beginning to deteriorate toward winter, so any invasion attempt will likely not occur until at least next spring. Thus, my guess is, they will press these night raids through the winter in hopes of cowing the British."

"Is that likely . . . or even possible?"

"In my opinion, sir, not a chance in hell . . . respectfully, Mister President. Mister Churchill has and continues to do a masterful job of inspiring the people and the military. They will take it, and while they take the senseless civilian attacks, Fighter Command will rapidly gain strength. By next spring, Fighter Command may well be stronger than they ever have been and may not be beatable. I have the feeling the Germans missed the only opportunity they will ever have to subdue England. The wild card in all this is the U-boat threat. The losses at sea are painful and hurting the British. They simply cannot sustain this loss rate. To be candid, Mister President, they need our help, especially with this horrific submarine threat."

"Are you familiar with the Tizard Mission, General Strong?" asked the President.

"I am aware the British government initiated a technical exchange program, but not much more, I'm afraid."

"Well, Secretary Stimson will correct me, if I get this wrong, but Sir Henry Tizard is leading a team of technical experts, who brought along quite a few physical examples of their technology. The first, full, joint meeting was held a couple of weeks ago. Van Bush tells me everyone has been overwhelmed by the technology and the exchange is progressing rapidly. As I understand the plan, after a few days as a joint team, they have since broken up into eight sub-teams and dispersed to various sites to complete the technical detail transfer. I tell you this, General Strong, because it is our intention to share our technology with our British cousins – Sir Henry's team – and to work together with the British to find solutions to operational problems like finding U-boats and intercepting night-bombers."

"Thank you for sharing that information with me, Mister President. We can only hope they find those solutions soon enough."

"Indeed. I am confident we will find solutions in short order. We will also help with the mass manufacturing of the developed equipment as quickly as humanly possible."

"Excellent."

"If I may, one of the items on the Tizard list was only generally covered. How is Frank Whittle's research with this new turbine engine technology progressing?" asked Roosevelt, catching Strong somewhat by surprise, not expecting the President of the United States to be aware of the experimental engine program.

"Slow but steady, would be my assessment, Mister President."

"When do they expect to fly this so called jet engine?"

"The best estimates are next year, maybe about mid-year."

"Haven't the Germans already flown a jet aircraft?"

"Yes, sir," answered Donovan. "The Heinkel factory flew a turbine-powered prototype monoplane last year . . . August as I recall."

"Correct," added Stephenson.

"When do we expect the Germans to have an operational version?" Roosevelt asked Donovan.

"Progress appears to be slow for them as well. A goodly portion of their production resources has been devoted to Messerschmitt and current fighter variants. Maybe late next year or the year following at their current rate."

"A jet-powered aircraft would give one side or the other a decided advantage, would it not?"

"Yes, sir, without question."

"What is the next move?" asked Roosevelt.

"Mister President," interjected Donovan sharply, "to address that question I would like to recommend a smaller audience."

Roosevelt nodded his head in recognition of Donovan's security concern. "Are there any other questions for General Strong?" he asked. There were none. "Thank you, George, for your precious time. Please keep up the great work over there for all of us."

"Thank you, Mister President."

"Harry, if you would excuse us," Roosevelt said, signaling Strong and Hopkins they were being asked to leave. The President thanked General Strong for his time and the briefing. They waited until the door was closed. "All right, then, Bill, what do you have?"

"First, let me say, 'Intrepid' has been most generous helping us expand the intelligence cooperative efforts. Second, Bill, please feel free to add your comments. We have an unprecedented view of the Axis Powers. Both

ULTRA and MAGIC have confirmed key elements. General Strong did not know where the other source of information came from, but ULTRA clearly established the invasion postponement two days ago. However, MAGIC yielded just this morning a strong indication the invasion would be initiated at 3 PM on the 23rd."

MAGIC was the classified code name for intercepts of Japanese government and military communications decoded by American intelligence groups. MAGIC was roughly equivalent to ULTRA for the British. The deciphering process proved more difficult and less productive for MAGIC than ULTRA, and ULTRA was quite daunting. Fortunately, for the British, the Germans were more structured, less suspicious and less cautious. The Japanese seemed to always worry about the integrity of their codes, while the Germans maintained a sense of invulnerability regarding their Enigma cipher machine and their code wheel sequences.

"Which one is correct?"

"We think ULTRA."

"Then, what is happening with MAGIC?" the President asked. He considered the consequences of losing the information from the Japanese Purple code machine, the counterpart to the German Enigma device. Purple and Enigma, MAGIC and ULTRA, would be their salvation, if their existence could be protected. "Have the Japanese figured out we have the diplomatic codes?"

"We don't believe so, Mister President. Although the MAGIC intercept does not refer directly to the invasion of Britain, it does leave everyone with that impression."

"Is it possible the Japanese may execute their own invasion on the 23rd?"

"It is possible."

"Where?"

"Could be Indochina, Singapore or Indonesia, we can't be sure."

"Then, let's get this MAGIC intercept to Winston."

"Yes, sir. I would suggest either Stephenson or me carry the message directly."

"Does your suggestion mean we still don't have confidence in the security of our communications link with SIS?" asked the President.

"We restructured the codes, Mister President, after the Kent affair," Stephenson stated. "We are trying to build confidence, but are not quite there according to 'C.'"

"That traitor Kent and the prostitute spy ring in London did more damage than we thought," commented Donovan.

"I agree with Bill." The President looked to 'Big Bill,' and then 'Little Bill.'

Stephenson took his cue. "I need to make the journey for other reasons. I shall be happy to carry the MAGIC intercept."

"Very good."

"We need a highly secure, closed, communications link to connect ULTRA and MAGIC for all of us to gain the most benefit."

"Agreed."

"Is there anything else on this evening's agenda, Henry?"

"No, Mister President, that should do it. Miss LeHand indicated before she left that Mrs. Roosevelt is waiting for a night cap with you."

"Then, I should not keep the lady waiting."

———

Saturday, 21.September.1940
RAF Middle Wallop
Middle Wallop, Hampshire, England
10:30 hours

The weather continued to improve as the sun rose from the eastern horizon. Only scattered clouds drifted toward the northeast. The only excitement in the morning was the cancellation of CROMWELL, the invasion imminent or underway condition, with Alert No.2 being reinstated. Invasion was still probable according to the intelligence community. The pilots of No.609 Squadron enjoyed the sun and cooler air. The whole situation seemed to be improving.

'Hunter' Drummond sat outside the dispersal tent with 'Harness' Kensington, 'Crazy' Kradilcek and 'Red' Burns. It was the first occasion in many weeks, if not months, they talked about things other than flying, the war or women. Oddly enough to Brian, they talked about the birds filling the air with the music of nature, and the deep, rich greens and browns of the trees, fields and hills surrounding the fighter base. RAF Middle Wallop sat at the southeastern edge of the famous Salisbury Plain. The wealth of history embedded into the region made everything seem mature . . . fully developed . . . deeper.

The telephone rang. An instant later, Squadron Leader Darling yelled, "Scramble Green Section." The three pilots grabbed their kit and leapt toward their machines. "Vector oh nine five, angels two. They have a low flier," he shouted to the backs of the running pilots.

They barely taxied halfway up the hill, just enough to give them plenty of takeoff room toward the east. The three Spitfires roared across the field.

Wheels came up immediately as they broke ground. They kept the fighters low grabbing speed instead of altitude. Jonathan started a slow climb to 2,000 feet.

"Bandy, this is Sorbo Green Leader," 'Harness' radioed. "We have the vector."

"Roger, Sorbo Green Leader. The Observer Corps spotted a single Junkers Double Eight. They estimate he is at cherubs two and heading toward Farnborough or Brooklands. Your heading is good for intercept. Good hunting."

"Roger, Bandy." Jonathan crossed his fist several times in front of his face signaling the two wingmen to take a combat spread formation. They needed three sets of eyes to find a solo, low-level attacker.

Small pockets of fog remained in low-lying areas. It was over a patch of what looked like white cotton the slate gray bomber revealed itself to Brian. "Sorbo Green Leader, tally ho. I've got him several miles just south of us."

"You are in best position, Green Three. Go get him 'Hunter.' We will cover your tail."

Brian rolled his Spitfire to the proper pursuit line, pushed his throttle through the emergency gate and pointed the nose down to gain speed. The Ju88 was the fastest of the three types of German medium bombers. He would need all the speed he could gather up.

The only German gun that could engage him was the dorsal gunner just behind the cockpit. A single, 7.9mm, MG15 machine-gun would be no match for the Spitfire's set of eight 0.303 machine-guns. The German gunner must have been very nervous; he opened fire far too early. The bomber pilot held his course, showing more discipline than his gunner. Brian adjusted his intercept line, as he closed with the bomber. The streams of tracers from the bomber's dorsal gun arced toward him. Brian placed his gunsight pipper at the left wing root where the wing joined the fuselage, checked his slip ball centered, and depressed the red firing button on his control spade. All eight Browning machine-guns burst to life. Impacts walked up the tail. The left engine burst into long, orangish-yellow, tongues of flames with a trail of black smoke.

Brian did not want to overshoot his prey. He barrel rolled over the top of the bomber keeping the now wounded aircraft in sight. He quickly settled his fighter into the right flank of the bomber. The dorsal gunner on the bomber did not have time to adjust his position when Brian opened fire. No more tracers came from the bomber's guns.

The German pilots quickly determined their diminished state. The pilot dove the bomber into a valley. He found a patch of fog and purposefully flew into it. The risk was incalculable. Brian pulled up and away from his target. He throttled back slightly to avoid overrunning the bomber. He fully expected a large fireball and explosion to erupt from the fog when the bomber impacted the ground or some unforeseen obstacle. He had only the hum of his Merlin engine and the rush of the air-stream past the canopy.

Brian throttled back a little more and gained a few hundred feet more altitude, as he searched the edges of the fog layer for the German bomber. Nothing. Just the white of the fog. Could the bomber have impacted without an explosion? Brian did several quick S turns to check behind him.

"Two o'clock, 'Hunter.'" radioed 'Crazy' Kradilcek.

Brian rolled back to the right sharply. There he was just above the top of the fog off to the right about a quarter of a mile away. Brian pushed his throttle back up to emergency power and quickly adjusted his flight path. The two other Spitfires trailed Brian by about a mile and another thousand feet of altitude. Brian watched the bomber. The bugger was either very foolish or a very committed man as he pressed toward his target. They were less than five miles from RAE Farnborough and about eight miles from the Brooklands Hurricane Works. This guy was really determined, and Brian had to stop him.

With the bomber's gun out of action, Brian moved quickly to the spine of the intruder. He watched the sparks of the multiple bullet impacts march toward the cockpit. In a flash, the bomber exploded scattering itself into thousands of pieces. Brian pulled up sharply, rolled and pulled harder. Fragments sprayed his fighter. The aircraft shuddered but kept flying.

"Bandy, this is Sorbo Green Leader calling. Splash one bandit," Jonathan radioed with a calm, solemn voice.

Brian looked back over his shoulder at the cloud of black smoke and brown dirt that marked the finale of the skirmish. "That one's for Roger," Brian said aloud into his oxygen mask. He rejoined the other two fighters. Jonathan looked him over. Brian appeared to be bleeding oil from the vicinity of the oil cooler. He throttled back to a slow cruise power setting, and Brian opened up his coolers. He watched the coolant temperature, and oil pressure and temperature. He made it back to RAF Middle Wallop without exceeding any limitations.

He inspected the aircraft along with Leading Aircraftman Bernard Gordon. The aircraft had multiple shrapnel hits, and there were numerous oil and coolant leaks. No serious damage had been done.

"You didn't really have to press in that close," Jonathan whispered to Brian, as they walked toward the tent.

"He was getting too close."

"Maybe, but you were incredibly lucky the explosion did not take you out as well. You need to be more careful."

"I suppose."

"It was for Roger."

Brian stopped and looked into his best friend's eyes. "You heard that?"

"Yes. I held 'Crazy' back. I thought you needed the victory for that reason."

"Thanks, Jonathan. It did feel good."

"Now, just don't do that again. You understand?"

Brian started walking again. "Yeah, sure."

The squadron had been released before Green Section completed their debriefing. 'Crazy' beat them to the Officer's Mess bar and had already begun to tell the story for the other pilots. It was the barrel roll repositioning maneuver so low to the ground that captured the imagination of the pilots. No.609 Squadron had another reason to celebrate. In addition to Brian's 17th confirmed victory, he had completed the exorcism of his demons and avenged the death of Roger Beamish. Brian enjoyed the moment, drinking beers with his friends and colleagues.

Saturday, 21.September.1940
Chequers Court
Ellesborough, Buckinghamshire, England
20:30 hours

Dinner was dispatched with quickly, which did not do justice to Mrs. Landemare's cooking, but Winston was eager for the comfort of the library's vintage, Georgian, wing-backed, leather chairs, a generous snifter of Hine cognac, and his favorite *Romeo y Julieta* Cuban cigar. Each of his guests partook of the evening's post-meal offerings. His audience this evening was unusually small.

The featured guest in Churchill's mind was Air Chief Marshal Sir Hugh Dowding. He wanted to get the venerable leader of fighter pilots away from his staff, away from the trappings of authority, away from Sir Hugh's entourage as well as his own, and into a more relaxed social environment. With the signs of the present battle finally beginning to turn positive, he wanted more personal insight into the man's thinking.

General Lord Gort, 6th Viscount Gort of Galway, John Standish Surtees Prendergast Vereker, VC, GCB, CBE, DSO, MVO, MC, added an intriguing complement to Sir Hugh, since he felt the sharp end of Dowding's compelling argument to not send ten, precious, additional, fighter squadrons to France as they fought valiantly to stop the German tidal wave last May. Lord Gort commanded a higher level of respect as the holder of the Victoria Cross – the kingdom's highest award for valor in combat.

Both men had been affable and cordial during dinner, but would they be so amenable when they jumped into the meat of this evening's intended discussion?

Of course, the ever-present Major General Sir Hastings 'Pug' Ismay would bring another interesting perspective. 'Pug' understood the military mind and was a first witness to the political decisions at the Cabinet level by his staff service from a time prior to the Munich Accord.

"Now that we are safely ensconced in this bastion of male conversation, I trust everyone enjoyed Mrs. Landemare's choice of cuisine and preparation," the Prime Minister began and took a healthy swallow after savoring the rich aroma.

"Exquisite, sir," offered General Gort. "Our compliments to Mrs. Landemare. And, thank you so much for inviting us to join you at Chequers."

"It is an honor to have you both here – the hero of Dunkirk and the hero of the Battle of Britain."

"Respectfully, Prime Minister, ascribing that label to an evacuation hardly seems appropriate."

"The Victoria Cross entitles you to that opinion, John," responded Churchill, using Lord Gort's given name. "Let me assure you, I think I have sufficient understanding of military affairs to say a fighting withdrawal is one of the most complicated and vulnerable maneuvers on the battlefield, so humor me, if you will."

Lord Gort bowed his head. "As you wish, sir."

"I am surrounded this night by knights of the realm, so it is quite apropos to offer my gratitude on behalf of His Majesty for your service to the Crown." Winston did not wait for acknowledgments. "What is your assessment of the air battle, Hugh?"

Dowding cleared his throat to buy himself a few moments consideration. "We are not out of the woods, just yet, but we see positive signs."

"How so?" asked Churchill, baiting the fighter commander.

"The replacement rate for our pilots has finally turned positive after two dreadful months of negative numbers. The 7th was the turning point, but it was the action of the 15th that gave the Germans a clear statement, we were not done."

"Well said, Hugh. I must say that is my assessment as well. I watched the daily Air Ministry reports with keen attention, as I have the Admiralty reports on this terrible Battle of the Atlantic. I can only imagine what those pilots had to face two, three . . . five times each day. The odds against them . . . the fatigue . . . the mortal danger . . . all alone in those confining cockpits . . . just them and the machine. Incredible."

"If you would allow, Prime Minister," interjected Lord Gort, "your broadcast on the 20th of August was most inspirational, not just to the air force but to the army as well."

"Thank you, John, but I was not being shot at and I could offer only words."

"You understate the importance of your words, Prime Minister," Ismay said.

Churchill waved his right hand dismissively. "I heard more than a few stories during the Dunkirk operation of the Army disparaging the Air Force. I would like to hear your perspective, John," but before Lord Gort could answer, he continued. "I can appreciate what those desperate hours and days must have meant to a Tommy on that beach. Even one German fighter strafing them as they tried to escape would have been too many. They faced terrible odds in the skies above those beaches. We," he nodded to Dowding, "could not spare more fighters. It could be readily argued we should have withdrawn all the remaining fighters, when the Germans began overrunning their airfields and forcing them to find new fields for their operations." Churchill stopped, took another sip of his cognac and looked to Lord Gort.

"I think all of our soldiers felt in their hearts the pilots were doing the best they could. However, they could not see our aircraft. They saw German aircraft. Seeing a British fighter or watching a burning German fighter going down would have done wonders for the morale of our soldiers. They just wanted to know they were not alone on those beaches."

"It is something to consider, Hugh, wouldn't you say?"

"Yes, sir. Sometimes perception is all the reality we have."

"I must tell you a little story of mine. On the 28th of last month, I dragged 'Pug' with me to Manston aerodrome to set things right, since the base commander decided to abandon operations, as a consequence of the losing battle he faced in repairing bomb damage and keeping the field open. The Germans were bombing the field faster than he could repair things. Regardless, I could not allow that action for the Germans to see. We could not give them that satisfaction. They bombed us while we were there."

"Everyone ran for cover, except our prime minister," Ismay said.

"And you, 'Pug.' You stood there with me."

"Only because I could not leave you alone, sir. My inclination was to seek shelter as well."

"Yes, well, we stood there and watched the battle play out before us. Our fighters set upon those damnable bombers. Scared the bloody hell out of them. They dropped their bombs early. Their bombs exploded in the sea or short of their target. I was there to inspire the Air Force personnel, trying to maintain that comparatively unimportant aerodrome operational. Yet, it was a flight of three Spitfires who inspired me. Once their work was done, they descended at high speed and passed very low over us in the middle of that cratered landing area. They pulled up, split and each of them rolled several times before they rejoined and returned to their base. I doubt those pilots knew it was me down there, and I know they could not see the tears of pride descending my cheeks. Magnificent display, I must say. Perhaps, if the Tommies at Dunkirk had seen just such a display, there feelings toward the Air Force might be different."

"I take your point, Prime Minister. Perceptions."

"Precisely. Now, I would like to hear your prognosis for the current air battle," persisted Churchill.

"We saw a marked change after the engagements of the 15th. Daylight raids have diminished dramatically, while nighttime bombing has increased. What we do see during the day are large-scale fighter-bomber sweeps."

"What does that mean?"

"Fighter-bomber sweeps?"

Churchill nodded.

"They have configured their fighters to carry a single 100-kilogram bomb. They fly at high altitude. If they are not challenged, they drop their bombs in the vicinity of a target. They are not particularly accurate. Their purpose is not to bomb ground targets effectively. Their sole purpose is to draw our fighters up and continue to deplete our fighter resources."

"I surmise from your choice of words, you have not been giving them the satisfaction."

"Correct, sir. We dispatch several sections or even a squadron to approach them, force them to drop their bombs early, and then stand off until they must return for fuel."

"Cat and mouse," said Churchill.

"In a form, yes, precisely. As I said earlier, the changes have enabled us to gain strength. Lord Beaverbrook has done a masterful job stimulating production. We have the fighter aeroplanes. We need the pilots to fly those aircraft he has provided."

"I would like to hear your perspective on this Big Wing concept we have been told about for the last month."

Sir Hugh showed slight clues of his irritation in his usual stoic expression. Churchill recognized the fighter leader had to feel frustration that a subordinate in his command created this controversy. He withdrew into his thoughts to consider just how much he wanted to present to the Prime Minister. After his moments of contemplation, Sir Hugh responded. "The Big Wing concept addresses an essential principle of warfare – mass. There is no question Big Wings would improve our counter-raid operations. However, the factor that overrides mass for Fighter Command is time. If we had another 20 minutes of warning, a Big Wing would be a very effective tool to deal with the enemy's massed bomber raids and their large fighter escort force."

"Explain to me your concern for time."

"In summary, our choice is to engage over the Channel and the coastline, or over London. The latter would yield far more bomb damage to the city. It takes time for a Big Wing to form as well as disperse for landing at multiple sites. To be direct, Prime Minister, if we knew when the enemy raids formed and exactly what their target was before they approached, the Big Wing concept would work. As I understand my mission, we must protect the people, to keep as many bombs as possible from causing casualties. The Big Wing concept serves a proper military purpose, but respectfully, the tactic will fail our primary mission."

This time it was Churchill's moment to think. He kept his eyes on Dowding, as he absorbed the air leader's words. Churchill would not press the topic. "What about these night raids?"

"They are far more problematic. The Radio Research Station at Bawdsey Manor has had some notable success, reducing the size of the transmitter-receiver unit to allow aircraft installation and practical operational use in the night interception task. The prototype unit has been installed on a Beaufighter. The crew achieved several victories so far. We need many more appropriately configured aircraft before we will have an effective unit."

"I am certain the night work is a high priority at the Air Ministry. Given Lord Gort's observations regarding the Army's perceptions during Dunkirk, I have waited to ask you this question, Hugh. I think the time is appropriate. I would like you to revisit our meeting of the 15th of May. Would you make the same recommendation?"

"I am not certain of the point of your query, Prime Minister."

"That was a critical meeting, Hugh. It was the day after the disaster at Sedan. I continue to rethink those moments. The question haunts me to this day. What would have happened if we had had sufficient fighter forces to protect those bombers and they had successfully destroyed those six pontoon bridges over the Meuse? Might we have had time to react properly?"

"We cannot rewrite history," protested Dowding.

"No, we most certainly cannot, Sir Hugh. Yet, those questions haunt me, as I said."

"Prime Minister, with all due respect, my primary duty was and remains the air defense of Great Britain. As I said at the outset of this evening's discussion, I stated we are not out of the dark woods. What we have endured for the last three months . . . well . . . we were within a mere few days of collapse, but that changed on the 7th, and that was with the ten squadrons we were ordered to send to France. If we had sent those squadrons to France, I think it safe to say, we would not have made it to the 7th without those ten squadrons. I regret not arguing forcefully for the withdrawal of the Advanced Strike Force after the loss at Sedan. We need those pilots we lost in France. We are not out of this, yet."

"Lord Gort, your perspective of this question?"

"Prime Minister, I am afraid I am with Sir Hugh on this question. I can tell you with some authority, we lost the Battle of France years before the Germans invaded Poland, even before Munich. We had insufficient forces in every category to stop the overwhelming, well-trained and well-equipped German army. The coordination between air and ground forces proved devastating. To be direct and blunt, our intelligence services failed to appreciate the breadth and extent of Hitler's rearming of Germany. By the time we witnessed the complete armed forces and their new, highly mobile tactics in Poland, it was too late. Even as we absorbed what happened in Poland, we could not adjust fast enough."

"Listening to your assessment, Lord Gort, perhaps you think we should have abandoned France before the fight," said Churchill.

"No, sir. Quite the contrary. We had no choice, based on treaty obligations, and as you articulated more than a few times, it was better to fight them over there than on British soil. My point is, we must learn from our mistakes. I was not a party to decisions by the government, so I am ill informed to ascertain what might have been done differently. What I can tell you is what we faced in the Low Countries.

"Our intelligence told us the force on the other side of the Ardennes was likely a reserve force, staged to support the assault into Holland and Belgium. The information on the enemy force in conjunction with the terrain and forest led me to belief the intelligence. The mass and swiftness of the Ardennes penetration and the crossing of the Meuse at Sedan staggeringly surprised us. We could not believe they could do it, because our tactics at the time told us we could not carry off that kind of highly mobile, armor assault in such terrain. Whether their intelligence was better and more precise, or they were simply lucky, the Ardennes attack split the seam between the French and us, bypassed the defenses of the Maginot Line, and complemented the northern attack perfectly. In hindsight, the timing of the main attack was about as perfect as any soldier could hope to achieve.

"I say all this to support Sir Hugh's statement. Ten additional fighter squadrons would not have made a difference in the outcome of the Battle of France."

"That is a rather stark point of view, John. As Hugh stated, we cannot rewrite history. None of the information available to you or to the War Cabinet even remotely anticipated the Ardennes attack, so I do not quibble with your assessment, John. And, I must confess, I wanted those fighter squadrons in France as much to bolster French morale and resolve. At the end of the day, I reluctantly agreed with Hugh's recommendation and took the unpleasant task of informing the French government of our decision. The question still haunts me, but both of you have enlightened me and assuaged my unease."

The four men continued their discussions well into the early hours of Sunday. Winston Churchill was not bashful, shy or reserved in his opinions regarding military affairs, but he was above all a pragmatist. He knew his generals were capable and up to the task. His Majesty's Government had to ensure the proper forces were applied at the most effective points, and their strategic and tactical intelligence had to improve. Churchill knew ULTRA had not given them timely information. He could only hope the performance of the GC&CS codebreakers would improve their battlefield information for future battles on the road to victory.

———

Chapter 7

Friendship, of itself a holy tie,
is made more sacred by adversity.

-- Charles Caleb Colton

Sunday, 22.September.1940
Cabinet War Rooms
New Public Offices
Westminster, London, England

Week 12

Churchill had departed Chequers immediately after lunch to return to London and the war. His guests had departed after their breakfast. 'Jock' Colville had briefed him on immediate messages as he ate his breakfast in bed. The journey had been peaceful and quiet, but comparatively slow due to fog and rain showers. Activity in the underground offices of His Majesty's Government remained incessant, and it was no different when Winston descended the stairway, passed the security room, and into the continuously lighted main passageway.

As Churchill entered his small, subterranean office, the haggard looking but professionally dressed William Stephenson stood. "Welcome back, Bill," said the Prime Minister. A silver tea service was left on the small table. The door was closed, leaving the two old friends alone in the small combination office and quarters.

"Good afternoon to you, Winston." Stephenson looked around the small room with its steel beams and heavy wooden planking. "Not exactly Number Ten, now is it?"

"This is your first time in the War Rooms, isn't it?"

"Yes. Rather dreary, I should say. Not exactly to my liking."

"Nor mine, actually. I would rather be above ground, but the War Cabinet insisted, quite strongly I might add. We must win this war as quickly as possible, so we may extricate ourselves from this hole."

"We shall do our best," Stephenson said with a strained smile.

"I assume from your exhausted appearance, it was not a pleasant journey."

"A serious understatement, of that I am certain." Stephenson paused to let the PM speak. When no further questions or comments came, it was understood he wanted more explanation. "The only available transport was a long range Hudson, for God's sake. The weather was absolutely abysmal. The crew withheld heat to gain the most range with the fuel on board. I was wrapped in blankets and still freezing. I am getting too old for that type of abuse."

"I am sorry your journey was that unpleasant."

"Not to worry. If I don't catch my death, I shall survive. I arrived at RAF Northolt," he checked his wristwatch, "two hours ago. I took a quick bath at the Savoy, which did refresh me. Hopefully, I shall remain awake long enough to complete my briefing."

"Then, let us waste not."

Stephenson shifted in his chair and nodded toward the teapot. Churchill nodded his head in response. With two cups poured, Stephenson took two sips, and then sat back in his chair. "Two days ago, actually it feels like last night, I met with President Roosevelt. Also present were Secretary of War Stimson, Harry Hopkins, Bill Donovan and a Brigadier Strong, George Strong of the Army, who was military liaison here, as I recall."

"I remember the name. He has led the American military observer contingent."

"Yes, precisely. The meeting was initially for Strong to report on the situation here."

"And?" asked the Prime Minister, somewhat impatiently.

"Suffice it to say, he reported on our progress most favorably. He has been impressed with the performance of Fighter Command. He reported to Franklin that we had won the Battle of Britain."

"Maybe a bit premature, but the future does look considerably brighter than a month ago."

"Nonetheless, Strong is a supporter."

"Excellent."

"Even Stimson seems predisposed to the cause."

"He has been. Does this mean they will be more forthcoming with fighter planes, destroyers and rifles?"

"We did not discuss the arms supply situation, but I believe Strong's report should help in that arena as well."

"Good, then the gamble will have paid off."

"Franklin recognizes your gamble as well as everyone else in the room. I suspect it is the fact you took the gamble so boldly that has accentuated the results. You believed in our ultimate success, as it were."

"Yes, but he gives me more credit than I deserve."

Stephenson chuckled and mumbled more to himself, "I think not."

"What else?"

"Strong and Hopkins were excused. The conversation turned to UL-TRA and MAGIC. The Americans have a MAGIC intercept that states an execution time tomorrow morning for what they believe could be the German invasion of Great Britain."

"Interesting."

"Quite. There was some discussion about the veracity of the message. They conceded it could be some other region. The decryption of various code words leads them to believe it is England."

"All our information indicates to the contrary."

"The President wanted me to make sure you received the message."

"I have."

"We also discussed the security of communications for the more active and prompt exchange of MAGIC and ULTRA intercepts."

"You should work this through Stewart."

"I am scheduled to meet 'C' at Queen Anne's Gate day after tomorrow."

Stephenson looked carefully to Winston and sipped his tea, as Winston considered the information. Churchill stood, went to the stand next to his cot and poured himself a cognac. The Prime Minister had his own meeting with 'C' later in the day, after the late afternoon War Cabinet meeting. The principal topic, this day, would not be the invasion, as it had been for the last month. Today, the primary topic was the pending Dakar raid, the risky effort to keep the Vichy French out of the war, and the decision to proceed. Lord Gort's words last night weighed heavily on his thinking.

"When do you head back to Washington?" asked Churchill.

"It depends on Tuesday's meeting with 'C.' Wednesday, maybe the week's end, I should think."

"I cannot tell you in glowing enough words what an absolutely exem-plary job you are doing, Bill. I sincerely doubt we would have progressed to the level of exchange we have without your personal involvement, and your linkage between Franklin and me. Recognition may take some time rising, but one day the world shall know of the truly significant role performed by 'Intrepid' in this sordid affair."

"Thank you, Winston. Recognition is not what drives me."

"Thank the good Lord for that," the Prime Minister laughed.

"I shall leave you to the business of state. Is there anything I should carry back with me?"

Winston considered the request. He thought about this evening's War Cabinet meeting. "Has the President been briefed on Operation MENACE, the Dakar raid?"

"In general."

"Permit me to give a brief sketch," he said, moving toward the enormous wall map of the world. "In July, about the time of the Oran incident, we were looking for aggressive actions to stop Nazi expansion through the Vichy French." He pointed to the Western most tip of the African continent. "Dakar, the former French West Africa, appeared to be moving toward reinforcement by the Vichy government and quite possibly the establishment of Nazi sub pens like Brest. Three Vichy French cruisers pressed the Straits of Gibraltar successfully on the 11th of September through some very unfortunate miscommunications by the Navy. We also tried to intercept the warships outside Dakar, but failed that as well. We received word this morning the Japanese crossed the border in three places into French Indochina, claiming authority by some recent treaty with Vichy France. Colonial French forces still in Indochina have mounted some resistance, but they are no match for the Japanese Imperial Army. The Vichy French remain a major concern and a serious threat. And now, the Japanese are much closer to Singapore than could ever be comfortable."

"Where does General de Gaulle fit into all this?" asked Stephenson.

Général de Brigade Charles André Joseph Marie de Gaulle, *le Armée de Terre*, had led a successful but in the end futile armor counter-attack at Montcomet upon the German penetration during the Battle of France. He argued vehemently against capitulation to the Germans and fled France on 17.June.1940. With other renegade French officers, de Gaulle established and led the Free French Forces in exile from his headquarters in England.

"He is becoming quite the thorn."

"In what way?"

"We planted the seed, and now he has made this operation his *raison d'être*. The bloody Vichy have reinforced the garrison, and the Battleship *Richelieu*, for God's sake, is in Dakar harbor under the guns of Fort Manuel. General de Gaulle insisted strongly the operation proceed despite the disadvantageous naval situation. He landed the Free French battalions at Rufisque this morning. Unfortunately, against all the meteorological forecast data, fog has enveloped the West African coast."

"So, what are they going to do?"

"We have a decision to make this afternoon or tomorrow at the latest. I expect a situation update from the Admiralty. Unless there is some overwhelming and compelling reason, I am afraid we must proceed with the operation and hope for the best. We must support the Free French, and we cannot afford to have U-boat pens at Dakar."

"What do you want me to tell the Americans?"

"Nothing at this moment. I thought you should know what was happening in case we need to discuss things with our cousins."

"I shall be in London for the next few days. I shall leave word with John Martin or Sir Edward before I head out across the great waters."

"So good to see you again, Bill. I shall endeavor to keep you informed of the Dakar situation. Once again, please convey my appreciation to President Roosevelt. I shall ask General Strong for a private chat when he returns to London."

"Excellent idea. Take care, Winston. I need to get out of this hole."

"I share your desire. Unfortunately, I do not enjoy your freedom. Safe journey, Bill."

Churchill returned to his preparation for the War Cabinet meeting. Sir Edward would appear shortly to confirm the agenda and any special arrangements. There was much to discuss.

———

Monday, 23.September.1940
RAF Middle Wallop
Middle Wallop, Hampshire, England

The long, late afternoon mission proved to be unique in many respects. The weather was nearly perfect for bombing and fighter operations – virtually no clouds or haze to obscure visibility. The squadron landed near sunset with their fuel tanks nearly empty and gun-port tapes missing. Aircraftman Colin Jenkins, the armorer for the 'PR-F' Spitfire, joined Leading Aircraftman Bernard Gordon, the crew chief, as Pilot Officer Brian Drummond jumped down from the wing, stretched his muscles and moved each of his joints.

"You had four rounds per gun remaining, Mister Drummond," said Jenkins.

"We had plenty to shoot at on this one," Brian responded.

"Can't find any hits," reported Aircraftman Jordan Toldson, the rigger on the crew.

"Any victories, Mister Drummond?" asked Gordon.

"No clear kills. The intel guys will have to judge the gun camera film. I know I had hits on several One Oh Nines, but I don't know if any went down. We were too busy with all the fighters."

"At least no damage," said Gordon. "We shall turn the bird around and get her ready for tomorrow."

"Met guys forecast several more days of good weather, so we should have plenty of business."

"We shall keep you up there."

Brian waved his hand as he walked toward the dispersal tent. The intelligence debriefings completed as dusk fell. No.609 Squadron was released from duty, as evening twilight disappeared. Brian thought of Charlotte, but Jonathan kept his attention close at hand. Supper remained lively and animated, but it was the conversation and twisting hands in mock aerial combat in the bar that occupied the pilots.

"We must be doing something right," said Flying Officer Jonathan Kensington. "Forty fighters escorting just six Junkers Double Eights. They are using their fastest bombers as bait to get us into a good old row."

"Unless they eliminate our fighters, they cannot safely conduct a Channel crossing and invasion," said Flight Lieutenant Robert Morrow.

"Maybe it's the last we'll see of those large raids a few weeks ago," Pilot Officer Frank Burns commented.

"No, but now they are using the night," Morrow said.

"They can't beat us during the day, so they hide in the night," added Burns.

Brian listened to the words. He knew the night-fighters were beginning to find more success. The local night-fighters, No.604 Squadron and their twin-engine, Blenheim fighters with the strange bulges and antennae, could be heard every night now. The din of excited and enthusiastic conversation in the Officer's Mess bar may have prevented hearing the drone of Blenheim engines taking off, but they all knew the night-fighters were busy. The major cities, London, Manchester, Birmingham, Liverpool and Coventry, attracted all the attention of the night raiders. Every report of bombs falling on London brought concerns and thoughts about Mary Spencer and the baby growing in her womb. Brian worried about Mary's safety with the nightly bombings and their very limited ability to defend the cities at night. Mary's pregnancy also brought the same question every time. Should he tell Charlotte about his baby with Mary?

The questions rushed at Brian. He did not want anything hidden between him and Charlotte. He wanted to tell her, but he was also afraid of what might happen? What would she think if she found out about Mary's pregnancy from someone else? Jonathan did not think he should take the risk, if he really felt she was that important. Mary was not likely to tell anyone. How could she find out? The night bombings plagued his thoughts as well.

"Are you with us?" asked Burns, touching Brian's shoulder.

Jonathan's curious expression told Brian his detached thoughts had become noticeable. "Sure," Brian answered.

Jonathan looked at him rather skewed, as if he knew where he had been.

"What did I say?" asked Burns.

"Don't play those games with me," barked Brian.

"My, oh, my, seems we touched a nerve."

"We were talking about the night-fighters," Jonathan interjected, to deflect the new course.

"What about the night-fighters?" Brian asked.

"It's kinda an unnatural act . . . flyin' at night," said Burns. Everyone in the conversation laughed. "I mean really . . . flyin' around in the pitch black. I'm more afraid of runnin' into the bastards than shootin' 'em down."

"The Six Oh Four lads manage to avoid running into their targets," added Morrow.

"I don't know how," Burns mumbled.

Brian remembered a quick conversation in passing a few days ago. "I was talking to a couple of the Six Oh Four guys, and they said the airborne RDF kit they use has improved. The controllers get them close with Chain Home, and then they take over with their on board sets."

"Maybe that will give them the edge."

"They'll need somethin.' The damn Germans have lost their taste for guns."

"They definitely have the advantage at night," Jonathan said.

"Maybe not for much longer," answered Brian.

"The sooner the better."

"They'll bomb London into oblivion, if we don't do somethin' to stop 'em."

"Why don't they come after us any more?" asked Brian.

"You mean bombing the aerodromes?" Brian nodded. Morrow continued, "They figured out they cannot beat us."

"I wish it were that simple."

"Worse. They finally figured out the importance of the Woolston Works," Jonathan said. "What if they cut off our supply of Spitfires?"

'Sparky' Morrow held up his hand like a traffic policeman. "Lord Beaverbrook pinched the Vickers chappies to step up Castle Bromwich. They are producing Mark Two's up there. Plus, I hear they have moved most of the heavy production tooling into small shops like that petrol station and garage between Salisbury and Southampton."

"I hope it works," Jonathan said.

"We all do," Morrow responded with solemnity.

Brian looked at his watch. It was nearly eleven o'clock. It was too late to call Charlotte, although the sound of her voice would be glorious and reassuring for him. "The Met guys say we've several more days of perfect weather. I don't know about the rest of you, but I'm going to get some sleep."

"Rather than drive to Winchester," jabbed Jonathan.

"Your dolly bird, aye, 'Hunter?'" asked Morrow.

Brian sneered at Jonathan for raising the topic among an intoxicated bunch of pilots. He swallowed the last inch of his warm beer and left the others. Several snide remarks followed him out of the bar. Brian resolved to give Jonathan a shot in the arm tomorrow morning. He did not mind being teased about other, more casual, female companionship they all sought on occasion. Charlotte represented something completely different for Brian. He wanted to keep it that way, not let the often crass and abrasive aviators, noted for their womanizing, lessen her image in his mind.

———

Tuesday, 24.September.1940
RAF Warmwell
Warmwell, Dorset, England

No.609 Squadron had redeployed to Warmwell at dawn. They had not done that in more than a month. The more frustrating part was, they had already flown two patrols, one over Plymouth and the West counties of Devon and Cornwall, and another patrol over Bristol and Cardiff. They found no targets, and none had been suggested. The use of precautionary patrols had become the exception in the last few months. So, what changed, Brian asked himself repeatedly during the day?

The third call came. The launch and vector took them almost due north back again toward Bristol, Bridgwater Bay and the River Severn. The Westland Works at Yeovil popped out of the green countryside like a cork on the water. The moderate altitude probably meant another patrol. No indication of a raid had been given by the controllers. They could see the mouth of the Severn pointing toward the Northeast. Then, just before they would normally be passed to Filton Sector Control, Middle Wallop Control called.

"Sorbo Leader, this is Bandy calling."

"Go ahead, Bandy."

"Vector one three five, angels five." Darling did not wait for the message to finish, as he turned the squadron. "Proceed at maximum available speed." Brian pushed his throttle through the emergency gate. With no engine power margin available, the flight's ability to maintain their formation positions would get a little sloppy. "We have a raid in progress, approximately 40 bandits."

"Roger, Bandy. On our way."

That call could only mean one thing, Brian told himself -- Woolston and the Supermarine Works. The Germans had definitely found the significance of the saw-toothed roof factory with its rather large seaplane ramp across the River Itchen from Southampton. The relatively low altitude told him it was probably another low-level penetration raid, maybe even the highly skilled, committed and aggressive guys from *Erprobungsgruppe* 210. These guys just don't quit, Brian said into his oxygen mask. The words made him consider taking off his mask, but the possibility of going up in a tangled fight without oxygen convinced Brian to keep his oxygen mask snapped to his leather helmet and his goggles down.

They could all see the smoke rising from the vicinity of Woolston. No words passed among the nine Spitfire pilots, but they all thought the same thing. Why had they been sent north? Was this some elaborate or sophisticated trick to open up an unopposed approach by the German raiders? What had gone wrong? As they moved closer, the scene became starker. The Woolston Works could not be seen through the smoke. This was not good, Brian told himself.

Just beyond the smoke from the Woolston Supermarine Works, or maybe it was around and through the smoke, a squadron of Hurricanes was fully engaged with a squadron or more of Bf109s. It had to be one of the Tangmere Hurricane squadrons. They received no identification. Why weren't there any Spitfires to deal with the 109s? While the Hurricane was a very capable fighter, it was not the best match for the Bf109E-4.

Darling positioned the squadron to join the fight. As they closed on the brawl, the markings of the Bf109s jumped out at Brian -- red propeller spinner in front of a solid black nose forward of the cockpit. It was *Erprobungsgruppe* 210. This time they flew the faster, more maneuverable single-engine fighters rather than the more cumbersome Me110s.

Brian chose an initial target as his colleagues were doing as well. As he began his intercept turn, closing with a German fighter chasing a Hurricane, Brian noted a light flash to his right. A quick look found the source. Four flights of four aircraft each were flying at nearly wave top level up the Solent from the Isle of Wight. More Bf109s.

Brian rolled his Spitfire sharply to the right, nearly on its back so he could clearly see the intruders. "Bandits, three o'clock low, coming up the Solent. We've got another wave inbound."

"Sorbo, disengage. Let's take the new guys," Darling radioed.

Brian pulled his nose down sharply in an effort to intercept the attackers before they reached the Supermarine factory. Realization came quickly. He was the closest RAF fighter to the second wave of Erpro210 attackers, and he had too much altitude to lose. The lead section of Germans would reach Woolston before Brian could approach firing range.

As he maneuvered for a diving, head-on engagement of the second section leader, he saw dark objects drop from the German fighters. Fuel tanks? The explosion below and in front of him startled him for just enough of a moment to spoil his line up. He had not expected bombs coming off Bf109 fighters. The first section of attackers pulled up underneath Brian. The second section was now too close. He shifted his aim to the third section. Brian was in a shallow descent, as he placed his gunsight reticle on the leader of the third section.

The machine guns burst to life. The stream of bullets reached out. The leader did not deviate from his bombing run at the Spitfire factory. The stream of bullets moved quickly to engulf the German fighter. The aircraft exploded in a fireball. Brian pulled up sharply to avoid as much of the shrapnel, and the engine choking heat and smoke.

He rolled left sharply into a very steep, level, high-energy turn to pick any of the attackers who might try to come back for a second pass. As Brian strained his neck to reacquire any targets, he saw bombs detonating beyond the Supermarine factory. At least they had spoiled the aim of the last few attackers.

Brian pushed harder on his throttle lever already against the forward mechanical stop. The Merlin engine roared at emergency power. There was no protest from the Spitfire. Brian scanned the sky and found one of the Erpro210 Bf109 fighter-bombers pulling up and turning to make his escape south. Several quick, precise control inputs reestablished a good intercept track on his target. Brian continued to scan for others trying to engage him as he kept returning to his prey. The German made no evasive maneuvers as Brian closed rapidly. The attack line was a little too steep on the German. Brian fired as he recognized his overshoot. A few bullets hit the target as it flashed by his right wing.

The 'PR-F' Spitfire shallowed the climb and rolled nearly vertical. He did not want to pull through too hard dissipating his airspeed and allowing his prey to escape. Brian looked over his right shoulder and slightly up out the top of his single piece canopy. The German attackers were running south. They were trying to extricate themselves rather than fight. Spitfires and Hurricanes almost instantly filled the sky around the Germans. It was now going to be a high-speed chase south. Brian was now too far back.

He scanned the sky looking for any trailers to cover several of the RAF fighters straining for the retreating shots. He watched the Spitfires press on with the chase. The Hurricanes pulled up and broke off the chase with the realization they did not have sufficient speed to continue and the Germans would not return for another try. Two of the last few Germans were hit. One was trailing a streamer of light gray smoke. They continued the chase past the Isle of Wight into the English Channel. The lead Spitfires pulled up to disengage. The Germans escaped, including the smokers.

Brian retarded his throttle, as he pulled up slightly. They rejoined as a squadron. Brian found Jonathan's 'PR-K' Spitfire and smartly maneuvered to the left wing position. Darling started them toward RAF Warmwell, and then a short time later, probably recognizing the time, altered course back to RAF Middle Wallop.

No words were spoken even to the tower for landing. All nine No.609 Squadron Spitfires landed. There was somberness to their actions on the ground. The ground crews scurried about reloading and refueling the fighters. There was sufficient daylight remaining for at least another sortie, if required.

The debriefing went quickly. They remained at Available status until the fighters were reported ready for launch, and then they were changed to Readiness status.

As was always the case especially when new things happened, the pilots compared information and tried to gather more clues. This time, even Flying Officer Royster helped gather information.

"What the hell was all that?" asked Frank Burns.

"It appears our friends with Erpro Two One Oh have been experimenting some more," answered Royster. "We heard reports from One One Group about these variants. From everything you lads reported, I think it is safe to say you encountered a fighter-bomber variant the Air Ministry has designated as an Edward Four slant Beer." Brian saw a clear image of the Bf109E-4/B fighter-bomber. "We believe it has the same armament as the fighter version, and is equipped with provisions to carry at least one 250-kilogram, high explosive bomb or other special munitions."

"Why did control send us up to Bristol?" asked Stockard.

"We will have to review the sequence," answered Royster.

"Could be many reasons," added Darling, "but, they caught us with our knickers down around our ankles."

"Putting a single Hurri squadron against One Oh Nines was not particularly bright either," Flying Sergeant Johnson commented.

"The bastards did everything right."

"Even two waves. The first, apparently, made an unopposed pass. Engaged the Hurris drawing them up and away from Woolston, made it look like they were the fight, and then that damn second wave came in underneath," Morrow said.

"If 'Hunter' hadn't seen that second wave approach, the second set might have been unopposed as well," Jonathan said.

"Rather smooth, if you ask me," Davies said.

The reports from Woolston began to come in, as they waited for another launch command and cogitated over the tactics. The Erpro210 raid began to take the appearance of a very successful attack. The entire factory had been damaged. Some reports said totally destroyed. One of the bombs penetrated the shelter protecting the employees. The emergency services were still counting the dead and wounded. A number near 100 had been suggested as the possible fatalities.

Although No.609 Squadron had done everything asked of it, the German attackers had apparently done everything precisely correct. Even with the damage, the pilots had a bizarre, grotesque admiration for the exceptional tactics and execution of the raid. They also recognized the significance of the attack on the Vickers-Supermarine Spitfire factory. The supply of the superior British fighters would be seriously affected for many weeks, if not months, and might not fully recover.

The evening meal and drinks in the Officer's Mess bar that late September night took on all the vestiges of a funeral wake. The usual levity and gaiety of the fighter pilots vanished as the consequences of the successful German raid sank into their thoughts. Strangely, the destruction of the Woolston Works made a more profound and deep effect on the mood of the pilots, more so than the bombing of London. Had the Germans just cut off their lifeblood? They all knew Castle Bromwich was in full production of the Spitfire Mark II, but Woolston represented the world speed record, the Schneider Trophy, Reginald Mitchell, and the birthplace of the vaunted Supermarine Spitfire. Many of the pilots, even those from the other squadrons that had not participated in the encounter, felt for the first time like they had failed in their duties. Of all the adversity, loss and damage involved in the Battle of France and the Battle of Britain, the reports that day from Woolston had a more grave impact on the Fighter Command pilots than any other. Their wellspring had been discovered and damaged.

Wednesday, 25.September.1940
RAF Middle Wallop
Middle Wallop, Hampshire, England
10:45 hours

The somber mood carried over into the next day. The weather remained nearly perfect for bombing and fighter operations. They all wondered, although none of them talked about it, whether the Germans would return to Woolston to finish whatever was left undone. They waited.

Reports from the East trickled in about reconnaissance flights and high fighter sweeps, but there was not much bombing through the morning hours. Pilot Officer Brian Drummond wanted another try at the Erpro210 guys. He felt an itch he did not mention to anyone.

The only telephone call into the squadron informed Squadron Leader Darling and notified Flying Officer Jonathan Kensington that an Anson would land just after noon to collect Jonathan for another visit to RAE Farnborough. They speculated what the reason might be, but did not guess. Jonathan would tell them what he could when he returned later in the afternoon.

Jonathan departed on time. The squadron remained at Readiness status all day. The only move had been a drop back to Available-in-30-Minutes status to allow the flight and ground crews a proper mid-day meal. No.10 Group saw no action. Even No.11 Group encountered isolated enemy activity. The near perfect weather had not been utilized to the extent it could have been used. The waiting time gave each of them more time to wallow in the mire of yesterday's loss.

The fires had been extinguished. The losses had been verified; 98 killed, 214 injured, the factory out of production for many weeks. They also learned 57 Spitfire Mark I's at various stages of assembly had been destroyed along with parts for many others. In one moment, the Germans achieved more impact on British fighter strength than any other single day in the war. The size of the loss staggered the Fighter Command pilots. They silently replayed each moment of yesterday's events trying to find their own reasons. How had the system succeeded for the months of pitched battle, only to fail so hugely when they actually began to use the word, victory?

They waited with their thoughts of yesterday for the Germans to come. They waited all day.

Corporal Warren walked over to the pilots to remind them their transport was still waiting for them. They hung up their flight equipment in the darkened tent, and returned to the Officer's Mess.

Wednesday, 25.September.1940
Cabinet War Rooms
New Public Offices
Westminster, London, England
16:00 hours

For the Prime Minister, the day's news continued to be quite bad. Lord Beaverbrook's report on the damage to Spitfire production from yesterday's highly successful, low-level raid on the Vickers-Supermarine Spitfire Woolston Works attempted to put as bright a face as possible on the grievous loss.

After the change in tactics by the Germans at the beginning of September, virtually all the tooling from Woolston as well as other key aircraft production factories had been dispersed to local automobile garages, shops and even private homes. While the losses at Woolston were enormous, Lord Beaverbrook believed they would return to full production within two to three weeks. Beaverbrook's foresight to disperse the production assets saved them from what could have been a fatal loss.

Also disturbing, German bombers dropped a string of incendiary bombs on Shepherd's Bush, West London, and managed to hit Wormwood Scrubs Prison. Prior to the war, as the threat of war had loomed larger by the day, the Security Service (MI5) moved their registry files out of central London, since the capital city was deemed a likely target for enemy bombardment. The MI5 files contained decades of information cards on suspects, leads, contacts, and all the other bits of information counter-intelligence operations turned up. The data bits were catalogued, classified and cross-referenced to the extent of their investigative knowledge. The bombs that hit The Scrubs ignited numerous fires. Several of those fires burned many of the file cabinets that held the MI5 registry files. As Brigadier 'K' Harker reported by early afternoon, most of the primary, original files had been destroyed before the fire brigade could extinguish the fires. The process of damage assessment would take several weeks to complete. Harker also reported that his predecessor, General Sir Vernon 'K' Kell, had been sufficiently apprehensive about the security and integrity of the registry files. Kell had the foresight to order the laborious task of creating a duplicate copy that had been completed prior to the Munich Accord. Harker asked for any discussion about a public statement to be held in abeyance until the damage assessment was completed and a plan developed. In addition to the loss of the original files, this morning's bombing accelerated plans to move MI5 from The Scrubs to the grounds of Blenheim Palace, Oxfordshire – farther away from London.

For the Prime Minister, the more significant bad news came from further south. The reports from the Dakar raid over the last few days remained quite negative. There were no bright edges to the reports.

The Royal Navy Battleship HMS *Resolution* along with the cruiser HMS *Cumberland* had been seriously damaged in action with the Vichy French Navy. The French Battleship *Richelieu* had been hit several times and the task force commander suspected the French capital ship might have been rendered immobile.

The vanguard of the Free French forces under the command of General de Gaulle had been ordered withdrawn earlier in the day. The poor luck with the weather seemed to be the largest factor. The persistent fog prevented any safe attempt at landing the remainder of the Free French troops, and prohibited any purposeful use of naval gunnery to support the ground forces. The naval engagements over the previous two days had been less than conclusive. The reports had left Churchill with the impression the losses far outweighed the successes.

The bitter and painful memories of the reports and decisions 25 years earlier when he was First Lord of the Admiralty gushed back to him. The ill-fated Gallipoli raid resulted in an obscene loss of exceptional young men and eventually led to his demise as First Lord. Churchill could not escape the comparisons although he refused to mention his thoughts to anyone, not even his closest advisors nor Clementine, his ultimate confidant. He wondered how the War Cabinet, Parliament and the public would receive the bad news from Dakar.

Winston Churchill also knew he had to reckon with the effect the defeat might have on their relationship with the Americans. Although support from the United States had grown in the last few months, American public opinion against any involvement in the war remained a formidable obstacle. Each and every setback would bolster the convictions of the American isolationists. He thought of every possible way to break the news to President Roosevelt. In the end, he knew there was only one way.

The Prime Minister reread the typed private message to his counterpart across the Atlantic Ocean. He tried to put himself in Roosevelt's place, as he read the short message. It was virtually impossible to make a defeat sound like a victory, especially placed in the context of the broad events during the summer. Operation MENACE had to be put in proper perspective. Not all operations would be successful. No one had expected the reduced visibility of the unusual fog.

MOST SECRET – PERSONAL AND PRIVATE

```
67
25.ix.40
FROM:  FORMER NAVAL PERSON
TO:  POTUS
I MUCH REGRET WE HAD TO ABANDON THE DAKAR
ENTERPRISE.  VICHY GOT IN BEFORE US AND
ANIMATED THE DEFENCE WITH PARTISANS AND GUNNERY
EXPERTS.  ALL FRIENDLY ELEMENTS WERE GRIPPED
AND HELD DOWN.  SEVERAL OF OUR SHIPS WERE HIT,
AND TO PERSIST WITH LANDING IN FORCE WOULD HAVE
TIED US TO AN UNDUE COMMITMENT, WHEN YOU THINK
OF WHAT WE HAVE ON OUR HANDS ALREADY.
WE CONTINUE TO CONSIDER REINFORCEMENT OF THE
FREETOWN GARRISON AS WELL AS OTHER ACTIONS
TO CONTAIN VICHY AND NAZI EXPANSION IN WEST
AFRICA.  FREE FRENCH FORCES UNDER COMMAND
OF GENERAL DE GAULLE SHALL BE REDEPLOYED TO
PROVIDE MOST LEVERAGE OF AVAILABLE FRIENDLY
FRENCH FORCES.
AS SOON AS APPROPRIATE PLANS HAVE BEEN CREATED,
WE SHALL ENDEAVOUR TO KEEP YOU INFORMED.
I MIGHT ADD OUR PROGRESS WITH CURRENT DEFENCE
OF ENGLAND HAS IMPROVED IN RECENT DAYS.  I
BELIEVE YOU HAVE ACCESS TO OUR MOST RECENT
INTELLIGENCE.  ENEMY APPEARS TO BE COMMITTED
TO NIGHT BOMBING OF OUR CITIES.    WE ARE NOT
THROUGH THIS TRIAL AS YET.  WE MUST THANK YOU
FOR YOUR CONTINUED SUPPORT.  WE LOOK FORWARD
WITH OPTIMISM TO BRIGHTER DAYS.
```

MOST SECRET – PERSONAL AND PRIVATE

Churchill considered whether he should mention the devastating news from Woolston. He shook his head even though no one could see him. The message said what it needed to say. There would be no purpose served to make it darker than it already was. War reports always had to be taken in a much larger context. From an American perspective, the failure of the Dakar raid had to be more significant.

Winston depressed the lever on his call box. The duty sergeant answered.

"Yes, sir."

"I believe Sir Edward is waiting for me. Would you be so kind to find him and send him in?"

"Right away, sir."

A minute later, a single knock on the door announced the arrival of Sir Edward Bridges, Secretary to the War Cabinet. "Yes, sir," he said.

"Ed, I have thought this through many times as well as from a variety of directions. Read this message to Franklin, and tell me what you think."

Churchill handed Bridges the single piece of typewritten paper. The Cabinet Secretary read through the note several times before looking back to the Prime Minister. "Very well put, I should think."

"I considered providing more information as well as a mention of the Woolston loss, but opted for simplicity."

"A more sound position."

Churchill considered the words. "Then, send this immediately via our most secure channel."

Sir Edward nodded his head several times. "Anything else?"

"No, I think not."

The Prime Minister was left once again with his memories of Gallipoli and other battles lost. He decided to downplay the results and not draw any further attention to the losses of the last few days. Maybe tomorrow would be a better day. He thought about calling Clementine, but decided against it. He needed to be alone with his burden.

———

Wednesday, 25.September.1940
RAF Middle Wallop
Middle Wallop, Hampshire, England
18:45 hours

Flying Officer Jonathan Kensington returned to RAF Middle Wallop, as No.609 Squadron was released from duty at dusk. They needed a diversion. They gathered in the Officer's Mess, collected up a pint of beer each, and surrounded Jonathan.

"What was it this time?" 'Boxer' Stockard was the first to ask.

"A One One Oh Destroyer."

"I'll be damned."

"We're starting our own bloody frigging *Luftwaffe*."

"So, how was it?"

"Just as we thought," answered Jonathan. "It is a fast bird, but it handles more like a bomber than a fighter. It is certainly not as agile as the One Oh Nine. I can also see why those Erpro Two One Oh blokes use it for low level bombing."

The group fell silent. None of them wanted to be reminded of yesterday. The image of the elite German squadron and their success came flooding back. Jonathan recognized the impact of his words. He cleared his throat.

"The Destroyer has an even worse view out the Perspex than the One Oh Nine. We should be thankful for our Malcolm hood canopy. I flew several short flights in the front seat as well as one flight in the back, gunner's station. I now know why I like the controls in my hands."

It took several more moments for the dark mood to be broken.

"Yeah, you like to hold all things between your legs," offered 'Red' Burns to the laughter of the others.

"Don't we all."

"So, you did not like the back seat, aye, 'Harness?'"

"Categorically not. The bugger I was flying with scared the bejeezus out of me with some of his shenanigans, and me without any controls. I don't know how those gunners stand it in the back seat."

"Easy," interjected Burns. "No problem. They put a pistol to their heads and say 'You vill like zis, or you can talk to ze Gestapo.'" They all laughed at the image. It fit with their thoughts.

"How come you are the only one who gets to fly these captured machines?" asked Stockard.

"Blessed."

"Maybe so."

"How did they get this One Ten?" asked Burns.

"What they told me was, the crew got both engines shot out and belly landed in a freshly plowed field near Wareham in Dorset. The pilot did a most impressive job of it. The gunner had been killed. The pilot was apparently dazed and did not have time to destroy the aircraft before the local Constabulary arrived to capture the bugger. They repaired the minor damage, replaced the engines and air-screws. They captured the beast two and a half months ago."

"Will others fly it?"

"Others are. The test guys made some quick flights. I was the fourth operational pilot to fly it, so they said. They decided, in our current situation, and in light of the two seats, to let us operational lads have a poke early."

"Lucky you."

"Quite."

The pilots stayed with the Me110C-4 topic and the litany of stories surrounding their collective experience fighting the twin-engine, two-seat, German fighter. Alcohol consumption brought more humor and levity to their camaraderie. Yet, they could not deny the fatigue brought on, even by waiting. Tomorrow would be another round of this dance, and their ranks began to thin with the march of time.

Thursday, 26.September.1940
RAF Middle Wallop
Middle Wallop, Hampshire, England

Dawn brought another fine weather day, and the prospect of more business for the pilots of Fighter Command. Pilot Officer Brian Drummond, as well as most of the other pilots of No.609 Squadron, did not find much pleasure in the high, wispy, lace work clouds that greeted them. The injury of two days ago remained with them. It would not affect their flying, but it still affected their thoughts.

Charlotte Palmer had seen the smoke from the Southampton area and worried about Brian. She waited for more than a day hoping Brian would call her. Charlotte's telephone call last night made the loss more graphic. Even the public knew what had happened. It was good to hear her voice although Charlotte told him about the strain of not knowing whether he was safe. Brian had been sorely tempted to make the run to Standing Oak Farm last night after the telephone call, however prudence won the moment. He hoped they would get a break soon. This was the fifth straight day of perfect or near perfect flying weather. Until the weather changed, there would be no breaks, but he did not tell Charlotte about this reality.

The squadron went directly to Readiness status shortly after Squadron Leader Darling placed the morning condition call. It was going to be a long day. They could only wonder what lay ahead.

The routine of No.609 Squadron continued as they waited through the morning mail delivery. Several dispatches and official communications were delivered to Squadron Leader Darling. Only a few personal letters came, none for Brian. Near the end of the stack, Brian's name was called by Corporal Warren. The envelope told him it was an official letter from Headquarters, Fighter Command. He opened the letter.

HEADQUARTERS, FIGHTER COMMAND

ROYAL AIR FORCE,

BENTLEY PRIORY, STANMORE, MIDDLESEX

Telephone Nos.: BUSHEY HEATH 1661 (6 lines)

BUSHEY HEATH 1646 (4 lines).

Telegraphic Address: "AIRGENARCH STANMORE."

Reference: -- FC/P.1164500

25th September, 1940.

To:

Pilot Officer Brian A. Drummond, DFC

No.609 Squadron, No.10 Group

RAF Middle Wallop, Hants

You are hereby ordered to stand detached from your present duties effective 30th September, 1940.

2. You shall proceed with no delay en route to RAF Church Fenton, North Yorkshire, for assignment to No.71 Squadron, No.12 Group.

3. You have been given a Priority 1 Transport pass. Present this letter to the proper authorities to ensure you are given the fastest available means en route.

4. Any difficulties with execution of these orders should be reported immediately to Wing Commander Douglas Higgenbotham, Logistics Support, this Command.

By order of the Commander,

H.C.T. Dowding

Air Chief Marshal,

Air Officer Commanding-in-Chief

Fighter Command, Royal Air Force

Brian reread the short, blunt letter twice before he looked up. His eyes found 'Red' Burns first. His red haired, American colleague held the same puzzled expression Brian must have displayed. Brian then looked for Jonathan who was outside since he had not received any mail. He handed Jonathan the letter. His friend only needed one reading.

"What does this mean?" asked Jonathan somewhat rhetorically.

"Hell if I know. I know what it says, but why? What is going on?"

"We had better talk to the Skipper. Maybe he knows something."

Before they could reenter the tent, Squadron Leader Darling came out followed by the other pilots as well as Corporal Warren. He held several pieces of paper. "I shall cut to the chase," he started. "As we were alerted earlier, orders have arrived, transferring 'Hunter' and 'Red' to Seven One Squadron, and 'Curly' and 'Crazy' are being transferred to Three Oh Three Squadron. Both are new squadrons created last week."

"What gives, Skipper," interjected Flight Lieutenant Morrow. "We are not at full strength, haven't been for weeks now, and they are transferring four more away from us. Are they disbanding the squadron?"

"No. They are not disbanding the squadron, as far as I know. It seems we have had a flood of American, Czech and Polish volunteers. The Air Ministry is forming the first of several squadrons with foreign pilots. We have had a Canadian squadron since shortly after the war began. I am told there will be additional squadrons standing up as more foreign pilots are assigned."

"That's what the orders say," Burns added.

"The loss of four more will put us less than half strength," Morrow protested.

"I think we are all aware of the numbers, Robert. Orders are orders."

"What are we going to do about it?"

"Obey our orders," Darling responded rather sternly. "According to the information in my cover letter, we have five, maybe six, new pilots coming to us over the next week or so."

"Now, isn't that just grand. We are in a brawl, we trade four exceptional fighter pilots for the opportunity to train a half dozen newbies," said Morrow.

No one spoke. Brian looked to Burns who could only smile and shrug his shoulders. Jonathan just shook his head. The sensations Brian experienced earlier in the year when he went through the tussle with the embassy guy over his violation of the federal Neutrality Act came back to him. The walls and bars of the cage began to close in upon him, again.

He had spent the last year of his life with this squadron. These were his friends, his mates, his brethren. Nothing against his fellow Americans, but Brian felt the enormous strain of the warrior's bond. His life had been dependent upon and supported by these men. He did not want to start over again with another squadron, even a new one. He also knew the numbers. Seven Americans had participated in the Battle of Britain. Four had already been killed in action. Most of the new pilots in No.71 Squadron would be brand new pilots, which was probably why they were forming the squadron at RAF

Church Fenton, well north of the principal aerial combat. North Yorkshire was nearly back to Scotland where he first joined No.609 Squadron.

"This must be the Eagle Squadron, one of the Two Three Eight lads referred to the other day," said Roland Stockard rather quietly and casually.

"Eagle, aye," snorted Morrow.

"Does have a nice ring to it, though," commented 'Fog' Johnson.

"Balderdash," Morrow barked.

"Misters Burns, Drummond, Kormer and Mansek, I would like to have a chat with each of you separately, if you do not mind," Darling said.

Frank Burns nodded and went with him into the tent. The others offered what amounted to condolences to the transferees. It felt like he was being banished from his family for some minor infraction. He had been with Jonathan since OTU7 advanced pilot training at RAF Hawarden, North Wales. Now, he would be on the other end of the country from his friends and compatriots. *Why do all the Americans have to be together? Why couldn't we just stay the way we are?* Brian tried to remember Malcolm Bainbridge's experience with No.43 Squadron in the Great War. He could not remember whether Malcolm had faced the exact same situation when the Americans joined the war in 1917.

Burns came out of the tent with a smile on his face. He motioned with his thumb over his shoulder for Brian to go in for his talk with Squadron Leader Darling. His leader waited in the small commander's compartment at the far end of the tent. A single, bare light bulb hung by a wire over his small field desk. Darling motioned to one of several folding chairs on the opposite side of the desk.

Darling cleared his throat. "Brian, I am going to say some things I do not want repeated to anyone." He paused to receive a head nod agreement from Brian. "I do not like these orders either. You are one of the best fighter pilots I know of in the Royal Air Force. You are the highest scoring pilot in this squadron. We cannot afford to lose you. However, the orders are clear."

"But, Skipper . . ."

Darling held up his hand. "No, buts. These orders are valid."

"Can't orders be changed?"

"We are at war, Brian. There are many things we must do in war that are not to our liking, but are usually in the greater good."

"But . . ."

Darling held up his hand again, but did not speak. Brian sat there staring at Darling. The walls of the cage moved closer and began choking off his air. There had to be a way out -- an escape route, or trap door. He wanted to stay with his squadron until the end. He wanted to feel the accomplishment with those he had shared the struggle.

"In my official capacity, I must tell you to stiffen up and follow your orders." He paused. Brian sensed there was something more he wanted to say. Brian nodded his head to acknowledge the words. "In a more private capacity, as a friend, I might choose to remind you of the events last winter and spring surrounding your possible extradition to America. You do recall those events, do you not?"

"Like a bad dream."

"Maybe I should remind you of your friendships -- the orders that were changed from Gladiators to Spits -- the extradition that was countermanded. Remember your friends."

Brian smiled. He knew exactly what Darling was talking about – Air Commodore John Spencer. "But, he's at Uxbridge, not Bentley Priory, now."

"Does that mean he does not have his friends and connections any more?"

"No."

"I might also remind you of the words from our group commander when he visited this facility last July."

Brian smiled more broadly, as he recalled the offer from Air Vice-Marshal Sir Quintin Brand to help in any way he could in recognition of their mutual friend, Pilot Officer Malcolm Bainbridge. Squadron Leader Darling was right. Why was this situation any different from the obstacles he had faced earlier in the year? "Are you saying you want me to stay?"

"I cannot say that officially. However, unofficially and within the confines of a private conversation between friends, I would say, emphatically yes, of course I want you to stay. There is no commander in his right mind that would not cherish your presence. On this one, though, I am afraid you are on your own. I cannot resist this change. I simply cannot assist."

"Thank you, sir."

"You must move quickly, Brian. You only have a few days, and two of those are the end of week. The 30th effective date is Monday."

"Yes, sir. I'll try to call him tonight once we're released."

"As you say, then. Now, get your focus and concentration back. We might take a scramble at any moment, and we need you at full performance."

"Thank you, sir."

"Thank you, Brian. Now, get out of here."

The bright sunlight hurt his eyes as he stepped outside the tent. He too smiled although probably for a different reason than Frank Burns. Brian walked toward his 'PR-F' Spitfire. He felt the urge to touch the aircraft, much as he had done at RAF Drem, a year ago. Jonathan caught up to him.

"What did the Skipper say?" asked Jonathan.

"I'm going to call Air Commodore Spencer tonight after dinner."

"So, you are going to try to have your orders changed."

"I'm going to try."

"I hope he doesn't know about your affair with his wife."

"Jesus, Jonathan," Brian responded sharply, and stopped to face his outspoken friend. "Could you have been any more vulgar?"

"Hey, Brian. I am not the one who was bonking my benefactor's wife and got her pregnant."

"Yeah, but you don't have to keep reminding me."

"Someone does. You seem to forget these things."

"Well, thanks . . . thanks a lot."

"What are friends for? Now, I suggest we keep walking before we raise any more suspicions by our observer crew," he said, nodding toward the tent and the other pilots invariably watching the two of them.

Brian started to walk. His thoughts did turn to Mary Spencer and the new life she carried. Jonathan was correct, as good friends should be in situations like this. He wanted everything to be healthy and safe for Mary, and selfishly, he did not want his relationship with Mary, which he truly hoped was in the past, to come between her husband and him. He liked Mary as a person, and as a woman and lover. She was fun to be around after he grew to know her. She could have found someone else to fill the void of her loneliness, but in a strange, bizarre way, Brian was glad she chose him.

"Do you think you can get them changed?" Jonathan finally asked, referring to Brian's orders.

"I don't know, but he got me into Spits, and he kept me from being sent back to the U.S."

"Maybe he can?"

"I sure as hell hope so."

"I do not want to throw another clot in the churn, but I heard someone say this Seven One Squadron is flying Brewster Buffalos, for Christ's sake."

"What!"

"I do not know whether it is true."

"From Spitfires to Buffalos . . . that just can't be."

Buffalo was the RAF name for the American-made Brewster F2A monoplane fighter. The aircraft was a short, stubby looking machine powered by a radial, air-cooled, Wright R-1820 Cyclone engine and armed with four

50-caliber, Browning machine-guns – two in the nose, two in the wing. The Buffalo had some advantages; one of them was not performance.

"At the end of the day, Brian, this Eagle Squadron . . . they are your countrymen."

Brian stopped again to look into his best friend's eyes. "Jonathan, I know that, but this is where I belong. We've been through so much together. I just can't imagine leaving you and the others, and certainly not Spits. I just can't. Anyway, you feel like my countrymen. I've become a man here with you."

"You might be ostracized by the others who join."

"Maybe, but not if they're smart, and not if they understand me and my reasons."

"Good on you, then."

They reached the 'PR-F' Spitfire. Leading Aircraftman Bernard Gordon joined them to ascertain what was happening. Satisfied there was no need for his presence, he returned to the ground crew tents. Brian walked around his fighter, touching the surfaces from the bare aluminum wing leading edges stripped of their paint and primer to the oil covered engine cowling. The machine held him in awe. It did not affect Jonathan in the same way. Brian told Jonathan again about the story of his first touchings of the Spitfire on the ramp at the Woolston Works with then Group Captain Spencer, the same day he arrived in Southampton. While his impressions had matured over the months of combat, being shot down three times and 18 aerial combat victories, Brian still felt a reverence for the sleek, curvaceous and powerful fighter. He simply could not stop flying the Spitfire. They had to let him stay.

Shortly after lunch, they launched toward the southwest. No.609 Squadron along with No.152 Squadron engaged a raid over Christchurch near Bournemouth and Poole. The best anyone could determine, the enemy raid was after the Bournemouth port facilities. It was a small raid, one squadron of He111P-2 bombers escorted by a squadron of Bf109E-4 fighters. The engagement was reminiscent of the early July encounters when the Germans attacked only shipping. The only difference was the vigor of the engagement. All the crews had more experience and steadier nerves. No losses could be documented on either side, and the bombs fell short of their targets into the Channel and harbor.

They returned to RAF Middle Wallop to more bad news. While they were engaged in the no consequences raid over Christchurch, a large attack went after Woolston again. This time it was a relatively high altitude attack by a large bomber force and an even larger fighter escort all around the bombers. The reports indicated something like 70 tons of high explosive bombs may have fallen on Woolston and the surrounding community -- more damage to the factory, virtually destroyed already, but fortunately, significantly less injury to people.

The No.609 Squadron pilots had become numb to the brutal assault of the last few days on the Woolston Works. They resigned themselves to doing the best they could at the tasks they were given. They could not fight everyone's fight. Maybe now the Germans would leave the hapless coastal city alone for a while.

After dinner, while the other pilots from all the squadrons retired to the bar for drinks and stories, Brian went to the converted closet, telephone booth. He asked for Bushey Heath 2471, the Spencer's residence telephone number.

"Bushey Heath two four seven one," Mary answered softly to the ring.

"Mary, this is Brian."

"Dear God above, he lives."

"I know it has been a few weeks since we talked."

"A few."

"How are you?"

"Other than beginning to show signs of pregnancy, I am actually doing quite well. Thank you for asking. I wish I had gotten pregnant sooner. I have actually seen more of John since the announcement than I have since the beginning of the war."

"Great."

"Did you call to tell me you wanted to see me, or that you wanted me to come to you?"

"A nice thought, but no."

"How disappointing, Brian. When will I see you again?"

"Maybe when things cool down," he lied. "The Germans found the Supermarine factory at Woolston, and they've bombed the hell out of it for the last few days."

"So, we heard." She paused. "When do I see you?"

"You don't give up, do you?"

"Not easily."

"Do you really think it's a good idea?"

"Brian, I am carrying our child, for God's sake. I wasn't after you for your sperm."

The words jolted Brian, as if he had been hit with an electric shock. Mary had always been straightforward and rather blunt. A lightning bolt hit him harder. *What am I going to tell Charlotte? What could I tell Charlotte? Would she ever understand?* Mary's persistence gripped him quite strongly. She had always been a formidable person.

"What did you call for?" she asked, when he did not pick up the conversation.

"I wanted to talk to you."

"And . . ."

"I am looking for Air Commodore Spencer."

Mary laughed hard, and then tailed off to a giggle. "You know better."

"You said he had been home more lately."

"No. I said I had seen him more. That does not mean he has been home. Anyway, he is still at work as far as I know. Do you have his office telephone number?"

"Yes."

"Anything else?"

Brian realized he had to tell Mary about Charlotte Palmer. She had to know where his feelings rested. He hoped she would not feel hurt by his love for Charlotte. He also hoped she would no longer expect him to visit her, to be with her. *Maybe we could be just friends?*

"I do want to see you. There are things I need to tell you."

"Good. I can't wait. You let me know when. Make it soon before I swell too large with child."

The distinct wail of an air raid siren filled the background from Mary's end. The Germans must be starting their nightly bombing of London, Brian told himself.

"They will cut us off shortly. I love you, Brian."

"I love . . . ," he said to a dead telephone line, ". . . you, Mary," he finished to himself.

Brian replaced the hand set into the cradle. He sat in the booth for a few moments lost in his thoughts. There was so much left undone. His focus changed to his transfer orders. He started to lift the handset to ask for the No.11 Group Headquarters number. With a bombing raid on London, the group controllers would be preoccupied. He replaced the handset one more time.

Brian joined his colleagues in the bar for a few nightly drinks. The others were well oiled by the time he arrived. He would have to stay sober, and try to call back later in the evening after the night raids were complete.

The bar room conversations were as they were most of the time. Brian joined in immediately without giving anyone an opportunity to speculate about his absence. He nursed his beer and accentuated his laughter to stay within the lines of the playing field.

The evening participants began to thin when Brian checked his wristwatch. It was 23:17. One of the No.604 Squadron night-fighter pilots joined the smaller group. Their aircraft had returned, and they had been released for the night. He waited for a few more minutes. The last two key observers, Jonathan Kensington and Roland Stockard, called it a night. Brian followed. Instead of turning right up the stairs to the bedrooms, Brian continued straight ahead into the sitting room and the empty telephone booth.

Pilot Officer Brian Drummond asked the operator for Uxbridge 2896. The woman answering the call did not identify her organization for security reasons. Brian asked for Air Commodore Spencer. Without making any commitments or indications, the operator asked him to wait.

"Air Commodore Spencer."

"Sir, this is Pilot Officer Drummond."

"Good evening, Brian" he said with a heavy, tired voice. "It is late for a line pilot. Anything wrong?"

"I know it's late for you as well, sir. I called your house earlier and talked to Mrs. Spencer. She told me you were still on duty. Our conversation was cut when the night air raid began. So, I waited until our night-fighters returned and were released."

"Yes, Brian, and it is quite late."

Brian felt a little silly babbling like he was. "Yes, sir. Sorry. I received orders today transferring me to Seven One Squadron."

"The Eagle Squadron."

"So, we heard. Anyway, I want to stay with Six Oh Nine."

"Seven One is an American squadron."

"I know, sir, but I've been with Six Oh Nine since nearly the beginning of the war."

The telephone line fell silent as neither officer spoke. Brian wondered what John Spencer was thinking about on the other end. *Maybe I called at a very bad time.* He certainly did not hear the usually friendly voice of his benefactor. Brian's heart jumped out of his chest. *Has he found out about his affair with Mary? Dear God Almighty, please do not let it be so*, he told himself, as the pounding in his chest focused his thoughts.

"I think you should follow your orders," Spencer said finally.

"Why?"

"Brian, you do not ask a senior officer why."

The junior officer swallowed hard. He had never heard John Spencer speak to him so sharply. *What had happened? Was it the discovery of the affair, or something else?*

"I am sorry, Brian. I am very tired. It has been a very long day. We have many Americans in the process of joining Fighter Command. We are forming a couple, maybe several, squadrons with the Americans, just as we have the Czechs and Poles. They are your countrymen, as well."

"Yes, sir. My apologies. I did not mean to question your position. I just want to stay with my squadron."

"You will do well wherever you are assigned."

"Then I should give up and go?"

John Spencer paused to consider the question. Brian thought the hesitation was a good sign. At least, he was thinking about it.

"All right, all right, you are a persistent sod. I shall look into the possibility. I can make no promises, you understand, but I shall give it a go."

"Thank you, sir. Thank you very much."

"Now, go to bed and get some sleep."

"Yes, sir. You should do the same."

"I am asleep already," John Spencer said, as he hung up the telephone.

———

Friday, 27.September.1940
RAF Middle Wallop
Middle Wallop, Hampshire, England
11:45 hours

Most of the pilots had drifted off into a mid-morning slumber, warmed by the sun through clear skies. Several others played a card game of some sort. Brian kept his thoughts on last night's telephone call, and could only wonder whether Air Commodore Spencer would be successful. He wanted to keep his positive thoughts on success, although as time passed, consideration of flying in an all American squadron nudged its way into his consciousness. He began telling himself that as long as he was flying Spitfires he could fly anywhere, but how could he possibly enjoy flying a Brewster Buffalo?

Any discussion between the pilots for the last few days centered on the success of the Germans in attacking the Supermarine factory. They all expected the call to come sending them south again to defend Woolston and Southampton. The poor Supermarine Works had been nearly bombed into oblivion, but the Germans kept coming back. The tenacity of the enemy seemed to well up from their recognition of the oversight regarding that very important target. The pilots expected the Germans to keep bombing the place until there was nothing recognizable at the site, perhaps in retribution for the success of the Spitfire.

The sun was approaching its local zenith when Brian decided it was getting too hot. He moved into the tent for some shade. Kensington, Kradil-cek and Johnson joined him inside. They just sat down in the folding chairs scattered about the interior and had not yet initiated any conversation when the telephone rang. Brian expected Corporal Warren to say the magic words, which she did.

"Scramble the squadron." As they ran out the door, she added, "It is Bristol."

No words passed among the pilots as they raced into the air. Darling headed them northwest toward Bristol. The check-in radio call confirmed their destination and kept them at low altitude. The possibilities disturbed Brian, as they probably did his compadres as well. The intercept vector and information had the earmarkings of what had become their nemesis squadron. Was this another repeat performance of three days earlier -- a feint toward Bristol with the real attack coming against Woolston or Southampton? There was a difference. The controllers must have sensed the need for reinforcing information. The Observer Corps had spotted the low-level raiders flying across Cornwall just west of St. Eval continuing north over the water. Four squadrons had been launched to intercept the skillful intruders -- two Hurricane squadrons from St. Eval and Exeter, a Spitfire squadron from Pembrey in Wales, and the Spitfires of No.609 Squadron.

'Spike' Darling's squadron was the last to arrive at the party. None of the pilots missed the red spinners and black noses on the mixed flight of Bf109 fighters and Me110 fighter-bombers. The more amazing element was the sight of Bf109s so far west. They were beyond the range of the German single engine fighter from any part of France. The same markings made Brian suspicious of possible modifications to enhance the service range of the German fighter. The clot of fighting aircraft moved up the mouth of the Severn estuary toward Bristol. *These guys are some tough puppies*, Brian said into his oxygen mask. The will and commitment of the Erpro210 pilots had to be admired no matter what side you flew on. Darling positioned the squadron to join the fight head-on in an attempt to break the concentration and aim of the bombers.

The Bf109s fought tenaciously to keep the British fighters off the Me110s. The scene looked like a mad swarm of hornets until the features of the aircraft popped out. They spread out to face the approaching swarm. Brian wondered if the risk of collision would be greater than the bullets, with so many aircraft engaged in the traveling fight.

Brian lined up on an Me110 pressing toward the Bristol Aircraft factory. The flashes from the nose of the German fighter-bomber signaled the beginning of the shoot out. Brian watched the German aircraft rapidly filling his sight. He kept the gunsight reticle high on the canopy and waited for the instant his opponent's wings spanned the inner ranging circle. Brian depressed and held his firing button as the two aircraft closed. Miraculously, the German's bullet stream missed Brian entirely while his danced over the engines, nose and canopy. The Me110 flailed for a moment, and then flipped on its back as Brian pulled up sharply and rolled to the vertical. He jerked his aircraft sharply to avoid several other British and German aircraft.

Once clear of the other aircraft, he kept his fighter in a moderate arcing turn. He saw a flash of a fireball below him. Green and brown parts came out one side while gray and black parts exited the other side of the fireball. Someone had collided and died.

Brian strained his neck muscles pulling his head back looking for a target. The sky instantly filled with Bf109s, Me110s, Hurricanes and Spitfires. The low altitude of the engagement compounded the intensity. A set of aircraft passed in front of him. The tail letters jumped out at him – 'PR-K,' it was Jonathan. Two Bf109s dogged him through the twisting, turning fight.

"'Harness,' pull back and through. I'll scrub 'em," shouted Brian.

Jonathan Kensington did as Brian asked. The range closed quickly. He moved his nose onto the closest German and fired. Several hits caused the attacker to break off. He fought with his bird, not responding quickly enough to the other attacker. Sensing Brian was close, the German pulled up and rolled out of the fight.

Brian gave chase.

Almost on cue, the Germans made a run at low level south trying to escape. They raced across Avon into Dorset. The Me110 rear seat gunners were more effective than usual. Then, Brian noticed the black dots coming toward them. *More Spits or Hurris*, he said to himself and smiled. *We've got them now.* In an instant, the angled features of Bf109s, these with yellow spinners joined to help their brethren. This had become a well-orchestrated and timed raid. They now faced fresh Germans.

The two Hurricane squadrons broke off due to low fuel, as did the Pembrey Spitfire squadron. The dozen German fighters made it through the earlier warning system without being intercepted. Too many unusual events made Brian question the condition of the air defense system. When he had time to consider the situation more carefully, he would be suspicious and concerned. At the moment, he wanted to face the new Germans.

The next phase of the engagement began with the head-on pass. Firing began early without consequence for both sides. The climbing turns started the two groups of fighters into the tangle.

The new Germans accomplished their mission -- engage the interceptors and allow their colleagues to make their escape. Brian did not take long to figure out that the new guys were also very good.

The fight wound its way up in altitude as each pilot sought the superior position into the next sequence. They spent most of their time protecting one another. As soon as Brian found a clear shot, a colleague in trouble demanded his attention. It was the same for all of them. As Brian rolled hard into another attack maneuver, he noticed the yellow spinners were now mixed with white spinners. The Germans were adding fighters.

The 'PR-A' Spitfire of Squadron Leader Darling dove past him just off his left wing. Two Germans followed him closely. He was in a very bad way. Brian rolled hard and pulled his nose through to engage his leader's attackers quickly. Brian fired an out of range burst to get their attention. It did not work. He pushed his throttle harder although it was already at the forward stop. The engine roared at emergency power, and the scream of the air-stream mounted quickly.

The crack of 20mm shells exploding on the skin of his fighter startled Brian. He rolled sharply in the opposite direction trying to spoil the aim of his attacker. He completed a large barrel roll and kept the nose coming up into a steep climb. Tracers passed his cockpit. The size of 20mm tracers so close to his cockpit sucked the breath out of his chest. He jinked back in the opposite direction, and jerked the nose down into a steep dive. His pursuer was very good and not easily shaken. His only hope now was a high-speed dive where his control forces would be slightly more manageable than those of the Bf109E-4. Several RAF pilots had been able to force Germans into dives they could not recover from before they hit the ground.

Tracers continued to fill the sky around him. Airspeed built up rapidly. The controls stiffened as the airspeed needle moved quickly clockwise across the face of the indicator dial. A few more hits and shudders. This guy was not going to give up.

Then, he saw and felt hits across the top of his right wing, into the engine and forward edge of the cockpit. Brian groaned into his oxygen mask with the intense, burning sting of shrapnel in his right leg. The engine burst into flame lapping frantically around the outside of the cockpit. Oil immediately coated the windscreen obliterating his forward vision. Brian used both arms on the control spade to pull back. He had to get out of the dive, or he would impact the ground.

Brian watched the attitude gyro slowly move toward the representation of the horizon and wings level. Fire continued from the engine compartment. The fuel tanks between the cockpit and the engine gave Brian a sickening sensation. Instinctively, he checked his fuel gauge. Only five gallons remained. That was better than full tanks, less fuel to burn, although the fuel vapor filling the tanks was more explosive. The black oil film smeared its way across his canopy. In minutes, he would not be able to see. Flames burst into the cockpit filling the space with smoke. Brian thanked God he kept his oxygen mask and goggles on. The flames disappeared as quickly as they had come. He had to use his hands to pat out the small flames on his legs. The engine started to miss a few cylinders in the firing order and began to backfire in protest of its injury. Large backfires rocked the aircraft, and then the big Merlin engine stopped.

Brian lowered the nose to keep some airspeed, holding around 100 mph. At least the fire was gone, but it could come back in an instant. Altitude -- 2,500 feet and descending rapidly. He had to get out of the stricken fighter. Brian pulled the canopy latch. It would not move. The fire was out as best he could tell. The holes in the cockpit skin allowed the smoke to dissipate quickly. He had to get out of the fighter, now turned glider. Brian checked the edges of the canopy. He found the problem. A 20mm shell, probably the one that sprayed shrapnel into his leg now soaked red with his blood, mushroomed the metal railing of the canopy and bent the canopy latch linkage. He grabbed the handhold, strained and screamed, but the canopy did not even twitch. He was entombed. 800 feet. Not much time left.

The entire right side on the canopy was opaque from the oil. A small patch at the top and a very small hole, about the size of one of his mother's country biscuits, remained at the aft left edge of the canopy. 300 feet. There was no time for other actions.

Brian concentrated on the hole. He saw trees, then the ground, and then more trees. He checked his airspeed. 80 mph. He pulled back a little more as the ground came up to him. The ground jumped at him, he pulled back more. Impact!

Friday, 27.September.1940
Romsey, Hampshire, England
13:05 hours

Peter Carrwood and his two young helpers had just finished plowing the barley stubble into the rich, brown, sweet earth, when the violent ballet began over their heads. They watched with removed fascination the maneuvering of the aircraft. The distinct chatter of machine guns added to the poignancy of the otherwise graceful dance in the sky. The magnetic attraction of the aerial engagement and the silent approach of a lone Spitfire trailing a long streamer of black smoke kept them from noticing the stricken aircraft until it was nearly upon them.

"What the . . . ," Carrwood grunted. The initial shock froze the three men, and then Peter recognized the situation and sprung into action. "Johnny, run to the lodge. Call the emergency services. Tell them we have an aeroplane crash. Darren, step lively to the lorry, lad. Fetch the spades, maybe the ax." Neither man moved immediately. "Move smartly. We have an airman who needs help."

Peter Carrwood started to run toward the area where he thought the aircraft would land. He recognized the shape of the Spitfire fighter. The large and nearly covered markings on the tail, 'PR,' stimulated him to say aloud, but to himself, "He's one of ours."

The frozen propeller, blackened nose and visible flames emanating from several holes behind the propeller added confirmation, not really necessary, that the pilot was in serious trouble. The movement of the aircraft as the pilot tried to land told Carrwood the pilot was still alive and struggling with his broken machine.

The aircraft flattened out just before the first impact. An enormous explosion of dirt marked the first hit, and then the aircraft emerged from the dirt cloud, fluttered several times and hit again. More dirt and a couple of additional impacts established a trail across his field. After the third impact, the nose pitched up slightly, and then nosed down sharply burying the nose in the dirt.

"Hurry," Carrwood screamed as loudly as he could.

He reached the partially buried aircraft with its nose in the soil and its tail thrust into the air like a photograph of ostriches in Africa. The aircraft continued to smolder. The heavy, choking smell of the dirt cloud and acrid petroleum smoke caused Carrwood to struggle for air. He used his shirt in an attempt to filter the fouled air. Breathing became a chore, but the possibility of the aircraft's fuel igniting dominated his thoughts.

Carrwood tried to find a handle on the canopy. The thick coating of oil made the entire canopy very slippery. He tried to find an edge, a latch, an opening of some kind to pull back the hood. He wiped feverishly at the oil on the transparency. The view of the interior was blurred and dark, but everything seemed black and covered with oil except for the bright red splatters of blood, and the dirty, abused torso of a man slumped forward in his harness straps.

"Dear Mother of God," he said more to himself.

Frustrated with his lack of progress, Carrwood stood up straight, stepped back a couple of steps, and shouted, "Darren, I need the ax now."

The young man was nearly halfway to the crashed aircraft. He had the ax and two shovels. The yellowish-orange tongues of flame burst up through a gaping hole in the nose. The cockpit began to fill with black, ugly, oily smoke. "Hurry, lad," he shouted, although he could see the young man struggling across the soft dirt.

Peter grabbed the ax. "Pitch some dirt into that hole," he commanded, nodding toward the flames. "Put those flames out." The young assistant sprang into action.

Carrwood hesitated, uncertain what might happen if he struck the glass, or whatever material it was, with the sharp ax. He looked into the cockpit once more. The pilot's head and face were down. He decided to first tap, and then strike the ax near the forward edge. The canopy fractured, not like glass, but more like pottery.

He checked Darren. He shoveled like a demon possessed, although there were no signs of flames any more. Carrwood used quick short strokes to cut away the remainder of the canopy.

"Help me pull the pilot out," Peter said. Darren responded promptly.

Carrwood leaned into the pilot's space. Soot, oil, blood and even dirt coated every square inch of surface. He grabbed the pilot's left shoulder strap and followed it down to the buckle, unlatched the harness. The pilot fell forward like a gush of pudding bursting from a large vat.

The two men struggled to grip the pilot's clothing and lift the large, limp man. Both men fought as much as their lungs would allow, demanding air for the exertion against the smoke, petroleum and dirt in the air around them. Carrwood gave up trying to be gentle. They had to get this man out and away from the broken machine. "Grab whatever you can," he commanded. "Ready, heave," he added, as both men strained against the pilot's weight and their mounting fatigue from insufficient oxygen.

The pilot came free. They held his torso. The man's head, completely covered by his oil-blackened helmet, goggles and mask, flopped around like a rag doll. He was only halfway out when chords from the man's headset and helmet drew taut.

"Cut 'em," Carrwood gasped.

Darren reached for the large work knife on his hip, drew it and swung it swiftly across the several chords. They parted causing Peter to lose his balance. He fell backward holding onto the pilot, pulling him down with him and on top of him. Carrwood scrambled to get out from under the larger man. Both men, gasping for breath, reached under the limp man's arms and dragged him ten or so yards away from the aircraft, and nearly to the tree line along the north edge of the field.

They lay the blackened man down. Carrwood stopped to clear his lungs of the foul air.

"Go . . . see . . . where . . . Johnny is," Carrwood said. Darren sprang to his feet and took off running.

Peter Carrwood looked down at the pilot. His entire head was covered by his helmet, goggles and mask, all covered with oil and soot. He tried to unfasten the slippery straps of the mask, and then finally in frustration nearly ripped them away.

The fresh, fair skin under the mask contrasted sharply with the black coating on the rest of him. Blood from his mouth highlighted the spot even more vividly. He pulled off the goggles, and then unstrapped the helmet. Blood covered the right side of his head and matted his light brown hair.

Carrwood stared at the young man's face. He was a very good-looking, almost aristocratic looking, man, and apparently quite young. He returned to the situation. Carrwood continued to open his clothing to assess his injuries. Several wounds dotted his chest, like ice pick wounds. His right trouser leg was torn just below the knee. Red dominated the area. The jagged edges of a white bone protruded from the torn cloth and flesh.

The haunting wail of the emergency vehicle sirens grew rapidly in strength. The fire truck and ambulance hesitated at the country lane probably trying to find a way into the field. Carrwood waved broadly to them. They continued down the lane to the edge of the field, and then drove right through the small fence and along the edge of the field where the soil was more compact.

The emergency services personnel moved quickly, professionally and precisely. They assumed the care of the pilot. The fire fighters tore open the engine compartment to ensure there was no fire. The medical personnel examined the pilot. One of them looked directly into Peter Carrwood's eyes and shook his head. Not a good sign, Peter told himself. However, they moved like blurred images to help the stricken man.

The ambulance was the first to depart with its siren blaring. The fire fighters finished their work. The senior man came over to Peter.

"You did right, sir."

"Thank you, although I don't know if we saved him."

"You have done the best you could. The surgeons have him now. They will do right by the man. If they can save him, they will."

"I hope so."

"Now, there may be unexpended ordnance on board the aircraft. I would strongly suggest you stand clear of the wreckage until the explosive ordnance disposal chappies have a good pass at her. The Air Force should be out here within a few days to clear the wreckage off your property. Do you understand?"

"Certainly."

"Any problem with that, sir."

"No."

"Then, if you will excuse us, we shall be off."

The fire fighters departed without their sirens. Johnny joined them. Peter and Darren sat down and simply stared at the Royal Air Force fighter. None of them said anything for many minutes. Johnny was the first to ask what happened. The three men compared observations and shared the story of that afternoon.

Saturday, 28.September.1940
Hampshire Central Hospital
Winchester, Hampshire, England
15:45 hours

Flying Officer Jonathan Kensington had been torn by the tragedy of yesterday's losses. The squadron lost its energy and will to fight when they landed and understood the losses. Pilot Officer Kormer 'Curly' Mansek, one of their two Czech pilots, died instantly when he collided head-on with an Me110 fighter. His best friend, Pilot Officer Brian 'Hunter' Drummond had been shot down, again. It had been two months since the last time. Jonathan felt a twinge of guilt that it was Brian who pushed so hard and took so many chances that he risked what happened yesterday, but he also felt the threads of superiority. He had not been shot down yet in the battle. Jonathan knocked on the wooden door frame outside the surgeon's office to avoid bad luck.

The losses for No.609 Squadron all of a sudden had become more staggering. Several of the pilots saw Mansek's collision and the explosion. It was a pilot's worst nightmare. They had lost eight pilots since the battle began. Squadron Leader Darling told them replacements were on the way, but none of them had arrived. They needed to gain strength.

As Jonathan waited in the medicinal smelling office, he silently thanked Squadron Leader Darling for releasing him from duty, and cursed Darling for asking him to represent the squadron. He also did not look forward to carrying the news to Mrs. Charlotte Palmer. This task was not going to be easy no matter how you cut it.

"Mister Kensington," the nurse called. Jonathan stood. "Doctor Morgan can see you now."

Jonathan did not answer other than he walked through a small, densely populated, administrative office to one of several peripheral offices. The room he entered was quite small. There was barely enough room for a desk, two guest chairs and a modest bookcase filled with medical books, papers and a few memorabilia.

"Good morning, I am Doctor Morgan," said the thin, nearly bald man dressed in a white overcoat with a white shirt and red necktie. He extended his right hand to Jonathan.

"Good morning, doctor. I am Flying Officer Kensington. I am here for Brian Drummond."

Doctor Morgan motioned to one of the chairs on the opposite side of the desk. He sat down, shuffled through a stack of folders. "Yes, now, let's see," he said, opening one of the folders and leafing through several papers. "I suspect you have been through this before since your colleague has been here before."

"All too familiar, I'm afraid."

"Yes, well, if you will permit me, I shall speak frankly." Morgan paused to receive a head nod agreement from Jonathan. "The bottom line is, your friend is quite lucky to be alive. He was brought in from . . . ," he paused to search the papers in the folder, ". . . from Romsey, yesterday afternoon, by the emergency services ambulance unit."

Jonathan swallowed hard against his sinking heart. "Will he survive?"

"Difficult to say, I'm afraid. We probably will not have a view until he regains consciousness. It could take days or weeks, for that matter."

"How serious?" Jonathan asked, and then regretted asking the question.

"Very, is the best adjective. He went into emergency surgery shortly after his arrival. He had a skull fracture, two broken ribs and a punctured lung, multiple shrapnel wounds through his torso and legs, and a rather nasty compound fracture of his right leg. In addition, he experienced serious burns on his legs, hands and neck. Your colleague took quite a beating."

"He was trying to protect our squadron leader from two Germans. I saw three Germans jump him." Jonathan swallowed hard, again, as he recalled the engagement. He felt a compelling need to explain the situation to the doctor, although he had not asked for the description. He cleared his throat. "I was not close enough to help him. We all had our hands full. They shot his engine out and apparently he could not get out of the aeroplane."

"Yes, most unfortunate. On the lucky side, a farmer and his two field hands happened to be close by, so they tell me. Anyway, your friend is lucky.

He was in surgery most of the night. He faces at least one or two more oper-ations to remove the remainder of the shrapnel, and it is too early to tell what therapy he might need for the burns."

"Will he be able to fly again?"

"Dear boy," the man said like he poked Jonathan with a stick, "as I said, he is lucky to be alive. Until we know how serious his head injury is, we can make no prognosis. Our most critical task at the moment is stabilizing his vital signs and slowly coaxing him back to consciousness. It shall be at least six weeks for his broken bones to knit."

"He lives to fly," Jonathan said almost to himself.

"He had better concentrate on living, full stop. He is in a most critical condition, Mister Kensington. Does he have family close?"

"He is an American. He has a woman friend. My family is about the closest, well, maybe there are a few others who are close."

"I would suggest you seek as much help from his friends as you can. He needs family and friends to bolster the fight in him. He needs moral and emotional support in his current struggle."

Jonathan looked at the physician with an odd expression of incredulity. He asked himself, *how am I going to give an unconscious man moral support?*

Doctor Morgan nodded his head in recognition. "We do not know if he can hear us, but we want to assume he can."

"I shall take that responsibility."

"Very well, then, are there any other questions?" the doctor asked, as he closed the folder.

"No, sir."

Jonathan left the doctor's small office with a heavy, cold stone in his gut. Brian had been banged up pretty good. The prospect of not flying again would not sit well with Brian, or any of them. He asked to see Brian. The nurse led him through the corridors to a set of special rooms. Medical equipment filled most of the room with a singular bed in the middle. Several nurses remained in the room. One checked heartbeat and blood pressure. The attending nurses nodded to the first visitor. No one spoke. The room smelled of death, and it sent a shocking chill through Jonathan's body. It looked like the inanimate mound of flesh in the bed was part of some grotesque medical experiment.

The man in the bed was rather large, but not recognizable. Nearly his entire head and neck were bandaged. His right leg was raised in a sling. A sheet covered his body except for his arms. Jonathan looked to one of the attending nurses, touched his right index finger to his lips, and then pointed to the body. The nurse nodded her head, and then gestured toward Brian.

Jonathan walked to the far side of the bed and Brian's right side. He thought for a moment about what he was going to say, and then leaned forward near Brian's right ear. He could smell burned flesh through the biting odor of antiseptic.

"Brian," he whispered. "It's Jonathan. I'm here with you, my friend. We need you back in the cockpit. You need to mend quickly. We will take care of everything," he said, not really knowing what that meant. "You just concentrate on getting better." Jonathan grasped Brian's right hand. It was cool, but not cold. He hoped for a response, but none came. "Hang in there, mate."

Saturday, 28.September.1940
Standing Oak Farm
Winchester, Hampshire, England
17:00 hours

Charlotte Palmer walked back from the barn. The afternoon chores had taken much longer than she anticipated. Her two elderly helpers, Lionel Bridges and Horace Morgan, while very well intentioned, were becoming less helpful. Lionel, nearly 72 years old, dropped and broke more items than he actually moved. The small wages she paid them made the limited work she received from them roughly a wash.

"Good day, Mrs. Palmer," said Mister Bridges.

"Yes, good day, madam," added Morgan.

"Thank you for your help, Misters Bridges and Morgan. We shall see you Monday. Enjoy the Lord's day."

The two old men rode away in a barely functional automobile. The drone of aircraft engines overhead stopped her halfway to the house and drew her eyes skyward. A half dozen small airplanes flew toward the southeast at a moderate altitude. She could not avoid the thoughts of Brian. A smile grew across her face.

It had been nearly five days since she had heard from him. The good weather of the last week must have kept the Royal Air Force very busy. She longed to hear his voice. She wanted more, much more, but would accept just his voice on the telephone to know he was safe.

Charlotte continued into the house. It was late afternoon, and her stomach had begun to growl. She opted for some hearty cheddar cheese with crackers, and a muffin she made this morning. A large pot of aromatic tea accentuated the flavors. The pinch of scarce sugar along with the richness of fresh cream from the morning milking added smoothness to the hearty tea. Brian liked the light, thrown together meals they often begrudgingly made time for among their more intimate interests. Everything seemed to remind her of Brian.

He was so young, almost innocent looking, and yet the hidden strength she felt around him defied his years. His youthful vigor and zest for life had rubbed off on her. Charlotte admitted to herself many times, she had never felt so alive, so full of energy, so optimistic about the future despite the ugly war around them. The life growing in her womb brought an even greater connection he might never know.

She leaned back in the chair and placed her hand on her belly, as if she might feel the baby move within her. She and Ian had never discussed having children, nor did it seem they would. With Brian, her pregnancy felt natural, normal and the right thing to do. It was his proposal of marriage that occupied most of her thoughts over the last few days.

He was so young -- seven years younger -- and yet, she felt no age gap between them. She wanted to marry him because she loved him. The swiftness of her attraction and connection to him startled her when she considered the events. Conversely, she did not want to marry him because of the baby, although she could not deny the safety and legitimacy the marriage would bring. Charlotte knew all too well the ostracism that comes to unwed mothers, even accomplished women like her. Standing Oak Farm gave her some degree of independence and protection, but she could not become a hermit. The struggle she faced at the moment was how to tell Brian of her pregnancy before she began to show signs, and not allow the pregnancy to become the reason for their marriage. Yes, she did want to marry him for the right reasons.

The sounds of an automobile engine, squeaking brakes slowing the car down the hill and the crunch of the gravel under its wheels quickened her heart. *Could it be Brian, coming to surprise me?* She smiled, as she walked to the front window.

A taxi carried a passenger she could not see at first. The first clue as the taxi turned in the drive was the medium blue of a Royal Air Force uniform. She sprang to the door wanting Brian to see her excitement.

A shorter officer backed out of the taxi, paid the driver and turned toward her. The man smiled meekly, and then warmth vanished. She recognized the face, but agonized over his name -- Brian's friend from his squadron. He walked toward her, and his expression turned more serious.

Her mind flashed back to a moment three months earlier when a Royal Navy commander arrived in an official motorcar. Her heart seemed to stop beating. Blood drained from her head. She felt faint, dizzy and nauseous. She knew why Brian's friend was here. Charlotte's knees buckled. She fell hard into a pile on her front porch.

The man jumped toward her too late to help. He placed a hand on her shoulder, as she began to weep silently at first. Her throat closed off the air, but she did not care to breathe any more. She had lost too many people in her life. *Why, for the first time since Ian's death, when I actually began to love again, did God decide to take Brian from me? Why was life so cruel?* She felt the man's strong hands under her arms lifting her to her feet.

"We should go inside, Mrs. Palmer," he said.

He knew her. *Why can't I remember his name?*

She tried to stand. Her knees nearly buckled again. She could not move. Charlotte fought to control her breathing and gather what strength she could find. She stiffened, sniffled back her tears, took a deep breath, walked very slowly into the house and turned to face the man.

"My apologies, sir. I do not like turning into a blubbering pile, but I have lost so much, and now I have lost, again."

"Not yet," he said.

Charlotte froze, somewhat shocked by his words. "What was that?" she asked, as if she had not heard him correctly.

The man smiled a warmer, more open smile than his nervous smile when he arrived. "Brian has been shot down, again, I'm afraid, but he is alive."

"Where?"

"He is in Hampshire Central Hospital at Winchester."

"That's where he was before."

"Yes."

Charlotte swallowed hard and clutched her sweater around her shoulders, crossing her arms over her chest, as if a chilled wind had just blown through the house. "How is he?"

"In bad shape, to be honest, but the surgeons are hopeful."

"I should go to him."

He held up his hand and took one step toward her. "No. Now is not the time. The surgeons advised me this morning that he needs another operation later this afternoon."

"Another?"

"According to the doctor I talked to this morning, they tended to him most of the night. He has some rather serious injuries, so they tell me."

"How serious?"

The RAF officer stared at her, as he considered her words and how much he should say. "They suggested that if you are able, you should visit him tomorrow afternoon. They tell me his best medicine at this juncture is a friendly, caring voice."

"Voice?"

The officer was puzzled, as if he did not understand the question. Several moments passed before he grasped her concern. "Yes, well, not to worry just yet. He is unconscious. He apparently took a rather severe blow to the head when he crashed."

"What happened?"

The man again considered how much he should say. "We were in a rather nasty row with a particularly skilled group of German blokes, and outnumbered as usual. Our squadron leader got jumped by several of them, Brian waded into the middle of them to help the skipper, and one of them managed to get some lucky shots. As best we can determine, his canopy must have been jammed shut, and he could not get out."

"Dear Merciful God above."

"He has the best care."

Charlotte turned without another word and walked toward the kitchen. "Would you like some tea?" She turned sharply realizing her *faux pas*. "I am so sorry. I have been terribly crass." She walked toward him. "Excuse my absent-mindedness. I know you are Brian's friend, but I seem to have forgotten your name."

The officer laughed. "Of all the things you must be grappling with, my name is unimportant. However, I am Jonathan Kensington, and indeed, I am Brian's friend. He is my best friend."

"Yes, of course," she answered noticeably blushing. "How stupid of me." She extended her hand to him. "It is good to see you, again. I only wish under better circumstances. Would you care for some tea and biscuits?"

"That would be lovely, if you don't mind."

"Not at all," she said, as she turned again toward the kitchen.

As Charlotte made a large pot of tea and gathered up some fresh cookies, they talked about other things, the weather, the farm and his family home, Carlingon Castle, near Newcastle-upon-Tyne. He was able to laugh, maybe in an attempt to bring some lightness to their conversation, but she remained somber.

Charlotte struggled to keep an amicable façade, but the conflict within her made the task quite difficult. She believed she truly loved Brian. She wanted Brian. This damnable war drained the little pleasure that remained. The prospect of losing another person she loved made her cold. She wanted to sink into a cave and blot out the world that brought so much misery to her life.

They sat, sipped their tea and Jonathan ate several cookies. Charlotte could barely drink her tea, and the thought of eating anything accentuated the nauseous waves rolling through her body. As he tried to carry on a conversation, Charlotte sank deeper into her cave.

"It will be all right," he said, finally acknowledging her morose state.

"No, it won't."

"Why do you say that?"

"Because I'm pregnant," she blurted out, and then felt mortal shame for her loss of control. "Oh, dear God, I am terribly sorry, Mister Kensington. Please accept my apologies," she choked out, as she began uncontrollably crying -- the emotional reaction she disliked so much, but found herself, at times, powerless to avoid. She saw his startled expression, and that took the last vestiges of control she held on to at the moment.

Charlotte covered her face with both hands and lowered her head to her knees. Her cries became more audible and choked. Jonathan moved to sit next to her and rubbed her back trying to soothe her. His recognition of her emotions made her fall deeper into the pit. She felt so lost, so alone. The more he tried to comfort her, the more alone, out of control and desperate she felt.

The war was a long way from being over, and Charlotte told herself she could not stand the threat of more loss and more pain. She told herself she would rather feel nothing. The pleasure was uplifting, invigorating and joyous, but the valleys that came with the peaks were unbearable. It was simply easier, and better, to feel nothing until this madness was past.

What am I going to do with a new baby? Tending a large farm with no help other than two elderly, well-intentioned, but barely functional men took most of her time without the added burden of a baby. On top of all that, the baby's father was a fighter pilot in the Royal Air Force with a brutal war all around them. And if that was not enough, he was seven years younger than she was and an American volunteer, to add the icing to the cake. He could be killed, or decide he had had enough and leave. Dear God Almighty, how did she get herself into this dreadful situation?

Charlotte Palmer fought the battle inside her to gain control. She took several deep breaths to calm her chest. She sat up slightly, retrieved a handkerchief from her sweater pocket and dabbed the tears from her cheeks and eyes. Charlotte stood, walked around the table, drew her shoulders back and turned to face Jonathan.

"I must apologize . . . ," she stopped her words when he held up his hand.

"No need to apologize, Mrs. Palmer."

"You simply must call me, Charlotte."

He chuckled. "Then, you must call me, Jonathan."

They both smiled and nodded their heads.

"I must apologize for my dreadful behavior." She did not stop with his hand this time. "I did not mean to burden you with my problems."

"It is quite all right. Does Brian know?" he asked.

Charlotte felt tears coming again, but strained to hold them back. "No."

"You must tell him."

"I can't."

"Why?"

"It is my problem. He has enough to worry about."

"Nonsense, if you would allow me to be so bold. He has told me he asked you to marry him."

Those words nearly took her over the edge, again. She swallowed hard several times and gritted her teeth against the urge to cry. "Yes."

"Well, then, there you have it."

"What do you mean?"

"Charlotte, I shall speak plainly, and I hope I do not offend you. I have lived closely with Brian Drummond for more than a year of this bloody war," he said, and then caught himself. "Excuse me. I did not intend to curse."

"I have heard much worse."

"Yes, well then, as I was saying, Brian is a good man. He is a decorated, accomplished, exceptional pilot. The country needs his skills. So, we have a mutual need. You need him to be a husband and father to your child, and we need him to help us defend this country."

"Since we seem to be talking about needs, I am certain you will agree, he does not need the added burden of a family to worry about in this war. He has told me, his concentration and focus are critical to his success."

"He has told you correctly. He is one of the best in that respect."

"Maybe it was a thought about me that got him shot down."

Jonathan laughed hard. Charlotte felt a little silly, as if she said something completely out of context. She wanted to confront Jonathan's reaction.

"His desire to help a fellow aviator in trouble was what got him shot down. By the way, I might add, his aggressive action in the air enabled our squadron leader to escape the deadly situation he was in, so Brian accomplished what he intended. It was a terrible price, agreed, but he did it."

"So, how does that change my situation?"

"Again, I hate to poke my nose in where it does not belong, but Brian needs my help, and maybe you do as well." Charlotte nodded her head and lowered her eyes knowing that he was correct. She did need his help. "Do you love Brian?"

"A rather personal question from a man I barely know."

"Yes, quite, but as I said early on, I am a plain speaker, and this war has removed much of my prudence and delicacy. Do you?"

"Yes."

"Then, let us take a look at the facts, as I know them. You met under rather unusual circumstances, grew to love each other, enjoy the company of one another and now have a child between you. A rather tidy package, if you ask me."

Charlotte walked over to the sofa, sat on the other end from Jonathan, poured herself a fresh cup of tea and freshened his cup. After adding a pinch of sugar and some cream, she took several sips, placed her cup down and sat back into the sofa.

"As you said, you are a rather plain speaker."

"My apologies, if I have offended you."

"No need. You are correct. I have enjoyed him more than any other man in my life. I want to be with him, but I am so worried."

"Let me tell you a little more about our mutual friend. He has an incredible will," *except when it comes to women*, Jonathan told himself, "and a very clear sense of who he is and what he must do. He needs us at this moment. He is engaged in a tough struggle, and he must maintain his will to survive. I truly believe, if he knows you are with him and especially if he knows he has a child to live for, he will do what he must do to recover from his injuries. He needs you, Charlotte. Go to him tomorrow, and tell him how you feel. Tell him about your child. If you truly love him, marry him. It will make him an even greater person."

Charlotte liked the words Jonathan offered to her. She considered for an instant, they might be just words he thought she wanted to hear, but the thought passed quickly. *He was right. He had to be right.*

"Then, so it shall be. I will go to him tomorrow."

"Smashing. Absolutely smashing."

The conversation returned to other topics. Charlotte took Jonathan on a short tour of the farm, as she had done for Brian many weeks earlier. Jonathan was finally able to draw laughter from her. Charlotte made dinner for them. As the taxi arrived, Jonathan kissed Charlotte on the cheek and wished her good luck for her visit tomorrow.

Chapter 8

We acquire the strength we have overcome.

-- Ralph Waldo Emerson

Sunday, 29.September.1940
Hampshire Central Hospital
Winchester, Hampshire, England
14:30 hours

Week 13

Charlotte Palmer found herself staring at the two-story, brick hospital. The queasy sensation in her gut confirmed what she began to realize -- she was growing to hate hospitals. She did not like the antiseptic smell barely masking the odors of death and disease. Since Jonathan's visit to the farm yesterday, she struggled with the words. *What and how will I tell Brian? What will I see in there?* From Jonathan's description, the sight of Brian, broken and bandaged, would not be pleasant. Charlotte prepared herself for what she was about to hear, see, smell and feel. She wanted to be strong, to help Brian recover from his injuries.

Charlotte took a deep breath and muttered to herself, "Be strong," as she walked toward the main entrance to the hospital.

She asked the elderly woman behind the receptionist's desk for Brian Drummond's location. The woman checked her large journal register, and then looked into Charlotte's eyes. The woman communicated, condolences, with her eyes, but she said, "One moment, please," as she lifted the telephone handset. "A woman visitor to see Patient Drummond, madam," she said. She received a set of instructions, as she nodded her head several times as she listened. "Are you family?"

Charlotte cleared her throat of nervousness. "Not exactly. I am his fiancée."

The woman repeated the message, and then looked to Charlotte again, this time with an apologetic expression. "So sorry, your name, please."

"Charlotte Palmer."

"A Miss Palmer, Charlotte Palmer," she said. The woman looked at Charlotte, nodded her head and returned the handset to the cradle. "A nurse will be down shortly to escort you to the doctor's office."

"Thank you."

Charlotte had just walked across the small lobby and sat down when a rather large nurse in a typical white uniform summoned her. She followed the woman down several corridors until she led her through a door into a smaller lobby, and motioned toward a particular door. A man in a white smock stood from behind the desk in the small office.

"Good afternoon, Miss Palmer. I am Doctor Morgan." They shook hands, and he motioned for her to be seated. "Flying Officer Kensington advised me yesterday you would most likely be stopping by to see Pilot Officer Drummond." Charlotte nodded her head, not wanting to speak. "We do have some good news, he is showing signs of improvement, but he still must recover from some rather serious injuries, I'm afraid."

"How serious?" she asked, although Jonathan had told her quite a bit of information.

"Well, I suppose that depends on your point of perspective. He has injuries to his head, chest and right leg. Our most immediate concern is his concussion, one or more strong blows to the head. He remains unconscious. We need him to regain consciousness before we can truly assess the extent of his injuries."

"Will he recover?" she choked out.

"Difficult to say, I'm afraid. Until we can fully evaluate the extent of his head injuries, there is no way to truly say."

Charlotte realized the doctor was not going to tell her very much, and she did not want to waste any more time. She wanted to see Brian, to be with him. "May I see him, now?"

"Yes, certainly. I must remind you, he is unconscious. We have no idea whether he can hear you or not, but it is very important that you remain positive and upbeat with your voice. We encourage you to touch him gently and give him words of encouragement. I must also tell you, he is heavily bandaged. He did have rather extensive injuries, you know."

"Yes, sir. I understand."

The doctor asked his nurse to take Charlotte to Brian. She thanked the doctor for his time, although she did not feel thankful. The nurse led her to a different part of the hospital. She stopped in the corridor near a set of rooms, and repeated the cautions and instruction as if they were rehearsed.

A single bed with a menagerie of medical equipment surrounding it and a large person under sheets and two attending nurses occupied the entire room. There was no way to identify the patient from the doorway. Her heart pounded in her chest. She gasped slightly and immediately covered her mouth to prevent any further sounds escaping.

His right leg was heavily bandaged and elevated above the bed. There was virtually no skin showing. *How will I ever be able to recognize him? Was this inanimate patient really Brian? What if they had the wrong man?* She swallowed hard several times choking back her nausea. Her legs turned to wet noodles. She reached for the door jam to brace herself. The large nurse grabbed her from behind as one of the attending nurses came to her. The younger woman looked into her eyes and gave her strength.

Charlotte gradually regained her composure. She straightened her dress suit coat, and then walked toward the left side of the bed. She looked down through the bandages covering his head. His face was noticeably swollen, his eyes were shut, but it looked like Brian. Charlotte felt herself weaken, again. One of the nurses guided her into a chair next to the bed. Now, Charlotte was becoming angry with herself. *Why am I so dizzy and light-headed?* She prided herself on her strength. *Why am I having so much trouble? Maybe this is love, maybe my pregnancy, but something had to change. I must get control of my emotions.*

Charlotte stood and leaned toward one of the nurses. "May I see him?" she asked.

"Are you sure you want to do this?"

"Yes."

"Very well, then, I must caution you, he was badly injured in the crash."

"I understand."

With one nurse standing beside her and holding her waist, the other nurse slowly pulled back the sheet covering him. Square bandages covered many spots across his chest. His right arm was nearly completely bandaged. Seeing the exposed juncture of his legs caused her to flush, not because she had not seen him before, but because he was exposed in front of other women. She tried to be strong and ignore her emotions. His left leg was the least bandaged of his extremities with only wraps above and below his knee. As Charlotte took in his whole body, there was no doubt in her mind that it was Brian lying still before her.

"Thank you," she said. The nurses covered him, again. "I need to talk to him." They nodded. "In private," she said with a modicum of insistence.

The nurses looked at one another. The older one said, "You must encourage him. We will be just outside, if you need us."

Charlotte nodded and waited for them to leave. She stood next to his head. She lifted the corner of the sheet to make sure she did not touch any wounds. She placed her left hand on his left shoulder, and felt his warmth and solid muscles. She leaned next to the place where his left ear would be and said, "Brian, it is me. I am here with you. You can feel me with you." She stopped to look for any signs of response. There were none. "The doctors tell me you shall fully recover," she fibbed.

Charlotte rethought the words she wanted to tell him. She might need to repeat everything once he regained consciousness. *I might need to repeat everything once he regains consciousness. What if my news is not what he wants to hear? Maybe he would be upset, as some men were when presented with such news? Maybe I should wait until I can see his reaction?* Charlotte recalled Jonathan's words to her yesterday. He seemed to be absolutely certain she should tell him, to help him recover, to give him something to live for through his struggle.

She kissed his shoulder, and then his nose and could barely reach his cheek through the bandages.

"Brian, I want you to know from the deepest part of my heart and my soul, I love you very much, more than I have ever loved another person. You are a very special human being. You asked me to marry you a week ago. I want to tell you I have thought about nothing else since then. I want to marry you, Brian, but I am so worried about you. If you have a family to worry about, you might not be able to concentrate. I do not want this dreadful war to claim you from me. I want you all to myself."

She paused again and stood up to search for a response. She clasped his left hand. It remained limp and without strength. Charlotte pulled the chair closer to the head of the bed and sat down. She leaned forward slightly nearly resting her head on the bed next to his bandaged head.

"I want you to know," she whispered to him, "you have so much to live for, and not just flying and me. I want you to know I am simply ecstatic that I carry our child. I am pregnant, Brian, and our child needs a strong father. I want to marry you and raise a family you can be proud of for all time."

Tears of happiness rolled down her cheeks as she stood again to take in a better view of Brian. She pulled the sheet down to his waist hoping to see his chest shake or his hands twitch. Nothing moved other than the very slow, barely perceptible expansion and contraction of his breathing. Charlotte dabbed away her tears and retained her smile as if he could see her.

She took the moment for a more careful observation of his bandages. Although the medical personnel had not told her, from the information Jonathan Kensington provided, there were probably jagged shrapnel wounds under the bandages along with the surgery to re-inflate his collapsed lung. She could only wonder and imagine the extent of damage to his head and right leg. She would have to wait to know more. Then, a staggering thought came to her.

What if his injuries give him a permanent disability, and he could no longer fly or serve in the military? The thought came to her like being dowsed with both scalding and icy water at the same time. Her head hurt from the pounding in her chest and head. She might have him all to herself, and they might live as a proper family. In contrast, she knew how much he loved and even cherished his flying, especially Spitfires. *If he could not fly again, would the limitation suck the life out of him?* The extremes of the facets to that central question became unbearable. She forced herself to think of other things.

Charlotte talked to Brian about any and every thing she could think of -- the farm, the weather, the approaching winter, the colors on the autumnal deciduous trees. She kept talking to him until the nurses motioned that they needed to return to perform their duties.

She watched the nurses performing their periodic checks. When they started to change his bandages, they asked her to leave. She did not protest.

"I must go now," she whispered to Brian. "I love you very much, Brian. Come back to me soon. I need you."

Charlotte nodded to the nurses to say thank you, and then left the room. Once outside, she checked the time. The late afternoon hour meant the evening milking time would be near. She would have no help this Sunday. She would have to milk the cows herself, as she did in the morning.

As she rode back to Standing Oak Farm, Charlotte actually smiled to herself. She felt the relief of an enormous burden being lifted off her shoulders. Maybe Brian could not hear her message, but she sensed that he did and she felt good. New warmth embraced her.

———

Sunday, 29.September.1940
Headquarters, Fighter Command
Bentley Priory
Stanmore, Middlesex, England
19:30 hours

The evening conference of Fighter Command's senior controllers kept Air Commodore John Spencer from his duties at Uxbridge, but also gave him an excuse to go home for supper with his wife, Mary. While the weather was not perfect with scattered clouds at various levels, the significant reduction in daylight bombing brought a mixture of blessings and lament. The night's dose of bombs on London still lay ahead. The Headquarters changed in the month since his assignment to Air Vice-Marshal Park's No.11 Group.

The mortal resolve of men in a death struggle had given way to a teeth-gritting, quiet determination. No one talked about victory or even winning the battle in open conversations, but between them in private, they knew they had been successful. Fighter Command had held the line and turned back the German invasion. Hitler experienced his first taste of a major defeat in this war.

There was no gloating, no celebration, and no victory dance. There was only the recognition of and commitment to the long haul toward ultimate victory.

The senior controllers grappled with the changes in tactics and the need for technological advancements to deal with the night-raiders. There were only two, maybe three if you counted the developmental units, squadrons of night-fighters in the whole of Fighter Command. Hardly an impressive force when placed against the enormous quantity of German bombers. The night raids of hundreds of medium bombers was better than the incredible masses of the daylight raids earlier in the month, but still many times the number of night-fighters the RAF could put into the air on any given night.

"Can't quite bring yourself to leave the monastery, can you?" said Air Commodore Herbert Maple, DFC, Deputy Chief of Operations, from behind John Spencer.

"No, Herbert. I rather like it here."

"Ah, but you have all the action down there at One One Group."

"I suppose you are right there, but I still miss the strategy."

"How is Mary doing with her pregnancy?"

"Other than needing me home more, she is doing rather well, actually. Thank you for asking, sir."

Maple held up his finger for a pause, and then tapped on his right temple, as if he was trying to shake out some bit of information. "As I recall, John, didn't you help a young American volunteer pilot? What was his name?"

"Drummond, Brian Drummond."

"Yes, that's it. I assume you have heard the stories."

"I am afraid not, Herbert. Has something happened to him?"

"In a manner of speaking," said Maple, motioning toward a couple of large chairs set to the side of the lobby. John Spencer did not like the implications, and he mentally kicked himself for not paying closer attention to the operations reports and casualty lists. "As we hear more, it appears we may have a *bona fide* hero on our hands in the form of your young friend. Do you recall the raid on Bristol and Filton by that damnable experimental unit?"

John remembered hearing several colleagues mention a low-level attack in the West Country, but he did not listen closely since it was in No.10 Group's area of operations. "Yes," he answered without embellishment.

"It seems your young friend with his squadron, Six Oh Nine, I believe," Spencer nodded, "took on the German fighter-bombers head-on at a hundred feet or so. One collision, so the reports say, along with another victory by your lad. The rub, though, he waded into a lopsided fight to help his leader and got tagged. The report states his canopy jammed, Perspex covered with oil, wounded, and the lad manages to land his crippled Spit with the Merlin stopped in a freshly plowed field. The good Lord has taken a liking to this young fellow."

"He has been lucky on several occasions."

"I can't recall the details."

"He was the pilot rescued by the Hampshire woman who the King awarded the George Cross to in August."

"Smashing."

"It is difficult to keep up with Brian, but I think this is his fourth knockdown."

"And, his 19th aerial victory according to the report. Doesn't he already have a Distinguished Flying Cross?"

"Yes, sir, he does."

"He will probably get another from the likes of this."

John's mind latched on to another aspect of the story Maple was telling. He was almost afraid to ask. He hoped the air commodore would tell him Brian's condition. Since he had not offered condolences, and he used the future tense relative to Brian's possible decoration, John surmised Brian was alive, but how badly hurt?

"What is his medical condition, if I may ask?"

"You may want to verify this, but . . . I believe he is bashed up a bit, but will survive."

"Thank God."

"Care for him do you?"

"Yes, as does Mary. He is rather like a son we never had, or we may have."

Both men laughed. Air Commodore Maple closed their conversation with his well wishes to both John and Mary, and then excused himself. John sat in the lobby as the last remnants of the day shift trickled out of the old converted monastery.

As John stood to depart, Air Chief Marshal Sir Hugh Dowding walked slowly down the west corridor toward the lobby. John smiled at his commander, and then felt his smile wash away as he noticed the gaunt, drawn, tired, and almost sickly appearance of Dowding. He had always been rather staid. They did not call him 'Stuffy' for no reason. John knew Sir Hugh took this battle, this moment in history, as a very serious and personal event. He had an attachment, a kinship, with his pilots, his 'chicks,' as the press coined the term. He had thrown his whole life force into giving his 'chicks' as much of an edge in battle as he possibly could and into carrying the awesome burden of protecting the United Kingdom from the invaders.

"Good evening, Sir Hugh," John said, as Dowding approached. He thought he even saw a slight smile appear briefly.

"Good evening, John. How have you been in the trenches?" Dowding asked in the vernacular of those who fought in the Great War.

"Reasonably well, sir."

"And Mary?"

"She seems to be doing quite well. The pregnancy is progressing normally, so the physicians tell us. Thank you for asking."

"Excellent."

John Spencer felt the urge to hear the opinion of his commander. "There is more talk in the last few weeks that we have won the battle."

"I am not quite ready to take that view," he answered, and then looked around the lobby. No one, other than the guards was in sight, but the room did tend to echo slightly. "Walk with me."

As they turned toward the large entrance doors, a female leading aircraftman walked smartly toward them. Dowding moved smoothly toward the door opening and holding the door for the young, handsome woman.

"Thank you, sir," she said in a delicate, soft voice.

"You are most welcome, miss," he responded, "and, thank you for your service."

The woman stopped, turned to face her commander, snapped her heels together and brought her right hand, palm forward, up to her temple in a precise, sharp, crisp salute. Tears sparkled on her cheeks.

Dowding returned the salute. John noticed him swallow hard several times, as the venerable warrior undoubtedly choked back his emotions. The Women's Auxiliary Air Force leading aircraftman pivoted and marched away smartly. Air Commodore John Spencer was equally moved by the brief, nearly wordless exchange of mutual respect.

Dowding cleared his throat, clasped his hands behind him, and walked slowly across the gravel driveway. His driver jumped out of the automobile, saluted and opened the door for his distinguished passenger. Dowding returned his salute and shook his head that he was not quite ready to leave. They walked a dozen yards before Dowding spoke.

"As you may be aware," he began, "the Germans have wisely chosen to abandon an invasion attempt, at least for now."

"Yes, sir, and they appear to have given up daylight bombing as well."

"While some may view that as victory, I am afraid I do not. As long as our cities are being bombed, we have not won the battle. The night intercept problem continues to plague me. We have not found the proper tools, as yet."

"We will."

"Perhaps, but I suspect my tenure is drawing to a close."

John was shocked, maybe even staggered slightly. Sir Hugh Dowding, nearly singularly, had been the father, architect, instrument and driving force behind the entire Air Defense System of Great Britain. From the Chain Home Radio Direction Finding network, the Spitfire and Hurricane fighters, the elaborately simple Observer Corps, to the integration of all the elements, Dowding had been at the heart of the growth of Fighter Command since the early '30's. John had been lucky enough to witness Dowding stand up and defy his uncle, the intractable and forceful Winston Churchill, in order to prevent additional, precious fighter squadrons from being squandered in the futile Battle of France. If there was one person who epitomized the aerial defense of the United Kingdom in the summer of 1940, it was Air Chief Marshal Sir Hugh Dowding.

"You can't be serious, sir."

"Not to worry, John. I am tired and old. I have served my purpose. I have done my duty."

"Why?"

"Well . . . interesting question, I suppose." He paused to consider his words. "Depending on your perspective, it is time for the old horse to rest."

"I should call my uncle."

Dowding held up his hand. "Please do not do that," he said. "The Prime Minister is under enormous pressure from a variety of sources. He does not need to deal with a tired, old mule. Lord Gort and I had dinner with your uncle at Chequers a week ago, today. He is still brooding over that very decision to withhold the ten additional fighter squadrons from being transferred to France."

"This is Leigh-Mallory's doing, isn't it?"

Dowding drew up sharply and faced his younger and shorter, former staff secretary. "Those thoughts will not serve you well, John. I strongly suggest you expunge them from your consciousness immediately. I have discussed this with Keith Park as well. The Air Ministry believes they need a change in leadership to counter the night bombing problem. The changes are pending, soon, I should think."

"Park, too?"

"Afraid, so."

"How soon?"

"I suspect, this month or next, at the latest."

"Dear God, what has come over them? Can't they see what has happened?"

Dowding actually smiled. "Your expressions mean a great deal to me, John, more than you will probably ever know. Thank you for your loyalty, as well. I will ask only one favor . . . when you do manage an opportunity to speak to your uncle in private, please convey my sincerest, humblest and most heartfelt appreciation and admiration for what he has done. If anyone has won this battle, it is Winston Churchill who carried the day."

John felt tears nearly escape his control. He swallowed hard several times to regain control. "I shall be honored to carry the message."

"Now, promise me, you will go home to your wife. She needs you more than we do at the moment."

"I will, sir," John said, and then crisply saluted his commander.

Dowding returned the salute, turned and walked proudly toward his staff car. John Spencer stood watching Dowding leave. He did not move. He saluted the car as it passed. John felt history envelop him in a warm blanket. He had once again been a party to history. The conflicting emotions of pride and resentment filled him as he watched the car drive away and disappear into the surrounding forest of evergreen trees.

———

Monday, 30.September.1940
RAF Middle Wallop
Middle Wallop, Hampshire, England

Flying Officer Jonathan Kensington cleared his thoughts, as his 'PR-K' Spitfire Mark IA rose gracefully into the calm morning air. He checked his right wing. The 'PR-S' Spitfire of Pilot Officer Janus 'Crazy' Kradilcek floated in perfect position. Jonathan glanced to his left, and then caught himself. Brian was not flying this mission.

The thoughts of Brian crept back into his awareness. The early morning telephone call to the Hampshire Central Hospital confirmed his friend's medical condition as unchanged. They were still waiting for him to regain consciousness. Jonathan felt good about Charlotte Palmer's feelings for Brian. She was older than his friend, but she certainly did not look much older. She had a quiet, subdued sensuality to her, even in her grief. Jonathan smiled underneath his oxygen mask. He saw what would probably be daily visits to Brian. She cared for him a great deal. She would be good for Brian.

Jonathan now regretted calling his sister, Rosemary, regarding Brian's crash. He knew she would not forgive him, if he had not called, but in a strange way, he felt as though he violated something special. Knowing his sister and the relationship she had with Brian, she would probably come down from Oxford to visit her lover. *Have I done the wrong thing for Charlotte and Brian?* He hoped not, after all they were friends.

Jonathan shook his head, trying to shake loose from a cobweb. Middle Wallop Sector Control pointed the squadron toward the southwest. He wondered if they were about to face the experts from Erpro210, again. They had been vectored to 26,000 feet. The last bunch of Erpro210 raids had been at very low altitude, so maybe it would be a different lot.

They found the inbound raid just off the coast, east of Plymouth. A relatively small group of Me110s configured for bombing and escorted by a large squadron of Bf109 fighters, or maybe two small squadrons. The engagement was brief and ferocious. The Me110 fighter-bombers did not seem to have the dedication, commitment and will displayed by the Erpro210 crews. As the Spitfires worked their way through the Bf109s, the Me110s dropped their bombs into the water and turned for home. While the Bf109 pilots had a penchant for the fight, they disengaged promptly when their charges turned tail. No.609 Squadron gave a short chase, but Squadron Leader Darling did not seem to have a taste for the fight either. Maybe the intensity of the last three months had worn everyone out.

Initially, Sector Control instructed them to recover at RAF Warmwell, and then changed their direction to RAF Middle Wallop. They landed without incident. None of the aircraft had been hit in the exchange with the small, German, fighter-bomber raid, and they had accomplished their mission. The pilots quietly completed their routine without the bravado and excitement that characterized so much of their actions over the last few months. Aerial combat had become the norm. It had become a job, not the adventure it used to be when everything was changing.

They waited through late morning, the mid-day meal at the Officer's Mess and a short nap in the afternoon sun. They waited as they had become accustomed to waiting, as though they were gardeners waiting for the shrubs and grass to grow, so they would have something to do.

Even when the telephone rang, none of them reacted. "Scramble the squadron," announced Corporal Warren. "It is London this time."

They were in the air in short order climbing due east and passing 12,000 feet when another radio call came.

"Sorbo Leader, this is Bandy calling."

"Go ahead, Bandy."

"Sorbo Leader, this is Bandy. Vector two four oh, angels three one."

"Roger Bandy," Darling answered, as he turned the squadron around to the right toward the southwest.

They seemed to be receiving more direction changes in recent days. Was the early warning and control system having some, or maybe more, difficulty locating enemy intrusions? They spent more time chasing around the sky like amateurs racing around trying to plug multiple leaks in the dike and missing all of them.

This time they found their targets, a squadron of Bf109E-4/B fighter-bombers headed for Yeovil or Filton. The odds were only two to one for this engagement. Half the Germans shed their bombs to nimble up their machines to meet the British fighters. The others pressed on for several miles until they decided the threat was greater than the mission.

The high altitude engagement in the bitter cold, thin air with condensation trails woven into a tapestry turned out to be mortal struggle. The Bf109s had the edge in overall performance at 30,000 feet. Several aircraft on both sides were hit. None were shot down in the 30-minute engagement. When the Germans broke off the attack probably with low fuel, the British gave chase.

Jonathan noticed his coolant temperature rising. One or more of the hits he had taken probably punctured his ethylene glycol coolant system. He had to get on the ground before the coolant ran out, his engine overheated, and then fused itself solid.

"Sorbo Leader, this is Sorbo Green. I apparently have a glycol leak. Breaking off."

"Roger Sorbo Green. Take 'Crazy' with you," commanded Darling.

No more words were needed as the two fighters rolled left toward RAF Middle Wallop. Jonathan pulled back the throttle to just above idle power to give the engine as much margin as possible. He allowed his airspeed to dissipate and began his descent. To give himself the best chance of saving the aircraft and avoiding landing where he did not want to, Jonathan did not push his descent. If he could get overhead RAF Middle Wallop, he would be able to glide to a landing at the airfield.

The coolant temperature continued to rise. Minutes later, the engine oil temperature began to rise as the oil pressure fell. He did not have much more time. He was passing 10,000 feet and probably within gliding distance of RAF Middle Wallop. Jonathan considered stopping the engine to save it, but did not have enough confidence in his ability to maneuver the fighter with a frozen engine into the shrinking space of the landing area. The airframe was more important than the engine.

As he made descending turns around the airfield, the Merlin began to sputter, kick and protest the abuse. He was as close to gliding as he could get as the large, three blade, propeller continued to turn slowly.

'Crazy' Kradilcek called the tower to let them know what was happening. Jonathan could see the emergency vehicles move into position to assist him should he need the help. The coolant temperature passed the redline. Time would soon be measured in fractions of a second. He did not like the line up. He felt he might land short.

Jonathan nudged the throttle open slightly. The big engine responded in defiance of its mortal wounds. As he crossed the perimeter tree line, the Merlin barked out an ugly, metallic grinding sound as it began to consume itself. The main wheels touched down. He bounced twice before firmly planting the aircraft on the ground. He promptly shut off the fuel, and the airscrew stopped abruptly.

The emergency vehicles surrounded him. Smoke wafted from the cowling edges. Jonathan unstrapped from his stricken Spitfire as though nothing unusual had happened and he was not in the middle of the landing area with his propeller stopped, smoke coming from his engine and surrounded by emergency vehicles. As he stepped down off the left wing, the fire crews opened the engine cowling. There was no sign of fire, so they waited. The strong, peculiar biting odor of boiled ethylene glycol soon dominated the oily smoke. Jonathan looked under the wing to see the coolant fluid dripping from numerous locations near the large coolant radiator under the right wing.

The ground crew quickly lifted the tail wheel onto the towing dolly to move the aircraft out of the landing area and toward the maintenance hangars. Repairs and probably replacement of the Merlin engine were required.

Jonathan declined the offer of a ride on the towing tractor. He felt like a walk along the southern edge of the airfield toward the squadron's dispersal area. He remained lucky. Brian had been shot down four times, had several more aircraft seriously damaged, and this was only Jonathan's third close call. He also had four confirmed aerial victories to Brian's 19 victories. Was that really the difference, he asked himself?

"You lucky again, 'Harness.'" said 'Crazy' Kradilcek.

"Better lucky than good, as we say old bean." Jonathan thought of his right wing-man's nine victories since the war began. He forced his thoughts to change. As he hung up his flight equipment on his assigned peg, and then looked to Corporal Warren, he said, "Any news about Mister Drummond?" he asked.

"Oh, yes sir, thank you for reminding me. The hospital called. Mister Drummond has not regained consciousness as of early this afternoon."

"So, I suppose there is no reason to go down to Winchester this evening."

"I suppose not, sir."

Jonathan went back outside to join Janus Kradilcek. They talked as they usually did about the engagement just encountered, as they waited for the remainder of the squadron to return. With Brian in the hospital, Kradilcek and him on the ground, there were only six fighters in the air. Jonathan looked across the line. He counted four new unmarked Spitfires at various spots in the dispersal scheme. At least he would not have to wait for another aircraft. They all wondered when the promised replacement pilots would arrive to fill the empty seats.

The vibrant hum of several Merlins at moderate power caught their attention. The six Spitfires arced across the southern horizon in a descending left turn toward RAF Middle Wallop. The flight had to be No.609 Squadron returning from the afternoon mission. *No other losses*, Jonathan told himself *. . . that's good.*

Corporal Warren joined Kensington and Kradilcek, as they watched the six Spitfires land, and then snake their way to their respective parking spots, turn and shutdown. They traded a few quick stories, completed their debriefing and were released a short time later. The earlier sunsets were a welcome sign of approaching winter.

After the evening meal and before the near nightly exchange of war stories over drinks, Jonathan called Charlotte Palmer. She had been to see Brian in the hospital. She confirmed the message the hospital passed to Jennifer Warren. The doctors remained hopeful according to Charlotte. She did not mention or even hint about her revelations of Saturday. She seemed to be in good spirits, and Jonathan left the conversation that way. He also considered calling Air Commodore Spencer or his wife, Mary, but did not. Calling Rosemary was enough. He did not need to complicate Brian's life anymore than it already was. He left those thoughts, his duty done, and turned to beer and stories with his fellow aviators.

———

Tuesday, 1.October.1940
Headquarters, No.11 Group
Uxbridge, Middlesex, England

"**A**ir Commodore Spencer," called the young communications officer.

"Yes," he answered.

"I was asked to relay a message from Fighter Command. Four Two One Flight is now operational." The man waited as though he deserved an explanation of what No.421 Flight was.

"Thank you," responded John Spencer. He was not permitted to discuss the mission of the special unit of Spitfires.

The duty controller waited until the small booth was cleared of inappropriate listeners. "Do you think these high altitude Spits are going to work?"

"I certainly hope so. It is about all we have in the short term."

"Any idea when they can modify the Chain Home RDF to pick up high altitude intruders?"

"None, to my knowledge."

"The higher they go, the harder it gets."

"The crews are getting frustrated with all the mis-starts and wasted time. Sir Hugh and Air Vice-Marshal Park think it will work, and I know 'Stuffy' is pushing as hard as he can to improve the performance of RDF."

"Being unreliable above 15,000 feet does not cut it with today's German tactics," he said and stood up straight.

John looked to his right to see Air Vice-Marshal Keith Park walking across the gallery toward the controller's booth. John joined his controller standing for the senior officer.

"I assume you have heard the news regarding Four Two One?"

"Yes, sir."

"Assign them to a separate radio frequency and try to keep two birds airborne at altitude during the daylight hours."

"That will stretch them rather thin," John said, as if the group commander might not have realized how many specially configured, high altitude versions of the Spitfire Mark II were assigned to No.421 Flight.

"Yes, of course. Lord Beaverbrook is trying to modify six more to be ready within a week or two, so the lads will have to carry the burden."

"Yes, sir."

"Would you be so kind to prepare a brief instruction to the sector controllers on the use of Four Two One?"

"I shall have it ready for your signature this afternoon."

"Thank you, John." There was finality to Park's intonation. "Oh, yes, I nearly forgot. There is more news as well." Park paused to receive John's head nod. "The King's honors list was published yesterday. Air Chief Marshal Dowding will receive the Knights Grand Cross of the Most Honorable Order of Bath."

"So, it is true?"

"What?"

"Sir Hugh stopped for a short chat when I was at Headquarters on Sunday. He said the Air Ministry would soon relieve him. I suspect the GCB is an effort by His Majesty's Government to soften the blow."

"He probably told you, I am to be relieved as well."

"Yes, sir, he did."

"Sir Cyril is to be promoted to Marshal of the Air Force."

John did not respond at first. He wondered how much Sir Cyril Newall had to do with the demise of Dowding and Park, the two air force leaders most instrumental in the successful defense of Great Britain and the thwarted invasion attempt were being turned out at the pinnacle of their achievement. John felt a passionate loyalty to Dowding. He wanted glory and accolades for the man who built Fighter Command and the Air Defense system of Great Britain despite the vast and significant obstacles during the pacifist pre-war decade. Dowding had been right in virtually every aspect. *What on earth could be the reason to relieve Dowding and Park? Sir Cyril knew the real story. This has to be a political move for some other reason.*

"Our first one."

"Yes, indeed, and he certainly deserves it."

"If I may be so bold, sir, what are you going to do?"

Park actually laughed. "I serve at the King's will. Actually, the Air Ministry is suggesting the air defense command for Malta and the Central Mediterranean."

"At least it is warm," John said matter-of-factly.

"That it is." Park looked to the situation map. "It does appear business may be picking up," Park said, motioning with his head. "I should leave you to your task."

"Good day, sir."

"Good day to both of you," said Park as he left.

The two controllers watched their commander depart the Operations Room, and then looked to the board. Several small high altitude raids approached the coast. There was nothing particularly exciting or unusual. Actions were taken.

"Is that all true?" asked the controller, as they waited.

John Spencer looked directly in the man's eyes. "First, you are not to repeat any of this to another living soul until it is public knowledge." John looked back to the situation map. "Second, to answer your question, it would appear so."

"They saved the country, and this is how they are rewarded."

"There must be much more to the story, I should think."

"Leigh-Mallory and Bader with his big wing theories."

John looked back into the man's eyes. "I shall ask you to keep these thoughts between us. I suspect so as well. We have been concentrating on the battle, not the politics."

Activity on the board began to pick up a certain agitation. The controllers turned to their duties. As questions were asked and directions given, John's thoughts bounced from the demise of Dowding and Park, to Mary and the baby, and even to Brian Drummond still unconscious in the Hampshire hospital. At that moment, John decided he would ask for tomorrow off to visit his wounded protégé. Hopefully, Mary would feel like traveling, and they could spend the day together. John Spencer began to look at life from a different perspective.

———

Wednesday, 2. October. 1940
Hampshire Central Hospital
Winchester, Hampshire, England
09:40 hours

For Pilot Officer Brian Drummond life stepped from a burning Spitfire cockpit to an almost painfully bright, light-green room filled with equipment and shelves of material. It took Brian several minutes to gain some degree of orientation. As he looked around the room, his location became obvious, and then he heard, "Nancy, fetch the doctor. Our patient is conscious."

An attractive woman, not beautiful and not homely, dressed in a white nurse's uniform came to him and stood next to his chest. Her warm hand touched the skin of his left arm. "Mister Drummond, I am Nurse Driffield. You are in a hospital." Brian nodded his head. The simple movement produced a noticeable pain in his head. "You have been in a coma for five days, and it appears you will fully recover."

"Where?" he asked softly.

"You are in the Hampshire Central Hospital in Winchester."

"Again."

The nurse giggled quietly. "Yes, Mister Drummond, you have returned."

Brian thought of his colleagues. "Squadron?"

"Not to worry, Mister Drummond. Several of your mates including your commanding officer have been to see you. Everything is well in hand." Brian wanted to ask about Charlotte, but did not know how to phrase the question. "Your fiancée has been here everyday as well."

Fiancée, the nurse said. How did they know I had proposed to her? Confusion occupied his thoughts. *Which woman? When the nurse said, fiancée, was she referring to Charlotte, or could it be someone else, like Rosemary using the title as a ploy? Who was claiming to be his fiancée?* He wanted to ask the nurse, what his 'fiancée' looked like, but quickly recognized how odd that would sound. Or, they might think he had lost his mind, or at least his memory.

"When?" he asked.

"She usually visits in the afternoons before she must return to the farm for the evening milking. She should be here in a couple of hours."

Before Brian could ask any more questions, two men, probably doctors, and another nurse joined them in the room. One of the men introduced himself as Doctor Morgan while the other man remained in the background. The doctor asked Brian numerous questions to ascertain any effects on his memory, orientation or thought processes. Satisfied with his initial checks, the doctor told Brian about his broken leg, fractured skull, broken ribs, several second-degree burns and multiple contusions. The doctor also said he would be out of action for at least six weeks. He would most likely remain in the hospital for a few more days while they evaluated his recovery progress, and then he would need to convalesce at a site of his choosing. The doctor impartially offered Brian a choice of his squadron, a special convalescent hospital in the North, or his fiancée's home since she offered.

There it was again, fiancée. Everyone seemed to know my fiancée except me. The nurse said the woman had to leave each day for the afternoon milking; that meant Charlotte. If she was referring to herself as his fiancée, then that probably meant she had accepted his marriage proposal, and had not been able to tell him. His heart beat faster. The nurse had said a couple of hours. He wanted the time to pass quickly. He wanted to hear the words from Charlotte's lips.

With medical examinations complete, they transferred him to a rolling table and moved him to the general ward. He recognized the large room. They put him in a bed in the same wing, just the other side, from where he was on his earlier stay. He could look straight out the windows this time. The ward held many more patients than his previous stay. At the other end of the ward, Brian saw an armed Army guard. The nurses noticed his focus and told him about an injured German pilot. Brian was curious, but his curiosity would have to wait.

Shortly before lunch, the smiling faces of Squadron Leader Darling and Flying Officer Kensington lit up the room, as they swaggered toward Brian.

"Good to have you back among us," said Darling.

"It's good to be back, although my headache gives me pause for regret."

They both laughed. "You took a bit of a pounding," Jonathan added.

"I remember getting shot up . . . ."

"Saving my backside," Darling interjected. "For which, I am deeply grateful and in your debt."

"Nonsense."

"Maybe to you, but not to me."

"I couldn't get out. It was like my dream, Jonathan. I couldn't get out. I couldn't see where I was going with oil all over the windscreen. I had a tiny spot of clear view at the left rear. I saw trees. I flared and hit the ground, then I woke up here, five days later, they tell me."

Darling cleared his throat nervously. "I do bring some rather good news, I should think. We received notice this morning. You have been awarded a second DFC."

"You are grabbing these things faster than they can be pinned on you," said Jonathan.

"Naw. Maybe they feel sorry for me because I've been shot down so many times."

"Hardly," protested Darling. "It is for 19 confirmed kills in the air, you twit."

"Skipper has some more good news."

"Ah, yes, lest we forget, the Air Ministry wishes to inform you that they reluctantly have consented to withdraw your transfer orders."

"You stay."

"Halle...lu... ," Brian started to say, and then abruptly stopped as the pain of his movement shot through every strand of nerve fiber. He groaned, but with a smile on his face.

"Sorry, old boy, didn't mean to excite you," said Darling. Brian moved his head very slowly in a nodding motion. "We thought you should hear from us directly."

"Flying?" asked Brian simply as he wondered why two pilots from the diminished squadron were at a hospital during daylight hours, and the sun was shining.

"Well, yes, of course," Darling answered once he figured out what Brian was asking. "The squadron has been temporarily designated a Class 'C' unit. We have four new pilots to train up quickly. We will move north to Catterick or Acklington in One Three Group in the next few weeks. 'Red' Burns left yesterday for Seven One Squadron. He still thinks you should be with the other Americans. Also, 'Crazy leaves tomorrow for Three Oh Three Squadron."

"I don't."

"For which we are most grateful. We need your experience."

". . . at being shot down?"

"No, you silly sod, at killing Germans. You are currently the third highest scoring ace in the whole of the RAF. You have much to teach the newbies. So, when you have mended, we need you back in the saddle to help the young ones."

Brian liked the sound of Darling's words. The words made him seem much older than he really was, and it was a good feeling. Brian nodded his head again slowly. Darling and Kensington continued to pass along news from the squadron as well as best wishes from the others and many from the outside. Brian heard some of the words, but his thoughts kept drifting to Charlotte.

He must have fallen asleep. When he opened his eyes, they were gone. More time had passed. He started to raise himself to find a clock. The pain forced him back down again. He also felt bad that Jonathan and Squadron Leader Darling had come down to see him, and he had fallen asleep. He would apologize to them when he saw them again.

The ward nurses saved him some light food. He enjoyed fresh tea, toast and jam, as well as some crackers and cheese. After his first meal in five days, Brian's next visitors arrived.

Mary and John Spencer walked into the general ward. Brian noticed them at the center entrance asking the desk nurse for his location. As they turned toward him, Brian could see the contrast in their faces and eyes. John smiled an optimistic, knowing smile from a pilot that knew what he had been through in this crash. Mary's face captured fear, uncertainty, resentment and concern. Brian waved his left hand and tried to smile to let them know he was all right, just a little banged up. This was the first time Brian had seen John Spencer since his promotion to air commodore. The broad black-light blue-black, joined strips on this tunic sleeves had replaced the four thin bands of a group captain.

Mary moved quickly to the left side of the bed, leaned over to kiss him, but could not find a cheek. She kissed his left shoulder. "I am becoming quite annoyed with your stunts, Brian," she said with some light-hearted irritation.

"No stunt, Mary," interjected John. "The lad is defending the country."

"Nonetheless, this is not pleasant."

John Spencer looked to Brian and smiled. "We met after the Great War, so she did not have to go through this aspect of war."

"Well, if it's any consolation, I don't want to go through this aspect anymore either," Brian said, producing laughs among the three until Brian groaned from the pain in his ribs.

"Poor baby," Mary said, as she touched Brian's shoulder.

"What do the surgeons say?" asked John.

"The doctor needs a few more days to evaluate my head injury, but other than that he says everything will mend."

The Spencer's had come to Winchester expecting to encourage Brian back from his coma. They discussed Brian's injuries as well as the events leading to his crash. Mary winced several times, but remained interested in Brian's story. John kept the discussion upbeat through the sequence of events and the inventory of injuries. Brian was thankful there was no mention of Mary's pregnancy. She never gave him a wink, a squeeze or other gesture of connection with him. She appeared to be the loving, caring wife of Air Commodore John Spencer. They also talked about the coming winter weather and even the anticipation of blooming flowers in the spring. There was optimism to their words, optimism Brian had not heard in many months. Brian sensed John Spencer knew more about the war situation, but he did not feel comfortable asking him with Mary present.

"I like the uniform change, sir. Congratulations on your promotion."

"Thank you, Brian."

"Now, I will see him even less now that he has become a general."

"Mary, please," John protested.

They stayed for a short time, until all, or at least most, of the pleasant topics had been discussed. John excused them. They wanted to return to London for supper. John proudly acknowledged his efforts to spend more time with his wife. Mary smiled. She looked back at Brian one more time, as they walked out of the general ward.

Brian felt better about his relationship with Mary. She seemed more like a friend, now, rather than a lover. There was not the slightest hint of any connection between them other than she cared about his health and well-being -- a caring friend. Maybe they were finally becoming what he had hoped they would be all along. He even began to rationalize that the life she carried was actually John's baby. Brian did not want to be between John and Mary. He wanted the child to be John's, or maybe just not his. Her presence made him less nervous and uneasy than it did months earlier. That was a good sign.

"Hello," came a soft, magnetic voice. At first, lost in his thoughts, he did not recognize the voice. His eyes were drawn to the source, and he felt a flush of embarrassment that he did not recognize Charlotte's soft, clear greeting.

"Hi," he answered, trying to sit up a little more. The pain instantly returned him to the bed with a groan.

"My dear, you mustn't move," she said. "Doctor Morgan just told me you had regained consciousness, and that everything seemed to be progressing normally."

"That's what he told me," he grimaced.

"He also said you would need some time to heal properly."

Brian nodded his head slowly. His eyes absorbed her unadorned beauty. His mind turned to one word, fiancée. He wanted to ask her, but he hesitated not wanting to scare her away.

"Are you all right?" she asked with a concerned expression on her face.

"I'm OK."

"Are you sure? You look distant. I thought you might be going into shock from the pain or something."

"No. I was just thinking."

"About what, may I ask?"

"About you."

"Oh dear," she said, as she flushed slightly, putting her hand to her face as if to hide a smile.

Brian decided to ask her directly. "The medical staff told me, my fiancée has been to visit me every day since my crash." She smiled, but did not answer. "Have you been calling yourself my fiancée?"

"I hope you don't mind," she responded. Brian tried to smile broadly despite the pain in his face and head. "I suppose I thought they would not object to me seeing you everyday, and after all, You did ask me to marry you."

"But, you never gave me your answer."

This time Charlotte smiled broadly with her full, pink lips highlighting her brilliant, white teeth and her sparkling blue-gray eyes. "Then, you don't remember anything I have been telling you each day?"

"I guess not," Brian said, his mind now racing through the possibilities.

Charlotte held her smile and looked around the room. Both adjacent beds held patients, one alert, and the other asleep. "Doctor Morgan tells me you will need several weeks of convalescence. I would like you to come to

Standing Oak Farm where I can care for you. We can talk there," she said, motioning with her eyes toward the adjacent patients.

"Take the offer, laddie," came the Scottish brogue from the alert patientin the next bed. "Won't be many of us to get an offer such as that."

"Thanks," Brian said, acknowledging the comment. He looked back to Charlotte. "I think I will, if it's no trouble."

"She's a fine one. Don't let her slip away."

Charlotte flushed noticeably and tried not to react. Brian understood her reluctance to talk about their relationship, but was annoyed that he had to wait. He started to get up, so they could go to a more private location. The pain kept him in place.

They remained together for several hours talking about the farm and the general activities of the last few days. They even engaged the Scotsman in the adjacent bed. Brian learned he was a seaman with the Sea Rescue Service. He had been injured during a difficult rescue of a German pilot, the one at the other end of the hall, and they both had been transferred to Winchester since the Southampton hospital was full.

The three of them managed to laugh about the stories around them. Even Brian's laughter, grimace, laughter cycle caused them all to laugh more. To Brian, the laughter and pain brought confirmation of life. He enjoyed the companionship.

Charlotte eventually forced herself to leave Brian. Her cows needed her attention more immediately than Brian. She whispered to him that she loved him and would return tomorrow. Brian thanked her for coming to visit. The Scotsman continued on, extolling her beauty and concern. Brian did not need the man's comments, but enjoyed them nonetheless.

———

Wednesday, 2. October. 1940
Cabinet War Rooms
New Public Offices
Whitehall, London, England
14:30 hours

The knock on the door of the prime minister's modest office broke Churchill's attention on the report before him. Assistant Private Secretary 'Jock' Colville did not wait for a response and inserted half his torso.

"Excuse me, Prime Minister. Secretary of State for Air Sir Archibald Sinclair has arrived for your requested meeting."

"Yes, yes, and 'Jock,' please no interruptions except for an emergency."

"Yes, sir."

Churchill shook Sinclair's hand and motioned for him to take the chair across the small, round, conference table in the crowded office-quarters. "How is the weather above ground?" asked Churchill.

"It is a fine autumn afternoon, Winston. Good flying weather, I'm afraid."

"For the hunters as well."

"Indeed!" Sinclair paused, but clearly had more he wanted to say. Churchill gave him the time. "Winston, I know you have wanted this day to go away, but I am afraid the day has come. With the threat of invasion diminishing by the day and the Germans transitioning to night operations, it is time to make the change."

"Are you absolutely certain this is the right thing to do and the proper time for this change?" Churchill did not wait for a reply. "Dowding built the fighter defense system that saved us from invasion. The pilots respect him for giving them the best chance against overwhelming enemy raids. Is this really how we choose to reward him for his accomplishment?"

"You have presented a compelling defense as well, I must say, Winston. However, this is the time. The system Dowding built did what we asked of them. He will be recognized for his contributions. However, we are clearly heading into a new form of aerial warfare. Our night intercept development has lagged."

"He had to win the daylight hours," protested Churchill. "We can forgive him his apparent neglect."

"And, he did what we asked of him. We need new thinking and new energy to win the night. Until we gain the upper hand, these night raids will punish our people . . . the people who depend upon us to protect them. Lastly, I must profess I have been very disappointed in his leadership of this nagging Big Wing debate that has distracted Fighter Command."

The Prime Minister took a moment to think about Sinclair's continuing argument. "Then, so be it."

"Will the War Cabinet support this change?"

Churchill stared at Sinclair and did not react. "Are you certain you want to go down this path?" Winston asked with stern, direct, unflinching concentration.

Sinclair stared back, but ultimately recognized Churchill most likely had done his political homework, especially for such a big personnel change in the middle of a battle. "I am only concerned that His Majesty's Government supports this necessary change."

"Do you wish to perform my duties?"

"No, sir. I . . . ," Sinclair pulled up on his own accord.

Again, Churchill did not respond or react, wanting Sinclair to say what he needed to say. Several long seconds passed. "The time for discussion and debate are over. We will reward these leaders for what they have accomplished, but the time has come for younger, fresh blood. We will make the following changes. Sir Charles Portal will be promoted to air chief marshal and will succeed Sir Cyril, who will likely retire from active service."

Air Marshal Sir Charles Frederick Algernon Portal, KCB, DSO & Bar, MC, served with distinction as Air Officer Commanding-in-Chief of Bomber Command, since April of this year.

"You must effect this change in three weeks, no later than Friday the 25th. Sir Cyril will be promoted to marshal of the air force on Friday – our first. I have recommended Sir Cyril for the King's honors to receive the Order of Merit and the Knight Grand Cross Order of Saint Michael and Saint George. If he wishes further assignment rather than retirement, I shall receive his wishes and your recommendation with favor.

"Thank you, Winston. When do you want to announce this change?"

"The principals should know immediately, to properly affect the change of command. Public announcement should be on the 24th or 25th, not before. This is going to be a controversial change. Best to make it a *fait accompli*, and not offer dissenters time for any protest."

"Very well. So it shall be. What about Dowding's and Park's replacement?"

"We must give Sir Charles some time to contribute to the selection of Dowding's successor, but the change at Fighter Command must be completed before the end of November. I have an idea for his successor, but I would like to know your suggestion, first."

"Since you raised this possibility, I have considered that choice. My recommendation will be Douglas."

Sir Cyril's number two, Deputy Chief of the Air Staff Air Vice Marshal William Sholto Douglas, CB, MC, DFC, had supported Leigh-Mallory and Bader in the Big Wing debate, but he had done so quietly without directly disagreeing with the Chief or making a public spectacle of himself. He was 11 years younger than Dowding and had a keen mind regarding the dynamics of modern air tactics.

"Actually, I agree. I have been impressed by Douglas's ability to walk the line. We will give Portal the opportunity to present his selection."

"Does Dowding retire?"

"If that is what he wishes. I have recommended to the King for Sir Hugh to received the Knight Grand Cross Order of the Bath. I expect the Central Chancery will announce Sir Hugh's GCB in the next few days with the King's investiture next week. Now, as you know, Sir Henry Tizard has been in the United States since August, executing the technical exchange program we set in motion two months ago. I have in mind Sir Hugh to be our air liaison officer to the United States. There is so much that he can encourage, not least of which is this airborne radar unit for night-interceptors. I would encourage him to take the assignment."

"He may well accept the position. Will he be promoted to marshal of the air force as well?"

"Not likely, Archie. I will agree to support a barony for him, depending upon his performance with the Americans. Now, Park must move as well."

"Winston . . . ," Sinclair protested and stopped when Churchill raised his right hand."

"Archie, listen to me, Sir Hugh has been our daylight air defense, but Park has been the face on the bombing of London. I recognize and acknowledge that Keith Park managed grossly insufficient resources and is largely responsible for stopping the German invasion. Yet, as I said, he is the face of the bombs that have rained down on Londoners. The people need to see a change."

"Winston, that is terribly unfair, and you know it."

"Perhaps it is, Archie, but it is what it is."

"Do we discharge him?' asked the Air Minister.

"Archie, please, Park has done a magnificent job. I do not want him pilloried. He is a very capable officer. He can continue to serve the Crown. If he wishes to continue his service to the King, we will find the appropriate assignment for him. I will recommend Park for Companion Order of the Bath."

"Who shall replace Park?"

"Intriguing question."

"Perhaps, I am inclined to challenge Leigh-Mallory. I have never admired snipes. He has been unprofessionally vocal with his criticism of Air Vice Marshal Park, the man standing the line, so perhaps it is best to give him the challenge Park has faced for the last three months – truth or consequences time. That said, I am comfortable with allowing the knock-on changes to play out after Portal's assignment."

Sinclair searched the Prime Minister's eyes. He was looking for sincerity. Churchill held his eyes.

"You are convinced this is necessary?"

"Yes."

"Then, so be it."

"I will inform Sir Cyril and Sir Charles. We shall move through the transition in an orderly and professional manner."

"Thank you, Archie."

"God bless them all for their service."

"Indeed. Quite so."

Sir Archie did not wait for further conversation. He stood and departed. Sinclair was clearly not happy with these changes, but they simply had to be done. Despite their accomplishments, the image of Fighter Command had to change. Churchill could easily classify these changes as political rather than professional, but they were necessary to move the Air Force and Fighter Command to the next level.

———

Wednesday, 2.October.1940
RAF Middle Wallop
Middle Wallop, Hampshire, England
17:15 hours

"**W**hat's the news?" asked 'Red' Burns, before Jonathan had stepped fully into the squadron dispersal tent.

Kensington nodded to Corporal Warren to record his return to duty. The visit to Hampshire Central Hospital had gone smoothly. The eyes of the pilots were on him, and the conversations and card games stopped.

"Our wounded comrade has regained consciousness." Words of relief, gratitude and hope filled the tent. "The doctors are still evaluating his head injuries and he's pretty busted up, but they believe he is likely to fully recover."

"When do we get him back?" asked 'Fog' Johnson.

"The doctors say at least four to six weeks . . . could be longer, depending upon how well he knits up and rehabilitates his leg. He only regained consciousness this morning after being in a coma for nearly a week."

"Is he going to be in the hospital the whole time?" 'Boxer' Stockard asked.

"No. I talked to his savior after I saw him. She has already talked to the doctors about him convalescing under her care, at her farm."

"Well, now," interjected 'Sparky' Morrow, "isn't that a tidy little package."

They all laughed at the image. Just then, they all noticed a different engine sound than they had been familiar with, in the last few months. One of the new pilots announced, "A new machine I've not seen before."

The pilots filed outside to witness and odd looking, twin-engine, tail-wheel configured aircraft taxi past their squadron area toward the No.604 Squadron area.

"Looks like the nighthawks got a new machine," 'Fog' announced. "Can we go take a look, Skipper?" 'Fog' shouted.

'Spike' joined them. "We've not been released, yet, lads." They all watched the new aircraft park and shutdown next to one of the Blenheim night-fighters. "That must be the new Beaufighter."

"What do you know about it?" 'Sparky' asked.

"Not much. They told us at Group the aircraft is supposed to have some new airborne RDF kit."

The distinctive ring of the alert telephone turned all heads toward the Dispersal Tent. Corporal Warren appeared. "The squadron is released," she announced.

"There you go, lads. Now you can go see the new machine, if you wish. The lorry transport should be here shortly, so if you are not here, you walk."

Jonathan and several of the other pilots secured their flying kit and headed off walking across the northwest corner of the landing area toward the No.604 squadron area. It was on their way to the Officer's Mess, so they would forgo the proffered truck transport on this evening.

The night-fighter pilots surrounded the new aircraft. Other pilots had joined them. Jonathan just listened and observed. The aircraft was indeed a Bristol Beaufighter Mark IF night-fighter, as 'Spike' had guessed. The two power-plants were Bristol Hercules Mark III, 14-cylinder, supercharged, radial engines that produced 1,400 shaft horsepower each. The most important piece of equipment on the aircraft had to be the new centimetric Mark IV Airborne Intercept radar unit in the nose dome. The aircraft carried four, drum-fed, Hispano-Suiza HS.404, 20 mm cannons under the nose. The aircraft had a conventional tail and a dorsal observer bubble. The radar operator sat in the tail under the bubble and also served as the cannon loader, changing ammunition drums and charging the cannons forward of his station. According to what Jonathan heard, the aircraft was fast, especially for its size, but not quite as fast as their Spitfire Mark IA fighters. It certainly had far more bite with the four cannons. Surprisingly, they were already anticipating upgraded variants with an additional six 0.303 machine-guns in the wings.

Apparently, no one other than designated flight crew were allowed inside the aircraft due to the nature of the special night intercept equipment installed. Jonathan could easily see the cockpit was not configured for day fighter operations; then again, the aircraft was designed to fly in the black of night and would only be dealing with targets directly in front of the aircraft.

The night-fighter guys had already been trained on the new intercept equipment at Bawdsey Manor. They were discernibly eager to get the aircraft and try out their new machine. Jonathan had always had a hard time imagining what night-fighter operations must be like. Day-fighter operations were hard enough with the confusion of so many aircraft turning and slashing at each other. Yet, trying to get within gun range of another aircraft in the inkwell of the night sky was just too much. They had flown enough at night in blacked out England to know just the navigation task alone was extraordinarily difficult, but purposefully mixing it up with other aircraft at night was simply mind-boggling. He sure hoped the night-fighter lads were correct in their expectation the new Beaufighter would be a game-changer, to at least seriously punish the German night bombers pounding London every night and other cities. They would need a lot more than one of the new night fighters, but they said more were on the way. There was hope.

———

Thursday, 3. October. 1940
Hampshire Central Hospital
Winchester, Hampshire, England

Brian was overwhelmed by the attention the staff and other patients showed him. Word spread that Brian was an accomplished fighter ace about to be awarded his second Distinguished Flying Cross. He was a genuine hero, and more importantly, his fiancée was the first recipient of the George Cross. The award for valor was the highest civil award, second only to the exalted Victoria Cross, they told each other. Everyone seemed to know that Brian and Charlotte had received their awards from King George VI himself.

At first, Brian enjoyed the recognition. He became quite self-conscious about it when the attention appeared to be endless. He especially but quietly objected to the word, hero. Brian could not bring himself to acknowledge or accept the word when used in reference to him or his actions. Everyone appeared to enjoy telling the stories they heard about his exploits, as if they had heard them directly from Brian, as though they were best friends. Brian nodded and smiled a perfunctory smile as the tales continued.

To break the monotony of the attention, Brian encouraged the nurses to help him get out of bed and move around the ward. He had to use crutches that were quite awkward and painful, but the movement actually felt good despite the pain. People began to sense his uneasiness with the attention, or maybe it was his struggle against the pain, as their words became gestures.

He started with small efforts assisted by the nursing staff and progressed quickly to greater efforts on his own. One of his first objectives was the German pilot at the other end of the ward. The eight beds on the far end of the ward were empty except for the German. His Home Forces guard, a private who appeared to be several years older than Brian, sat at the foot of the bed with his Enfield rifle across his lap.

As Brian approached, the guard stood holding the rifle across his chest in a ready position.

"It's all right, private. I'm not going to spring him."

"Sorry, sir. He is a prisoner. No one, other than the medical staff, is allowed to talk to him."

Brian looked into the German's focused blue eyes. His brown, curly hair made him look young, maybe about Brian's age. The man nodded his head with respect to Brian. Pilot Officer Drummond returned the gesture. Brian wanted to talk to the man. He decided to use his fame for a purpose.

"Do you know who I am?" asked Brian, feeling a little self-conscious about the words.

"Yes, sir."

"I would like to talk to the man, pilot to pilot. No harm will come."

The private hesitated, and then nodded his head and decided to stand at parade rest with his rifle at his side.

"Do you speak English?" Brian asked the German. The man shook his head, no. "Does anyone speak German?" he asked rather loudly to the ward.

"I do," answered one of the younger attending nurses. She walked over to him without being asked. The ward charge nurse started to object, and then decided not to interfere. She nodded to Brian to go ahead with his questions.

With the nurse translating both ways, the two pilots learned about each other. He was a second lieutenant, an *unterleutnant* in his words, and flew a Bf109E-4 fighter. He had been shot down three days ago and rescued from the English Channel. He had been flying fighters since the spring and had four victories. This was his first time being shot down. Brian told him about his victories and escapes. The man stiffened as he heard of Brian's accomplishments. The young nurse took it upon herself to add to Brian's words. The German said his squadron had been briefed on several of the British aces. The names seemed so odd coming from the German fighter pilot -- Tuck, Malan, Townsend, Lacey and Drummond. Being known by your adversary did not seem like a good thing to Brian. The stories coming back from the German made him feel very uncomfortable. Most of the information was precisely correct. The young nurse's translation of the German's words added to the buzz that filled the general ward.

To keep the conversation from becoming one sided, Brian recognized the names of the German pilots their intelligence folks briefed them about -- Galland, Weir, Mölders, Rubendörffer. Hearing the names of the German aces stiffened the man even more. Brian suggested that he relax. He had nothing to prove. The German told Brian that if he were flying for the *Luftwaffe*, he would surely have been awarded the Knight's Cross, the highest German military award for gallantry and valor. He would be revered in the German Air Force.

The two pilots shared friendly reminiscences about their early days of flying before the war. Brian learned for the first time about the Versailles Treaty restrictions on flying for the Germans. The *Luftwaffe* pilot spoke in reverent terms about Hitler's defiance of the restrictions, and the pride Hitler returned to the German people. Brian thought about confronting the German with Crystal Night and the invasion of innocent, neutral countries, but he chose to keep the conversation about flying. The man talked about learning to fly in gliders and then trainers and transport aircraft, until one of his mentors saw his talent and found him a seat in fighters.

The similarities in their flying backgrounds were eerie. They shared a common aviation heritage and love of flying. The mounting pain in Brian's head and chest convinced him to end the conversation and return to his bed. The German thanked Brian for the conversation and wished him the best of luck flying. Brian wanted to do the same, but knew the man's flying days would be over for a while, probably until the war ended. Brian told him to take care, and then thanked the nurse for translating and the guard for looking the other way to allow the conversation.

The relief of his bed sent Brian nearly into sleep. The adulation of others around him mounted again as Brian's conversation with the captured and injured German fighter pilot was passed among them. Brian found his relief in the person of Rosemary Kensington who arrived before lunch.

She kissed him on the cheek.

"Thank God you're here," Brian said.

"I did not think you would want to see me," she responded. "It has been too long since I have seen you or heard from you." Brian felt his face flush thinking back to her last visit. "You simply must show me more attention."

Brian looked over to the Scotsman who winked at him and did not say a word. He looked back into Rosemary's blue eyes. They were nearly identical to the German's eyes. Her eyes smiled at him.

Rosemary leaned near him and whispered, "Maybe I should sneak back in around midnight for another special treatment." Brian flushed again, but found himself remembering the pleasure of her efforts. He felt the familiar ache growing within him. There was an almost irresistible attraction to relieve the ache. The image of Charlotte's face came to him as well. Rosemary did not know about Charlotte Palmer. He wanted her to know.

Brian braced against the pain, trying to keep it to himself. They walked to the visitor's lounge. No one else occupied the room. They sat.

"Thank you for coming, Rosemary. It is good to see you."

"My brother called me to tell me of your crash. He called a couple of days ago, and you were still in a coma. I am pleased you are doing better."

"Thanks."

"How come you have not called me, and please do not tell me it is the war."

Brian knew he had an obligation to tell Jonathan's sister about his relationship with Charlotte Palmer. "I have always enjoyed your energy."

"So, you have found another woman."

Rosemary's bluntness stopped him for a moment. "In a manner of speaking," he said. "I'm going to marry the woman who saved my life last July."

"Is this hero worship, or is it serious?"

"It's serious," Brian protested, feeling a little offended.

"And, you didn't even give me a chance."

Brian did not know whether to apologize or change the subject. He chose the latter. "Your brother came down to visit yesterday. He was also the first to visit me after the crash."

"I do not want to talk about my brother. I want to talk about us." Brian nodded, not really wanting to talk about their relationship. "I thought we had something together."

"Sex," Brian reacted.

"Now, you are being as blunt and abrasive as me," she said, and then smiled.

"Sorry."

"No need. What is good for the gander, is good for the goose, or something roughly equivalent. And yes, we did have some great sex. Maybe we can have some more."

"I think not."

"Why, because you are betrothed to this woman savior of yours?"

"Yes."

"Brian, we have been through this before. No need to be so quaint on my account."

"Maybe, but it is how I feel, Rosemary. If you recall, I was a reluctant lover at first . . . ."

". . . Who became an enthusiastic lover," she interjected with a smile.

"OK. Yes. We did enjoy each other."

"Yes, we did, Brian. Yes, we did."

Rosemary Kensington stared out the window, probably at nothing in particular, although there was a nice stand of evergreen trees. Brian watched her, wondering what was going through her mind. *Did she resent me for my crude representation of their relationship? If their relationship deteriorated, would it affect my relationship with her brother?* Brian knew he enjoyed her energy, her zest for life, and he actually found her sharp wit stimulating. He wanted to tell her he never wanted their relationship to be physical, not that he was not attracted to her, but his connection with her brother was too important to risk. He had given into her seduction and relentless pressure at a weak moment. Brian liked Rosemary, perhaps even loved her in a special way, but she did not possess the attraction that Charlotte Palmer did for him.

Rosemary turned back to Brian, impervious to his stare. She looked into his eyes for several moments, and then smiled at him broadly. "I suppose I used you as much as you used me."

"I didn't use you," he protested.

"Nonsense, Brian. It is nothing to be ashamed of between a man and a woman. I was attracted to you from the very first moment I saw you at Carlingon Castle. I sensed something special in you, and I wanted to share it. I do sincerely thank you for what we have had together. I was right, you know," she paused, although Brian did not know how to respond. "You are something special. What do you want from me, Brian?"

He had not considered the question. He continued to look deeply into her eyes as his mind grappled with the question. Brian knew what he wanted from her, but hesitated to ask her. His concern rested upon her reaction. He decided candor and truth had to prevail.

"I think of Jonathan as my brother," Brian said. Rosemary began to shake her head in disagreement before she heard the words. She knew what he was going to say. Brian forced himself to continue. "I would like our relationship to be like brother and sister."

"Why are you doing this to me?"

"What do you mean?"

"Brothers and sisters do not make love together. Are you telling me our sex was not great, that you do not want to enjoy the pleasures of the flesh with me?"

Brian swallowed hard, not knowing where this conversation might lead, but determined to state his mind. "Rosemary, I'm trying to be as honest as I possibly can. I am not about to lie to you now. The sex was great, and Lord have mercy on my immortal soul, I do enjoy the pleasures of the flesh, as you say, and especially with you. There has always been a certain purity and innocence to your unbridled enjoyment, but I . . . ," he hesitated, struggling with his conflict of conscience. He could not tell her honestly he would not have sex with her. Brian found that part almost too hard to say. "I want to devote myself to my wife, and I have proposed to Charlotte, if she will have me."

Rosemary chuckled. There was a peculiar sparkle in her eyes, a mischievous, troublesome sparkle. "You are indeed quaint, and I suppose that is one of the reasons I think I love you." Her frivolity turned into a more cunning grin. "Does your Charlotte know about us?"

"I think so."

"I mean really know about us?"

"If you mean, have I told her all the intimate details? No."

"You told me about your other women, probably trying to scare me off, I should think. If your Charlotte can accept me close to you, as your sister, you say, then maybe it can work. I just like being around you, Brian. I may be able to accept no intimacy, if I do not feel jealousy from her."

Brian smiled for the first time since their conversation began. He wanted Charlotte to know he talked to Rosemary, and he certainly did not want Charlotte to feel threatened by her. It was worth a try. He had several other confessions to make to Charlotte before they made their commitment to each other.

"I would like to try."

"Then, so be it. I look forward to meeting your Charlotte. I think I do love you, Brian, and if it must be a sibling love, that is better than love unrequited. Now, the time has raced by, and I really must be back up to Oxford this evening. I am so happy you came through your latest little escapade without permanent damage. I care for you, Brian, and I wish you nothing but the best. I look forward to meeting your Charlotte as soon as it is convenient. Please stay in touch with me. Do not let time slip by any more."

"I'll do my best."

Rosemary stood and helped Brian to his feet. She embraced him gently, and then kissed him several times on the lips. She opened herself to him. Brian responded like an alcoholic faced with a full and open bottle of whiskey. Rosemary pressed herself against him as their tongues danced like serpents. Brian pushed back abruptly causing electric shocks of pain to various sites in his body.

"No," he gasped.

Rosemary smiled that mischievous, cunning smile of a fox. "Sorry. The temptation was too great. I just had to know if the passion was still there. I think we have answered the question. I mean no harm, Brian. I promise I shall be a good girl from now on."

Brian could only nod his head against the pain.

"Now, I must go," she said, leaning forward to kiss him again.

Brian drew back nearly falling over a chair causing more pain.

"Oh, stop, Brian. I am not going to bite. I just want to kiss you good-bye properly . . . as a sister."

Brian resisted no further, more from the mounting pain than reticence on his part. Rosemary kissed him on the cheek, smiled at him and left. Brian sat in the chair he nearly stumbled over to regain his composure and control his pain.

He looked at the clock. *Charlotte will be here shortly*, he said to himself. He had so much he needed to say to her. This was not the place. Brian committed himself to his rehabilitation to advance his release. The possibility of having some private time with Charlotte at Standing Oak Farm became his singular focus for the time being. He feared how she might take the confessions he had to make, but it was important to him they start off at a common point. He wanted nothing hidden from her. Brian knew he would have to work hard to hasten his release from the hospital as well as to avoid poking Charlotte's curiosity. Brian committed himself to the task.

———

Friday, 4. October. 1940
RAF Middle Wallop
Middle Wallop, Hampshire, England
11:30 hours

The day seemed like so many others for Flying Officer Jonathan Kensington. The major exception was the lack of machine-guns firing. They had flown two training sorties already, and it was just approaching lunchtime. The ground crews worked as feverishly as ever to fill the fuel tanks, check the guns, and evaluate the other aircraft components and systems. The aircrews waited, as it seemed they always did. There was not sufficient time for another sortie. They would miss lunch if they flew again. They would probably fly one or perhaps two more training sorties in the afternoon. The highly successful No.609 Squadron had been designated a Class 'C' unit the previous Sunday. None of them had flown a combat mission in nearly a week.

The four new pilots, two Englishmen, a Scot and an Irishman, were young, inexperienced, scared and tentative, but they were full of bravado and they knew how to fly – they were trainable. Between them, they barely reached 300 flying hours. It was almost sinful forcing such young, inexperienced pilots into the cockpit of a Spitfire fighter in combat.

The principal blessing came in the form of the dramatic reduction in combat intensity from the nearly intolerable levels of August and September. The last daylight-bombing raid had been carried out on the 30th of September. The remaining Class 'A' squadrons contended with the high-altitude fighter sweeps and the reconnaissance flights. Despite the cooler, but still great flying weather, the skies were relatively devoid of enemy aircraft, especially compared to earlier weeks.

As they waited for the lunch hour, the telephone rang. The veterans froze with their eyes toward the tent and their muscles tense ready to spring into action. The four new pilots continued to discuss something, not recognizing the significance of the distinctive bell.

Squadron Leader Darling emerged from the tent. "The lorry should be here shortly to take everyone up to the Mess for lunch. I have been called to Group. I shall return in a few hours."

Jonathan kept quiet and to himself through lunch. The newbies engaged anyone who would listen or talk in the conversations of flight. They sounded like they had all sounded months earlier, before the Battle of France and especially the mind-numbing, bone-aching fatigue of the Battle of Britain. They had so much to learn. Perhaps, the reduced intensity might give them a chance to learn before they found themselves immersed in the ball of twisting, turning fighters that proved so deadly to so many of their own.

They had the basic aviator skills. They could handle the aircraft. What they did not have was the air sense – combat savvy – they needed to stay alive among so many others trying to kill them. Somehow, Jonathan told himself, he had to tame the bravado and temper the edge for them to become effective warriors.

The pilots returned to the dispersal tent. Squadron Leader Darling had not returned from Group, yet, and Corporal Warren had no messages from him. Flight Lieutenant Robert 'Sparky' Morrow, acting as the second in command, reported the squadron ready for training. The squadron took off for an hour training sortie over the mountains of Wales. Morrow ran them through the variety of attack tactics they currently used, and then returned the squadron to their home base.

Darling's 'PR-A' Spitfire landed with a Hudson trainer. Squadron Leader Darling waited for another squadron leader to disembark from the Hudson, and waited for the twin-engine training aircraft to take off, and then walked with his guest to the tent. He called for the attention of all the pilots.

"I have some announcements to make," he paused to ensure he had everyone's attention. "It appears the Air Ministry has seen fit to promote me to wing commander." They all cheered for their leader. "I am to assume command of RAF Exeter." They congratulated him, but knew there was more to the announcements. "With that said, I would like to introduce Squadron Leader Jason Long-Roberts who will assume command of the squadron effective today. Further, on Monday, the squadron will return to One Three Group for rest, training and refit. We . . . well, actually . . . you will fly north to RAF Catterick . . . North Yorkshire. "

Jonathan wanted to be happy for the commander that led them through the many difficult months of combat, from Scotland to Hampshire, from the Phony War through the Battle of Britain. He also felt the uncertainty of new leadership. *What will happen to the squadron next?*

"There is more," he said to continue. "Air Vice-Marshal Brand has asked me to inform you, in case you have not seen the King's Honors List for this year, that Air Chief Marshal Sir Hugh Dowding has been recognized by the King and honored with the Knights Grand Cross of the Most Honorable Order of Bath, the highest military order of knighthood, for his leadership of Fighter Command."

"'Stuffy' deserves it, too," said Pilot Officer Roland Stockard.

"Hear, hear," added Flight Lieutenant John Davies.

"Also . . . ," he said, raising his hand for attention, "also, Air Chief Marshal Sir Cyril Newall will be promoted to Marshal of the Royal Air Force. Sir Quintin wishes me to add that we should expect, adjustments, I think is the word he used. We should expect these adjustments in the next few months."

"What does that mean, Skipper?" asked Flying Sergeant Miles Johnson.

"I am not really sure, actually. I suspect he knows more than he can tell us at the moment." Darling looked back across the small group. "Since we are on the topic of awards, our own Brian Drummond has officially been awarded his second Distinguished Flying Cross for his actions in aerial combat."

"Most deserved as well," said Flight Lieutenant Robert Morrow.

"Yes," added Pilot Officer Janus Kradilcek in a rare comment.

"How is he, Skipper?" asked Johnson.

"He has regained consciousness, finally. The surgeons want to evaluate him for a few more days, and then they expect him to be released for convalescence. I am afraid his injuries will keep him out of the air for a month or more. But . . . ," he said to stymie the grousing, "but, since not all of you may have heard, his transfer orders have been rescinded. He will remain with us, at least for the time being."

"Good-oh for him."

"Better than Burns," said Stockard.

"Easy, now," Darling cautioned.

"But, Skipper, 'Red' was rather quick to jump at the American squadron," Stockard responded.

"Maybe, but he had proper orders from the Air Ministry. He was ordered to go."

"He could have resisted."

"Enough of this," Darling commanded, wanting to terminate the course of the discussion. "If every American decided he did not want to be in the American squadron, there would not be an American squadron."

"Not all the Canadians are in the Canadian squadron, and not all the Czechs or Poles are in the Czech or Polish Squadrons," pressed Stockard, testing Darling's tolerance.

"I said, enough!"

"If I may be so bold," interjected Squadron Leader Long-Roberts with an obvious attempt to dissipate some of the tension, "I would also like to add to Wing Commander Darling's earlier words. He has been awarded the Distinguished Service Order for his exceptional service and leadership of this squadron. I know Wing Commander Darling would probably add this award reflects upon the performance of this squadron and all of you."

"Squadron Leader Long-Roberts is precisely correct. This is your award, and I thank you for it."

"Then, we must celebrate tonight," said Flight Lieutenant Morrow. Darling smiled and nodded his head. "Tonight it is, then. In deference to our flying sergeants and any of our enlisted rank who might care to join us, we shall have a go at the Mucky Ducky. Festivities begin at nine sharp."

Although the squadron remained at training status, the pilots fanned out among the aircraft to inform their crews and extended a rather rare invitation to the pilot's favorite pub in Andover. They respected Horatio Darling, and they wanted to show him. *He will be missed,* Jonathan told himself. The evening would also provide the social opportunity to measure up their new commander. Jonathan saw it coming. Tonight, they would let loose. He just hoped no one would be injured and nothing would be damaged. They were done . . . at least for a while.

———

Friday, 4.October.1940
Cabinet War Rooms
New Public Offices
Whitehall, London, England
15:45 hours

Sir Henry Tizard entered the prime minister's office in the underground complex. He had arrived yesterday from his mission to the United States. He had not seen the underground government offices of the War Cabinet and its support staff. A Home Forces sergeant gave him a very brief walking tour as they made their way down the main corridor of the bunker.

"Welcome to my humble abode, Sir Henry," Churchill said.

"This is different."

"Yes, the Cabinet War Rooms were completed shortly after your departure in August. We occupied these offices when the Germans attacked London a month ago."

"I have seen the horrific damage on my way here from Northolt. We have taken such a pounding."

"Quite true, and I am afraid we are a long stretch from the other end of the tunnel. By the way, Sir Henry, I truly hope someone told . . . you will most likely spend the night here, as the nightly bombardment may begin at any moment and certainly before we are done with our little chat."

"Yes, sir, they did. I brought a small bag with me."

"Excellent. I wanted a personal briefing on your mission as soon as you returned and before you became immersed in your next assignment." Sir Henry nodded his head. "Let us begin with how you left things in America."

"Doctor Cockcroft has ably assumed daily supervisory responsibility for the team. With the assistance of Lord Lothian and Bill Stephenson, we have established offices in Washington and New York. The eight teams have all completed their transfer tasks, and they are deep into the true exchange work, as we had hoped. I believe we will look back on this time and see that we were successful beyond our wildest imagination. The Americans have shared some extraordinary developments with electronic fire control. Together with our cavity magnetron, the electronics team feels they will have a workable system to improve efficiency of naval gun fire and especially anti-aircraft fire."

"We certainly need that now."

"The whole team, including the Americans, recognizes the immediacy of the need. I can assure you, they are working extra hours to produce operational systems. I must say, the American side has been most impressive in their engagement of industry and their parallel preparations for mass production."

"This is precisely where our decision to execute the exchange program will pay big dividends," Churchill said. "We need their manufacturing capacity most of all. The rest will sort itself out in due course."

"Then, I must say, you would be ecstatic at their commitment, especially within the electronic group. The Americans have created an independent defense research group with a large, expandable facility near the Massachusetts Institute of Technology, MIT, which they have called the Radiation Laboratory, or Rad Lab. The Americans have made more progress with what they call radar than we imagined?"

"Radar?"

"A convenient acronym for RAdio Detection And Ranging – RADAR – which I must say is a more descriptive term of where we are headed than our RDF contraction. They developed a key component that has eluded us. They call it a multiplexer. It is a sophisticated electronic switch that allows a single antenna to send and receive signals in a very efficient manner. As I said, between us, we are going to do very good things, very quickly."

"Excellent. That is most encouraging news. What about the airborne kit problem?"

"Yes, well," Sir Henry paused to collect his thoughts, "another bit of encouraging news, I must say. This is another area they are well advanced beyond us – minaturization, as they called it, of the electronics to support radar operations. The Rad Lab has made the airborne radar unit their number one priority, in deference to our present situation and pressing need."

"Good show. None too soon, I must confess. The Germans could not subdue our lads in the daylight hours, so out of frustration, they have taken to the night with considerable success. The Air Ministry is making progress with the night intercept problem, but it is agonizingly slow, and our people and cities are paying a terrible price. You recall our discussion in June on their radio bombing aid."

"Yes, sir. Now, that you mention it, I must confess my underestimation of Doctor Jones and his assessment of the German *Knickebein* system. I offer my most humble apologies for my mistake."

Doctor Reginald Victor 'R.V.' Jones was a comparatively young, 28-year-old scientist assigned to the Air Ministry Intelligence Branch research unit. He was the first person in His Majesty's Government to raise the alarm about the German system of using intersecting radio beams along with electronic equipment and an indicator instrument installed onboard some German bombers to rather precisely establish bomb release points at night and bad weather for reasonably accurate bombing of remote targets. Sir Henry had discounted Jones's assessment to preserve scarce research funds for other on-going development programs. The intellectual confrontation reached the War Cabinet and culminated in a rather volatile discussion in the Cabinet Room of Number 10 on the 21st of June that precipitated Sir Henry's resignation as the chief of the Air Ministry's research and development efforts. Within a couple of weeks of that milestone meeting, the *Knickebein* beams had been confirmed by independent means.

"Water under the bridge, Sir Henry. No apologies necessary."

"Thank you, sir. Nonetheless, the Americans were genuinely surprised by our disclosure of the German radio-assisted bombing system. They have numerous good ideas for countermeasures, which we immediately passed back to the experimental establishment. The Rad Lab joint team is also working on their version of a counter-measure system, since ours is more advanced toward an operational unit."

"Good. We need that now."

"The evidence is all around us. Seeing the destruction on the automobile journey into the city from Northolt is quite disturbing, to say the least."

"Yes, well, I can only say, the northside has not had it as bad as Canning Town, the Docklands and the south side."

"It is worse?"

"Yes."

"This would not be good news for our lads in America."

"Then, I must say, we should not distract them from their endeavors."

"Yes, sir."

"Now, that is the electronic group. What of the other seven groups, as I recall you mentioning?" asked Churchill.

"With only a few exceptions, all of the items on our list have progressed through the teams and been absorbed within industry. Professor Lindemann's aerial mine concept was rejected outright."

"The Professor will not be pleased."

"I presume not, but we did our best to present the work we have done. We should see U.S. produced items within a year for most of our items. I must note one problematic area – the uranium explosives matter. John Cockcroft is handling that area. The discussions have been frank and open to the best of our knowledge. You may be aware of Mark Oliphant's rather outspoken criticism of the Briggs Committee's . . . energy, shall we say. John has now added his voice to Oliphant's observations."

"What is the problem? I thought their physicists agreed with ours regarding feasibility."

"To be blunt, Prime Minister, the problem appears to be Lyman Briggs, who comes across as a plodding bureaucrat, with a limited budget, and a paucity of willingness to rock the boat."

"Sometimes, the boat must be rocked."

"Quite so. The complication is, Briggs is a political appointee by President Roosevelt himself."

"Ah, well, now we see the crux of the matter."

A dull but discernible series of thuds startled Tizard. "What was that?"

Churchill chuckled modestly at the noviceness of his guest. "Those, my dear Sir Henry, are exploding German bombs on this side of the Thames . . . perhaps a mile or less distant. I am afraid you shall have to suffer the discomfort of the War Rooms this evening." Sir Henry Tizard nodded his understanding. "You were talking about this fellow Briggs."

"My counterpart, Mister Vannevar Bush, and I have discussed the Briggs problem several times. Van feels it unwise for him to use his political capital to raise the matter to the President at this stage, meaning with the uncertainty of no final report from our MAUD Committee. Van and I have agreed the MAUD Committee final report may be sufficient impetus, unless you choose to raise the issue directly with President Roosevelt, as you have some insulation from the politics."

Churchill laughed hard. "I am politics, my dear Sir Henry. Nonetheless, I take your point. When I address the matter of uranium explosives with Franklin, it will be eye-to-eye. I must be able to see his eyes."

"Yes, sir. The Briggs Committee did agree to perform several minor experiments, such as the Frisch fission event to confirm the neutron levels first measured by the Germans."

"Do the Americans feel the same as we do about the German threat?" Churchill asked.

"John and I both believe they understand the threat and agree the threat is real. However, I think it best to say they do not feel the immediacy of the threat, or the urgency of the engineering and manufacturing tasks."

"Getting conventional armaments from the Americans has been more difficult than I imagined it would be. President Roosevelt has the unenviable burden of trying to help us without triggering public outrage by the powerful isolationist segment of the population, or at least the political activitsts in the government, the press and politically influential circles. I shall look for the opportunity to discuss the matter with Franklin, but I suspect such an opening will not come until after the threat of invasion has been substantively removed."

"I recognize it is outside my charter, but is the invasion still looming?"

Prime Minister Churchill held Sir Henry's eyes, as he considered how much he wished to discuss of the invasion. "Let it suffice to say, the threat has diminished for now and probably will remain modest through the difficult winter weather season. Thus, the answer to your question is, yes, it is still looming. Spring will bring our next major threat window, but I can assure you, we shall remain vigilant through the ensuing months."

"Thank you for your indulgence, Prime Minister."

"Is there anything else," more thuds again startled Tizard and caused Churchill to hesitate, "I should know about your visit to America."

"I do not know how anyone could be accustomed to that," he said, motioning over his shoulder.

"It gets worse," chuckled Churchill, "when they come closer. Anything else, Sir Henry?"

"No, sir. I think that is a reasonable summary."

"Very well. If any should come to mind, please do not hesitate to contact my office. The sergeant of the guard will assign you to a cot, ensure you are fed, and inform you regarding the rules of the house. Thank you for spending the time with me and especially for your able leadership of the exchange mission."

"You are most welcome, Prime Minister. It was an honor."

Sir Henry Tizard left the Prime Minister's office and 'Jock' Coville immediately entered to continue the matters of state. The thuds became more frequent and some closer, shaking loose dust that descended from the massive beams and girders above them. Churchill would spend this weekend in London rather than leaving for the peace of Chequers.

———

Saturday, 5.October.1940
Standing Oak Farm
Winchester, Hampshire, England
15:00 hours

Pilot Officer Brian Drummond had made a point of saying good-bye to each of the staff that helped him as well as several of the patients, including the German, before he departed Hampshire Central Hospital. An Air Force automobile had transported him and a young nurse to the idyllic farm of Charlotte Palmer. She stood at the door, as if she sensed his approach. The medium brown wool pants and richly knit, light brown sweater gave her a strong, confident, country appearance. Her hair actually appeared to be blond in the afternoon sun, rather than the premature silver he remembered.

He felt both excitement at the prospect of being with her for a period of time without duty commitments and uncertainty regarding the imposition of his convalescence. The sight of her waiting for him made his anticipation grow and his apprehension diminish, as the car descended the hill toward the farm buildings.

The three helped Brian from the car and into the house. The nurse offered the instructions for his rehabilitation. He was to walk and exercise as much as he could. He was expected to return to the hospital once a week for evaluation. A car would be sent, if they did not have access to transportation. She also left the telephone number for the hospital in case any complications should arise. With her instructions complete, the nurse left with the driver. Charlotte and Brian were alone.

She made tea for them and chattered away about recent events on the farm. Charlotte seemed very nervous, like a very young schoolgirl on her first outing away from home. She said as much several times although it did not have much of an effect. Her nervousness became humorous and cathartic for his apprehension.

"What is so funny?" she asked finally.

"I think you are more nervous than I am, and I didn't think that was possible."

"I am glad you are enjoying my uncertainty."

"I didn't mean it quite like that. I have my own uncertainty."

"Like what?"

"Like whether I am imposing upon your graciousness and charity."

"Nonsense, Brian," she said, as she placed the tea service on the sitting room table in front of the sofa. She poured the tea, added some cream and scarce sugar, and handed a cup and saucer to Brian. Charlotte chose to sit in the armchair at the end of the table. "We have so much to talk about."

"And, I haven't even received my hello kiss."

"Oh, my dear," she said, as she stood and moved toward him. "Heaven forbid." Charlotte leaned forward, grasped his face in her hands and gave him a warm, passionate kiss. "There, how was that?"

"Delightful. Thank you very much."

"Now, as I was saying. We have a great deal to talk about."

"As you said when I woke up."

Charlotte looked into Brian's eyes, smiled, and then returned to a more stoic expression. "I do not know where to start, so I suppose I should simply wade into the pond."

Brian nodded. He hoped her words would be good. He thought they would be, or she most likely would not have agreed to take care of him.

"Several weeks ago you asked me to marry you. I have been nearly distraught struggling with your proposal."

"Why?"

"I shall get to that. You must let me say what I must say."

"Only if it's good."

Charlotte grimaced at him in mocking protest, and then took a sip of tea. Her eyes did not return to his. "Now, you must let me get my feelings out." She hesitated to find her starting point. "I lost Ian in early June. Propriety dictates a year of mourning. I want to mourn him. He was a good man and deserves my respect. As I have told you before, I was not looking for love, and I thought you were a brash, pretentious, arrogant, young man. I was confused by your attention. For some time, I thought you were drawn to me because I saved your life, or maybe even worse, because I was a much older, widowed woman . . . and vulnerable." Brian started to react, but Charlotte held up her right index finger like a teacher expecting her students to be quiet. "Since those early impressions, I think we have come to know each other better, probably not as good as might otherwise be the case, given this dreadful war."

Brian studied her facial features and admired her elegance. Her simple clothes hid her other physical attributes, but it did not lessen his admiration.

"I was flattered by your attention and interest in me, but I am torn by so many conflicting elements, like our age difference, our citizenship, this war, and the dangers you face each day you are flying. I can truly say I believe you love me for the woman I am, not some image in your mind. I have come to love you more than I ever imagined possible, especially so soon after Ian's death.

"I think it is obvious you do not recall what I said to you while you were in a coma." Brian shook his head in agreement. "I said some things to you as encouragement for you to regain consciousness, but I also said them because I felt them and had to say them. In a manner of speaking, I am sorry you did not hear me, or remember what I said. It was easier when you could not respond or react, but I must say what I need to say, again."

Charlotte stopped to take another sip of tea and look over her cup at Brian. He wanted to know what she was going to say, but he was also afraid of what it might mean to their relationship. He had searched his memory many times since Wednesday, but to no avail. He waited patiently for her to continue, holding all his feelings behind a neutral exterior.

"You must know that my feelings for you have nothing to do with what I am about to say," she continued.

Brian swallowed hard fighting his emotions.

"I am pregnant."

Brian's heart nearly burst. He tried instinctively to leap to his feet. Pain shot through every nerve fiber in his body like a bolt of lightning. He laughed in relief and groaned in pain. Charlotte stood wanting to do something, but could not figure out what she would be able to do to help.

"I am sorry, Brian," she said, holding her face as though she had committed some social error.

Brian waved his left arm, as if signaling another pilot about to land with his wheels up. "No," he grunted between the shots of pain. "No."

"I am sorry."

"No," he grunted again.

"Brian, what can I do?" she asked in frustration.

Pilot Officer Brian Drummond mustered up his inner strength, sucked in a deep breath through his clenched teeth, and tried to sit up straight. He had some difficulty focusing his eyes on Charlotte's concerned eyes. He held up his left hand several times, trying to indicate for her to give him some time to compose himself.

"I must apologize to you," he said finally, not quite fully in control. "I should have been more careful with my enthusiasm. When the pain subsides, I think I can do more than I can or should. The pain reminds me that I have a long way to go. I must apologize to you. I know a pregnancy is not what you need."

"Not to worry. Ian and I had tried for several years, and then gave up."

"I am sorry I have added this complication, but it makes my commitment to you even greater. It also gives us the reason to get married."

"Wait, Brian. I will not allow my pregnancy out of wedlock to be the reason or even a reason to marry. I failed to take precautions."

"Bullshit," Brian said forcefully, and then caught himself. "I'm sorry. I didn't mean to curse, but we did this. We created a new life between us. It was not your fault or mine. I can understand this baby coming at a rather inopportune time, but we seem to rarely get a choice these days." He looked deeply into her eyes. Furrows of worry creased her forehead. "I want to marry you even more, Charlotte. I want to . . . no, wait," he said, realizing he had something he needed to say. He thought of Rosemary's visit and her request as well as the other pregnancy he was responsible for. "I know you have been honest with me, and there are some things I must say to you."

Charlotte stood and turned to look out the window. "If your revelations are about your previous romantic episodes, I would rather not know."

Brian wanted to see her face, her eyes. She kept her back to him, as she stared out the window. He wanted to tell her about his relationship with Rosemary and her request as a condition of their friendship. He felt an obligation to tell her about Mary Spencer's pregnancy, so he could be free of the burden. He wanted nothing between them that might cause conflict. He also knew the likelihood of Mary's baby coming between them was very remote. *Maybe she is right. If I said nothing, now, with the preface I have given, what would she think? Would she always be thinking about what she did not want to hear?*

Brian considered the things he might say to get out of the box. He had to say something even if it was not about Mary's pregnancy.

"I have always been told I was too young -- too young to fly, too young to go to war, too young to love, and too young to marry." She turned to look into his eyes. "My friends, especially Jonathan, have told me I am too young for you, or you are too old for me. I am concerned about my age, that I am too young for you, that I won't be able to make you happy." This time, Charlotte wanted to protest, but he stopped her. "I have felt a connection with you since the first time we met . . . when I was conscious," he added, causing them both to laugh as tension relief. "I know you are the one I am supposed to be with . . . and, it has nothing to do with your saving me. I am not a schoolboy. I may be young in years, but I have grown since I made my decision to go to Canada and join the RAF. I am a fighter pilot, a good one, I think, and I have done my share of living in the past year. I want you to know, from the bottom of my heart, I love you. I love you more than I thought possible. I want to be a part of your life, and I want you a part of mine. I want to be with you. Having a baby is like icing on the cake. It may not be the best time to get married or have a baby, but it is meant to be."

"What about the war?"

"The war is life, Charlotte. It simply is. We can't stop living our lives because there is a war on around us. There may never be a good time. We owe it to the child growing within you to make our lives the best we can make them, no matter what happens around us."

"What about Ian?"

"Your loss was tragic. I understand your commitment to his memory, but would he really want you to stop living for a year or more? Would he want you to not feel, to not love again? I know if I were in his place I would want the most happiness you could find. I am not asking you to forget Ian. Quite the contrary, especially as a naval officer who gave his life in the defense of freedom, we should all remember him, and never forget him."

"You speak wise words for someone so young."

"What did I do wrong?" Brian protested in a joking way. "What is it with all this young talk?"

"I am sorry. I meant it as a compliment."

"I know. I was just kidding."

Charlotte walked toward him, held his face again, and kissed him deeply. Brian felt the urge to lift her in his arms and carry her to the bedroom. He resented his injuries, but knew he would have to restrain himself for several more weeks.

She sat down next to him on the sofa. Brian shifted his position toward her against the pain he caused and kept to himself.

"Let's see if you can get this right, this time," he said. She waited for him. "Charlotte, will you marry me?"

"Yes, I will, and I will have your child."

Brian reached his left arm around her shoulders and drew her toward him. He held her close. He knew in his heart they would have a long and glorious life together no matter what lay ahead. The problems of yesterday would not become obstacles to their happiness.

———

Saturday, 5.October.1940
Cabinet War Rooms
New Public Offices
Westminster, London, England
23:45 hours

The Prime Minister stood as his chief of the Secret Intelligence Service entered his office/quarters. "Good evening to you, Stewart," Winston said, greeting his chief spy with a greater than usual cheeriness to his tone.

"And, a very good evening to you, Prime Minister."

"I trust you have the evidence?"

"Yes, sir. We do, finally."

Winston rubbed his hands together with a youthful sparkle in his eyes. "Let's have it."

Colonel Stewart 'C' Menzies retrieved a small stack of papers and photographs from his case. He motioned to the small desk befitting the diminutive size of the Prime Minister's private office and living quarters in the underground bunker. Winston nodded his head in agreement. Menzies spread several of the photographs. He could only recognize a few of the landmarks in the array of aerial photographs.

"These were taken over the last few days. This set," he pointed toward the half dozen closest to them, "just came in a few hours ago. If you will notice," he paused to point at several areas of the photographs, "the motorized barges are heading east in large numbers. The enormous number we saw in late August and September have nearly all been dispersed, returned to Germany, Belgium and Holland. The large numbers of bivouacked tents in these photos taken three weeks ago are now completely gone as you can see in these photographs from yesterday."

"Excellent."

"In addition, we have numerous intercepts, so many I chose not to bring them, from ULTRA and other sources that confirm conclusively the invasion has been postponed until next year . . . the earliest indication is the spring . . . April or May."

"Smashing, absolutely smashing."

"Their invasion fleet has vanished from the Channel. The marshaling of army units near the coast has been totally reversed. They have abandoned daylight-bombing. I think it may be safe to say, you have won the day, at least for the winter."

"They, Stewart, The Few, they have won the day. As the Lord is our witness, those young men stood to the mark and did not flinch or waver. They have beaten the most powerful air force yet assembled by man, and they beat them with skill, determination, prowess and perseverance. Mark my words, this summer shall go down in the history books as one of the most glorious triumphs of the human spirit . . . the will to be free."

"I do believe you are correct, Winston."

Churchill's expression turned serious. "While I suppose we should celebrate our success in stopping the vile, black tide that has washed across Europe, we cannot rejoice, as long as our citizens are being bombed every night." The series of dull thuds and descending dust punctuated Churchill's words, as if on cue. "We must now resolve to shoulder the long, tortuous task of reclaiming the lost lands and ridding Europe of this foul monster."

"Indeed."

"Since it is your business, what is the assessment of MI6 with respect to Hitler's next moves?"

"We have been asking that very question since the dissipation signs began to show up several weeks ago. Clearly, I must caveat our opinion with a healthy dose of caution, as we have little definitive evidence to support our current assessment. That said, we see the Germans have worked with characteristic efficiency to consolidate and solidify their gains of this year. They will most likely leave Vichy France alone and work through diplomatic means to draw Spain closer. Their dispersal of the invasion fleet would seem to be a pragmatic one – reduce the targets for Bomber Command – and not likely a statement of abandonment of Operation SEALION. We shall remain attentive to the intelligence we can collect regarding their intentions. Thus, the obvious primary choice is, they will renew their invasion attempt in the spring. They will most likely be unrelenting in their bombing campaign through the winter in an effort to soften us up. We have signs of his interest in other directions, but we find it hard to imagine how he could turn his attention elsewhere while we remain a viable threat to his security. As a consequence, the Battle of the Atlantic is likely to get worse before we can hope to find improvement. This is going to be a very difficult winter, I am afraid."

"I share that assessment. We will endure the bombing, but it is those damnable, bloody U-boats that are the most grievous and gruesome threat to the homeland and our sustainment." Churchill paused in dark contemplation. Menzies chose not to disturb his thoughts. "Are there any other options for the Germans?"

"Yes, to be succinct and direct. They need oil. They will be hard pressed to sustain their massive war machine without fuel and lubricants. Despite their curious non-aggression pact with the Soviets a year ago and their collaboration with the Soviets in consuming Poland, Hitler has a long and well-established animosity toward the communists. The Germans will most likely press hard on the Italians to take Cairo and the Suez. However, we believe it is the oil of the Persian Gulf and Baku that will most probably command the greatest attention. Hitler might actually attempt to subdue Stalin's Soviet Union for the double benefit of eradicating communism and obtaining the oil supply he needs to keep his armed forces running."

"Do you seriously think he might try that without beating us?"

"The oil has to be an enormously seductive objective for him. He is also above all else an ideologically driven man. While our analysts cannot imagine Hitler being so foolish to turn his attention eastward with the British Empire remaining his biggest threat, the other factor just might override any sense of prudence he might otherwise possess. Hitler is quite inflated with his sense of destiny from his striking success in the Battle of France and our embarrassing withdrawal from Dunkirk. His ego will find ways to discount the *Luftwaffe's* failure to beat our air force."

"Mark my words, Stewart, that vile man shall come to regret his failure to beat us here. Until he turns his attention elsewhere, we must concentrate on rebuilding Fighter Command. They have taken a brutal pounding over the last three months. We will continue to prepare for the inevitable spring offensive. Please keep me informed as we move through the winter.

"Yes, sir, by all means."

"Would you be so kind to return in the morning? I should think the War Cabinet and Defense Committee will want to see this latest information. We have a long road ahead, but now we know there shall be a brighter tomorrow."

"As you wish, Prime Minister."

"Excellent, then we shall see you in the morning. Pleasant dreams, Stewart, and thank you."

"Yes, sir. And, congratulations once again."

"We must thank The Few."

Chapter 9

But this momentous question,
like a fire bell in the night,
awakened and filled me with terror.

-- Thomas Jefferson

Sunday, 6.October.1940
Standing Oak Farm
Winchester, Hampshire, England
04:10 hours

Week 14

"**A**aaarrrrhhh," Brian groaned loudly and thrashed in bed.

The suddenness and intensity startled Charlotte. "Brian," she said, gently touching his shoulder.

"Nooooo!" he screamed, as he shot up to a seated position, ignoring the pain jolting his body and mind. Brian felt the scream an instant before he heard it. The stabbing pain in his head, chest and right leg followed his recognition by only a flash. Brian Drummond sat there staring into the dark room, frozen in time and by his fear.

"It was just a nightmare."

He did not move other than the uncontrollable shaking of his entire body, as if in a perpetual shiver of cold. His breathing remained deep and heavy as his body fought against the massive blast of adrenaline coursing through his veins.

"What was it?"

Brian just stared and barely shook his head.

"You are soaking wet, Brian. The sheets are wet." Charlotte reached over to switch on the small electric lamp behind the bed.

She looked at him shaking, his chest heaving, and staring off to some distant place. Charlotte moved to reassure him, wrapping her arms around him and pulling him gently to her bosom. Brian felt the softness of her unbridled breasts and inhaled the earthy sweetness of her skin. He could hear the pounding of her heart. His violent awakening had scared her. She tried not to show it, but her rapid heartbeat betrayed her. Charlotte wanted Brian to feel safe. He needed to know that much.

"What was it?" she whispered, again.

Brian could barely shake his head.

"My God, Brian. It must have been something . . . rather serious."

He shook his head, once more.

"I have never heard of a nightmare being so bad. It must have been terrible." Charlotte held him until his shaking ebbed. Brian's crash had been 10 days ago. He was released from the hospital just yesterday. His head bandaged from a skull fracture. His upper torso wrapped for two broken ribs and a punctured lung, and multiple shrapnel wounds across his torso and legs. He also had a below the knee cast on his right leg after surgery to repair a rather nasty compound fracture. In addition, he experienced serious burns on his legs, hands and neck. Brian certainly had more than a few reasons for nightmares. "Please tell me about it."

"I can't," he murmured, still staring to some distant point.

"Why? It will help to talk about it."

"You don't want to know," he murmured.

Charlotte held him more firmly, as if to reassure him that she was not going to leave him or turn away. She knew it had to be something serious, something he had probably lived through as a young fighter pilot in this dreadful war. She also knew that if she was going to help him recover and demonstrate her love she had to accept the horror he has undoubtedly lived through in his young life. "Yes, Brian, I do. I need to know what has hurt you so much."

"Can I have some tea?" he asked.

"Yes, yes, of course." Charlotte released him, got out of bed, and put on her heavy robe and slippers to lessen the chill of the night.

Brian fell back on the bed with a thud. Charlotte watched him for a moment. He stared wide-eyed at the ceiling. She left to make a small pot of tea for both of them. It was probably going to be some time before they could return to sleep.

Charlotte Palmer returned to Brian, carrying a tray of tea and a plate of cookies. He was just as she left him, uncovered and staring wide-eyed at the ceiling. Charlotte placed the tray on the nightstand beside him. She felt the sheets. Both top and bottom sheets were damp. She found a soft, heavy, spare blanket, pulled the covers and damp sheet off him, and then covered him with the blanket. The various bandages, bruises and burns covering virtually every inch of his body made her shudder every time she saw them. He must have been through so much to be so badly injured. Charlotte poured some tea for both of them. She added some cream, in abundant supply, and some precious sugar, an infrequent extravagance in war-ravaged England.

Brian began the process of sitting up. As he moved, he winced in pain and cycled his way up. Charlotte helped, using some extra pillows, to prop up behind him. She got him settled and replaced the warm blanket over him tucking it behind his shoulder. He reached both arms out to take the proffered cup of hot tea. She sat down beside him with her cup of tea. They took several sips to feel the warmth.

"Now, tell me about your nightmare."

"Charlotte, trust me. You do not want to know. Anything I say will not make you feel better."

She shook her head in disagreement. She knew he was probably correct. Whatever it was it would not be pleasant. But, she also knew she had to help him.

"Brian, if I am to help you, I must share your pain. Something truly dreadful has happened to you. So much so, that you have had a most bizarre nightmare and sweat-soaked the sheets. Your body is broken, cut, burned, bruised and otherwise tortured. Now, please tell me about it. Tell me, what could cause all this damage?"

Brian took several more sips of his tea, and then searched her eyes. He saw her sincerity and the love she felt for him. He realized he had to share the bad along with the good.

Brian tried to take a deep breath and winced with pain. He let it pass. "It was a recurring nightmare I have had several times before."

Charlotte waited patiently for him to continue. She raised her eyebrows as if to signal him to proceed.

"At first, I thought it was just a nightmare caused by the bloody war."

She waited some more. "But, it wasn't," she coaxed him.

Brian looked deeply into her eyes. "Are you really sure you want to hear this?"

"Yes, Brian. I am. I must be there for you."

He stared into her eyes for the longest time.

"It is the same every time.'" He looked at his cup of tea, now resting in his lap. "I was in a call box in London. Bombs started exploding all around. I tried to get out, but the door was jammed. There was fire. The buildings were burning and beginning to collapse all around me."

"Oh, Brian," Charlotte whispered, as she touched his cheek, and then stroked his damp hair.

"The harder I tried to get out of that booth, the closer the fire got and the more debris piled up . . . beginning to bury me." Brian closed his eyes. He shivered several times. "I usually wake up. This time it went further." Brian put his cup on the tray and looked back to Charlotte.

Charlotte continued to stroke his hair. "It is all right. You are safe."

"This time the rubble had piled up halfway all around. The heat was so intense. The booth is now beginning to burn. My clothes were on fire. My skin was burning."

"It was just a nightmare."

"So, I thought. When I got shot down this last time, it was that dream. It was a premonition. That dream was what happened." Brian began to sob. Tears streamed down his face. "I am going to die."

Charlotte embraced him. She kissed his head as he cried. "You are safe, Brian. Everything will be all right."

"I am dead," he cried.

"No, you are not. You are very much alive. It was just a dream, Brian."

"The canopy was jammed. The engine was on fire. Burning oil covered everything. The fire came into the cockpit. It was that burning telephone booth."

Now, Charlotte shuddered at the horror that did all this damage, that reduced this strapping young man to an infant awakened and scared from a nightmare. Charlotte felt pain, threads of anger over this ridiculous war, sympathy for the agony that consumed Brian, and regret she did not know anything better to do for him. All she could think to say was, "You are safe, now."

Charlotte wanted to ask him more. She wanted to know what Brian and the other RAF pilots were doing in the skies above them. Her curiosity nearly commanded her, but she backed away. She refreshed his tea and handed it to Brian. The injured pilot sipped, but kept his eyes lowered, lost in his thoughts.

"How do you feel?"

"Sick," he answered.

"Oh, Brian." Charlotte touched his cheek. "What can I do to help you?"

"Just stay with me. I will be OK. It was just a nightmare. I'll be alright."

Charlotte considered Brian's words and drifted into her own contemplative state beside her charge and bedmate.

Brian knew he should talk about what was rumbling around in his consciousness. The particular nightmare that haunted him and caused Charlotte such concern this morning had been with him since the previous winter at Drem, before the end of the Phony War and the invasion of France. The nightmare was the closest he had ever come to a premonition and the recent crash the closest he had come for fulfilling the premonition. Brian needed to get his mind off the morning's episode.

Charlotte must have sensed restlessness. "What do you want to do?"

"I'd like to go for a walk."

"It is still dark, Brian."

"I need some fresh air. I need to do something else."

"As you say, then. Would you like me to help you?"

"No. I need to do it myself."

Charlotte quickly moved to his side of the bed, to assist him should he need help.

Brian rolled slightly toward the edge of the bed and simultaneously pushed with his arm, as he swung his legs out. He sucked in a breath and winced against the pain shooting through various parts of his ravaged body. Brian on the edge of the bed, breathing quick, shallow breaths to let the pain subside and recover from his exertion. Charlotte started to reach for him several times, but he held up his hand to stop her. He reached for the crutches given to him at the hospital.

"Are you going to put something on?"

"Why? It's just us out here, and you've certainly seen all of me there is to see."

"It is a mite chilly outside. We are into autumn after all."

"The cool air will feel good . . . until I get cold."

"As you wish."

With the support of his crutches, Brian stood, fighting against more pain. He slowly moved through the house to the front door with Charlotte beside him. She opened the large, heavy, front door and stood aside to allow Brian to hobble through. After several steps, Brian stopped and slumped on his crutches.

"Are you alright?"

Brian hesitated quite some time before he responded. "Yes. The cool air feels good on what exposed skin I have left in all this."

"Do you feel better?"

"Yes . . . yes, I do, actually."

"Excellent."

Brian stood up as straight as he could manage. "Twilight is here. The sun will be up in 40 minutes." He looked around and above them. "Although, we may not get to see it. Another dreary day in Southern England." Brian hobbled along the path toward the lake. A light mist became a drizzle.

"Brian, we need to get you inside. The doctors cautioned you not to get your bandages or leg cast wet." He did not alter his movement. "Please, Brian, it is going to rain soon."

The wounded, young pilot stopped and turned to see Charlotte's beautiful but concern face. "I'm sorry. I do not mean to worry you." He looked to the sky, again. "Nearly milking time."

Charlotte laughed. "Indeed. Not your concern."

"I need to earn my keep."

Charlotte laughed, again, harder and deeper this time. "No, my dear man. You have defended us on our little island. You have done your duty and more than earned whatever keep can be provided. So, get your bum inside the cottage before serious rain arrives."

"Yes, ma'am." Brian moved back, stopping as he passed her, leaned toward her and kissed her on the lips. "You are a good woman. Thank you for caring."

"Nonsense. Now, get on with you." As Brian moved, she patted his bare buttocks.

"Ooowww!"

Charlotte gasped and clasped her hand over her mouth. "Brian, I am so sor . . . "

"I was just kidding," he laughed.

"You are an incorrigible tease." He smiled at her and continued his ambling movement to the open door. "Can you manage?"

"Yes."

"Dawn is nearly upon us. I need to get along with the morning milking. Mister Bridges and Mister Morgan should be here shortly to help."

Lionel Bridges and Horace Morgan were local, elderly men beyond conscription age, at least for now, who had worked for her family since they were teenagers. While they did not move fast or were they as strong anymore, they were helpful nonetheless.

"Would you like me to make you a fresh pot of tea?"

"No, thank you," he responded without looking around. "I can manage."

"I left the biscuits out. You can have those until I fix breakfast for us."

"Thank you, Charlotte."

As she went off to the barn, Brian made his way to the kitchen. He moved awkwardly with frequent pauses to catch his breath and allow the pain to subside, and after a frustratingly longer time than his mind accepted, Brian sat heavily in a sturdy chair at the kitchen table. The tea felt good. The butter cookies in the open tin actually had very little sugar due to the war shortages, but had just enough to recognize and taste.

Brian's thoughts gravitated to the morning's events. *I have got to get control of that damn nightmare. I am scaring the hell out of Charlotte. I do not want to scare her away . . . damaged goods. I have got to get back in the air just like that day two years ago, when I crashed after hitting the Seaver's dog on landing. Malcolm insisted I had to get back in the air immediately. He actually trusted me with his treasured Sopwith F.1 Camel — the airplane he flew and enjoyed the most during his service with the Royal Flying Corps during the Great War. Malcolm was correct then, and I am sure he would be insisting now, but I am in no shape to fly. Maybe I can convince someone to take me up in one of Six Oh Four Squadron's Blenheims or their new Beaufighter, or even a Lysander or Hudson . . . anything that flies.*

As he finished his second cup of tea, Charlotte entered. "I'm afraid you will need to put some clothes on, or at least a robe. Mister Morgan and Mister Bridges will be joining us for breakfast."

Brian did not argue or hesitate. He made his way to the bedroom, while Charlotte busied herself with food preparation. The sounds of a chef's implements in action brought back memories of his mother. Those were days seemingly long behind him, but the memories were still vivid. He heard the voices of Lionel and Horace added to Charlotte's effort. Brian struggled with his uniform trousers and even the pajamas someone had packed for him. Nothing would fit over his leg cast. He gave up, found the large, heavy robe Charlotte had given him, donned it and tied it off as best he could to cover himself.

"Ah, there you are, Mister Drummond," said Bridges. "We heard you had been fairly well bashed about. How are you getting on?"

"Great to see you again, Mister Bridges, and you Mister Morgan." Horace nodded his head. "About as well as can be expected, but I know I will recover quickly under Mrs. Palmer's expert care."

"No need to overstate it," added Charlotte. The two men chuckled in a rather odd, perhaps tense, manner. "Please be seated, gentlemen, so we can get Mister Drummond off his feet and onto his bum." They laughed more humorously this time. "Breakfast is nearly ready." The three did as they were requested.

Charlotte served each of them a large bowl of oatmeal, two fried eggs, cottage cheese, two pieces of toast, and a glass of milk. She also placed a bowl of butter and a large pot of tea on the table. Their entire breakfast had been grown and prepared on her farm. They ate well. There were no words between them, only consumption. Sated, their bellies full, Bridges and Morgan thanked Charlotte, and excused themselves to get on with their chores and assigned tasks for the day. Charlotte remained quiet for several minutes, lost in her thoughts. She pushed her dishes back and took another sip of tea.

Charlotte looked into Brian's eyes for several seconds before she spoke. "What are you going to do?"

"Today?"

"No, after your recovery?"

"I've got to get back to flying."

"Why, Brian? Flying has nearly killed you at least a couple of times I know about, and I imagine there are others as well."

"Flying has not hurt me. The Germans have."

"You damn well know what I mean."

Brian held her eyes. Do I really want to get into this? It has to come, so the sooner the better, I guess. "Charlotte, I am truly and terribly sorry I scared you this morning."

Charlotte held up her hand for him to stop. "Don't be peripheral with me, Brian. How can you even think of returning to flying after it has done this . . . ," she said motioning to his various wounds, ". . . to you. I have lost my father and my uncle to the previous war, and my husband to this war. And now, I have another man in my life who wants to continue flying bloody fighter aeroplanes. I fear I may not survive losing another man I love . . . no matter how necessary or righteous this war may be."

Brian swallowed hard. What the hell am I supposed to say to that? Brian reached for her hand, which she withdrew from him. "I love you, Charlotte. I know I have no reason to expect you to understand what flying means to me. I have no way to help you understand."

"Damn it, Brian, you are a bloody American volunteer. You have not been conscripted to serve. You are in this war simply because you want to be. This is not your fight. You can choose not to be."

"Yes, flying got me into this fight, but surely you must see this is not war for the sake of war. The Germans have overrun all of Europe. All that remains is England. We have been under the threat of German invasion since spring, and to my knowledge, that threat has not gone away. I can fly. I am good at what I do. My brothers depend upon me, as we depend on each other. We have been dreadfully short of fighters and pilots to fly them . . . to keep the Germans off this island . . . and, to keep you safe, Charlotte."

Tears descended on her cheeks. She quickly wiped them away but more followed. Charlotte covered her face with her hands and quietly sobbed. Brian placed a hand on her shoulder, and gently stroked her neck and shoulder.

Charlotte regained control, wiped the remnants of her tears away and looked into Brian's blue-grey eyes.

"Fate brought you into my life. I violated my mourning for my husband, because of you. My dear God above, I am pregnant with our child, because of the connection I felt for you."

"All the more reason to keep you safe," Brian quipped.

"Damn you, Brian! I have never felt so bloody conflicted in my life, and perhaps even worse, you have me cursing like a bloody, fucking sailor," she said, slamming her hand on the table and rattling all of the dishes.

"I'm sorry," Brian said rather meekly.

"You damn well should be."

"I have no intention of letting them get me."

"Don't be daft! None of them do. It just happens."

"This may not help, but let me tell you how I got here."

"I know how you got here."

"Yes, but not what led me to leave the only home I have ever known for Windsor, Canada, to join the RAF. As far back as I can remember, I have watched machines that fly. They call Wichita the aviation capital of the world because it has quite a few companies that build lots of airplanes. I met and began helping Malcolm Bainbridge, who owned several airplanes and his own air transportation company. In exchange for my work around his place, he took me flying one day when I was about nine years old. I was hooked. I loved it. I loved everything about flying. He began teaching me to fly. I made my first solo flight when I was still nine; that was in 1930, as the Great Depression gripped everyone. Through the worst days, Malcolm kept going somehow. I have no idea how, but he did. After Roosevelt became president in 1933, things began to turn around. I was right there with Malcolm. His best friend from the Great War was John Spencer, Group Captain Spencer, now Air Commodore Spencer. His uncle is none other than Winston Churchill. Anyway, as Hitler began to take over Germany, Malcolm, and indirectly for me, Group Captain Spencer, began warning all those who would listen about the dangers the Nazis presented in Germany."

"Churchill has been very singular and direct on that subject."

"Yes, quite so, and quite rightly it seems to me. I listened very carefully. I understood what they were saying. Shortly after the Munich Accord, I told Malcolm I want to fly the new fighter airplanes and defend freedom. He tried very hard to talk me out of the notion. I persisted. With the help of Group Captain Spencer, I found a place in Fighter Command, flying the best fighter aircraft in the world. Flying that machine gives me an indescribable sensation of freedom, of power, of satisfaction and even contentment. I am not too keen on the Germans shooting at us, but that is the price we pay for doing what we love."

"Do you love flying more than me?"

"Charlotte, that is not a fair or reasonable question. You love this farm."

"Yes, and I would give it all up to keep you safe and with me."

"I truly love you, Charlotte, as I have loved no other in my whole life. I will give up flying, if that is what I must do to hold you, but you may well break me as a man in doing so."

Charlotte stared at Brian for the longest time. She did not blink and just bore into him. Finally, she shook her head. "I cannot ask that. I can only hope you understand and appreciate what you are asking of me after all the sacrifice I have endured in my life."

"I believe I do, and I do not ask you for more."

"It will not be up to you."

Brian nodded his head. "Quite so. All I do is promise to love you as best I am able, and to do my best to survive all this, so we can live the rest of our lives together."

"I suppose I cannot ask for more."

Brian leaned across the table against the pain in his chest, and kissed her lips slowly and gently.

———

Sunday, 6.October.1940
RAF Middle Wallop
Middle Wallop, Hampshire, England
06:30 hours

A light rain kept everything damp. The low ceiling meant they were not likely to be launched. The weather forecast predicted light rain and low cloud ceiling all day. Flying Officer Jonathan Kensington occupied one of the many open lawn chairs inside the squadron's dispersal tent. They were down to less than half of their full strength number of pilots. All of their allotted Spitfire fighters were parked in a staggered, somewhat random fashion to reduce vulnerability from potential enemy attack, which they had fortunately not experienced in a few days short of a month.

Squadron Leader 'Stack' Long-Roberts sat at his field desk in the partitioned back corner of the tent shuffling the perpetual paperwork that passed across his desk. Squadron Operations Clerk Corporal Jennifer Warren, Women's

Auxiliary Air Force, sat at her table with the various telephones that were the tools of her job description.

Jonathan closed his eyes as the other veteran pilots had already done. He listened to the soft patter of the rain on the tent canvas. Perhaps today would be a restful day. With virtually no wind the sounds focused on only the rain. Tranquillity pushed the realities of war just a little farther away, at least for the moment.

The squadron had been designated a Class 'C' unit a week ago, which meant they were in a training and refit status, although for reasons unknown to the pilots, they had not yet been moved north out of harm's way, for the most part. On a day like today, they could not conduct useful training for the four new pilots assigned to them right out of the training unit. The squadron was finally scheduled for the move tomorrow to RAF Catterick, North Yorkshire, and into No.13 Group – the northernmost fighter group where the tattered squadrons like No.609 were allowed to regain their strength and readiness.

"The bloody bastards bombed London again last night," said 'Sparky' Morrow.

"Bombers at night and those damnable fighter-bomber sweeps during the day," 'Fog' Johnson contributed.

"The Six Oh Four night-fighters bagged another two of the intruders last night," added 'Waggle' Davies.

"We passed the autumnal equinox," 'Boxer' observed, "so at least the nights are getting longer than the days."

"At least we get a break from what we had a month ago," 'Fog' Johnson offered.

"Hear, hear."

Jonathan listened to the casual banter with distant awareness. His thoughts drifted to his friend and brother-in-arms. *Charlotte sounded a little frazzled yesterday.* She had just settled Brian in at her farm, after he was released from the hospital into her care for his convalescent leave. *I need to get down there to see him before we move north. He has to be better. The last time I saw him he was still unconscious . . . lucky that crash didn't kill him. I can't wait to hear his story.*

"Harness!" brought him back to the dispersal tent. Jonathan opened his eyes to see all the lads staring at him. "You must have been asleep again," 'Sparky' noted.

"No . . . just thinking about 'Hunter.'"

"Well, that is precisely what we were asking. Care to let us in on what you know?"

"I don't know much more than I did from my last visit, not quite a week ago. The last time I saw him was in the hospital and he was still unconscious. I talked to Charlotte last night on the tellie. He was released from the hospital yesterday. He's still pretty banged up. According to what she told me, the doctors expect him to be out of action for four to six weeks, perhaps longer."

"That's a shame."

"At least he's alive, and will return eventually."

"Indeed. Charlotte has agreed to be his caretaker during his convalescence."

'Boxer' chuckled, "Oh, I'm sure she will take very good care of him."

"No need to debase a good relationship."

Corporal Warren's alert telephone rang. They all froze, tensed like some Pavlovian response, ready to spring into action, and all eyes were on her.

"Six Oh Nine Squadron," she said. She listened, shook her head, and then nodded toward the squadron leader's office. "Skipper," she shouted, "Group on Red." She listened until she was sure their leader had picked up the phone, and then she returned her handset to the cradle.

The pilots relaxed.

"And no need to take offense, 'Harness,'" said 'Boxer,' "after all, he is quite the swordsman."

They all laughed at that image, including Jonathan.

"What is all this frivolity?" 'Stack' asked, as he appeared in the main part of the tent.

"Just complimenting 'Hunter' on his success with women."

"He will probably be out of action on that front as well as the flying business. Medical report from a couple of days ago indicated four to six weeks, maybe more. He will have to join us at Catterick. I have just been informed that Group expects to release us for transfer tomorrow morning. We will depart as soon as we are released. The advance team departed yesterday. The remainder of our personnel will depart by lorry immediately after we depart. Our young chicks can manage a Spit, but there will be no playtime enroute. Are we clear on that point?"

"Yes, sir," they said in unison.

"As such, a lorry has been parked in front of the Mess. It will remain there, under guard, until the caravan departure tomorrow. So, once we are released today, I suggest you pack up your kit. It must be on that lorry before you report to Dispersal tomorrow. Am I clear?"

"Yes, sir," they answered again in unison.

"Clearly, we do not have sufficient pilots to fly off all of our fighters, and the Ferry Command is saturated with other work; so, a Hudson has been laid on to bring several of you back to retrieve whatever aeroplanes remain. We will determine that tomorrow."

"Skipper, I was hoping to get down to Winchester to check up on 'Hunter' and keep him informed about our orders."

"Not today. Perhaps tomorrow or the following day. The Met Office is forecasting inclement weather for the next few days."

"Very well, sir. Thank you."

—

Monday, 7.October.1940
RAF Middle Wallop
Middle Wallop, Hampshire, England
07:40 hours

Everything was ready for the departure and transit flight to RAF Catterick. They had their maps, communications frequencies and callsigns since they would be crossing through No.12 Group and half of a dozen sector areas. The Skipper briefed them. They would fly fully armed, just in case, and the flight to North Yorkshire would take nearly two hours, so they would use moderate throttle settings and spread their formation once clear of the clouds to conserve fuel. They would be near their range limit, and he did not want to drop off any stragglers due to poor fuel husbandry.

The squadron had been reported to Sector Control as Available in 30 Minutes – the lowest level above Released -- an hour and a half ago, which should have been a simple formality.

The telephone rang. No one jumped this time.

"Six Oh Nine Squadron," Corporal Warren answered. She listened, shook her head, and then nodded toward the squadron leader's office. "Skipper," she shouted, "Group on Red."

The call lasted two minutes and was notably one-sided. Squadron Leader Long-Roberts joined his pilots. "Well, lads, a slight change in plans." They waited for him to continue. "We have been raised to Readiness. Group has an inbound raid. They are short ready units. Our birds are loaded and ready. We are a bit fragmented, and we are going to have to wing it. Our new replacement pilots will remain behind."

"Skipper!" protested one of them.

'Stack' raised his hand. "This is not a democracy. I acknowledge you eager little tadpoles want to have at Gerry, but you are not ready. 'Waggle,' you and 'Fog' will be with 'Sparky.' 'Harness,' you will be with 'Boxer' and me.

The bad guys appear to be headed for Weymouth, Yeovil or Bristol. We will be brought to Standby shortly and can expect the launch command in about 15 minutes. Group wants to place us in a loiter patrol orbit. Any questions?"

There were none. The veterans slimmed down to their combat kit and mentally prepared for combat. As they had been informed, the telephone rang. The veteran pilots stood and began walking to their respective aircraft.

Corporal Warren announced, "Standby."

Jonathan looked over his shoulder as he mounted the right wing of his 'PR-K' Spitfire Mark IA fighter. The four replacement pilots stood outside the Dispersal tent along with Corporal Warren, this time. He knew how they felt, but 'Stack' had probably saved their lives.

They strapped in, made all their connections – earphones, oxygen mask, parachute harness, lap and shoulder straps. Their ground crews stood ready. They waited. None of them heard the telephone, but they did hear the bell being rung by Corporal Warren. The crew chiefs gave the start signal.

Jonathan's big Merlin engine fired off on the first attempt and lustily roared to life. He got everything switched on, as the engine warmed up. He uncaged his attitude gyro, checked his heading set, and dialed in the current altimeter setting.

'Stack' signaled for taxi. The six Spitfires moved up the slight rise to the north end of the field, completed their engine checks. Blue Section launched first. Jonathan took up the left wing position. 'Boxer' had been the Blue Section right wing for as long as Jonathan had been with the squadron. 'Sparky's Red Section took up their position right and slightly below Blue Section.

"Bandy, Sorbo with you."

"Roger, Sorbo. Vector two five zero, climb to angels one five for now."

"Sorbo has it."

"Expect hand off to Lunar shortly."

"Roger, Bandy."

Jonathan trimmed up his aircraft for the moderate power climb, armed his guns, and checked his gunsight reticle was illuminated and set for the desired range, and then placed his gun camera in Ready position, which meant it would switch on when his trigger was depressed and would remain on for five seconds once released. They made their way through broken clouds that gave him spits of rain but nothing particularly turbulent.

"Sorbo, Bandy calling. Vector remains good. Switch to Lunar on one two six decimal four."

"Sorbo has it, see you when we're done, switching."

"Good hunting," they heard, as they all switched to the new frequency.

"Lunar, Sorbo with you on two five zero, passing angels nine for one five assigned."

"Roger Sorbo. Lunar has you. Continue climb to angels two five."

"Roger Lunar."

Jonathan rechecked all his armament switches. He also checked to make sure his oxygen mask was properly in place and snug with his darkened flight goggles in place and his oxygen system on. Jonathan kept scanning the sky around them, just as each of the other pilots was doing in his cockpit. The radio was surprisingly quiet. They leveled off at 25,000 feet and throttled back for maximum endurance. 'Stack' began a lazy zigzag pattern to slow their closure and to give each pilot a different view of the area. Jonathan thought he saw the fight below them, but the sightings were fleeting below and among the clouds. Then, he saw perhaps a squadron of 109s peeling off for their descent into whatever fight was there.

"Tally ho," Jonathan radioed. "Fighters two o'clock, just below the horizon, rolling for their attack."

"Lunar, Sorbo. Tally ho the fighters. We are engaging."

"Roger, Sorbo. Good hunting."

The 'PR-A' Spitfire pulled away from them. 'Stack' had gone to full throttle. Jonathan throttled up as well, feeling the emergency gate wire. He contemplated breaking the wire to catch up, but the opened spacing helped him scan the sky for other fighters. It took them several minutes to reach the other fighters, and the German fighters were coming up for their second perch. One of them must have spotted the six closing Spitfires, as they immediately altered their flight path, climbing straight toward them.

"Pick your target, lads," broadcast 'Stack.'

This would be a nearly head-on engagement, which meant a crazy tangle of fighters after the first pass. Jonathan pushed his throttle full forward, breaking the emergency gate wire. His Merlin engine was giving him all the horsepower it could produce. The white nose 109 grew rapidly in his gunsight reticle. Jonathan saw the cannon flashes on the nose and wings of the 109 just as he depressed his firing button and felt his guns erupt. Tracers filled the sky. They flashed past each other. Jonathan pulled back hard to climb, glanced over his shoulder to make sure he did not interfere with another Spitfire, and then rolled hard right, as he strained his neck looking back over his shoulder to find his target. Got him! Jonathan kept the nose coming around until the 109, who was trying for the same re-engagement was nearly on his nose. He rolled wings level with his nose slightly down, as his target was a few seconds behind and began filling his reticle. This time, Jonathan opened fire first. Impact flashes of molten metal appeared all over the nose and wings. The German never got to fire again. Large pieces of engine cowling flew off and an explosion of flame burst from his exposed engine. Jonathan rolled hard left to avoid the debris and the now flailing enemy fighter. Jonathan turned hard, straining his neck to look back. His target was spiraling toward the ground with a major fire nearly enveloping the entire fuselage. Jonathan immediately scanned for another target. The other Germans must have reached their bingo fuel level or run out of ammunition, as they all broke away heading south. Several Spitfires gave chase. Jonathan did the same. He found a trailing 109 and adjusted this flight path for a decent engagement angle. The German had to be very light. The closure rate, even with his descending engagement angle, was very slow, but it was still positive, so Jonathan pressed his attack.

"Sorbo, Lunar calling. Break off your engagement at the coastline and return to primary."

"Roger Lunar. Sorbo flight break it off and join up for RTB."

Jonathan winked at his lucky target, scanned the sky, again, to make sure he was clear, and throttled back. Within five minutes, Jonathan returned to his assigned position off the left wing of 'Stack's 'PR-A' Spitfire. 'Sparky's Vic of three Spitfires were formed up behind and slightly below 'Boxer's right wing. The Return To Base proved uneventful and routine. They landed in order, taxied to their Dispersal tent in the southwest corner of the aerodrome, and shutdown. The diminished ground crews took to the task of checking, refueling and rearming the Spitfires. Flying Officer Royster had already departed for Yorkshire, so they would have to debrief later.

Squadron Leader Long-Roberts made several telephone calls to establish their situation. No.10 Group released them. They would depart for the new base as soon as their aircraft completed the turnaround work. While the work would take longer than usual, since half the maintenance personnel left yesterday for RAF Catterick, 'Stack' led them back to the Mess for an early lunch. By the time they had eaten, their aircraft would be ready, and they would depart as a squadron for their renewal base. This time, the new guys would fly with them in their assigned flight positions. The veterans had done their job. It was time for rest and refurbishment. They left three aircraft behind due to insufficient pilots to fly them.

———

Monday, 7.October.1940
RAF Catterick
Catterick, North Yorkshire, England
14:15 hours

Unknown ground handlers directed them to their parking spots. An RAF wing commander, a flight lieutenant and a sergeant were standing just outside a green, clapboard hut that had to be their new Dispersal building. Squadron Leader Long-Roberts was the first to reach the welcoming party. Jonathan took his time shutting down his aircraft. As was their practice, he took his flying kit with him and left the cockpit open. The normal beehive of ground crew swarming over their aircraft was missing. The pace at this rural North Yorkshire aerodrome was decidedly slower than they had been accustomed to in Hampshire.

The airfield facilities were less in quantity and substance than they had at Middle Wallop. The grass landing area was well maintained, mowed regularly and with a single filled-in bomb crater. Everything was green – the buildings, the grass, the distant trees surrounding the base. In many respects, RAF Catterick had an idyllic quality.

The wing commander was in fact the base commander, whose name he did not hear. With his welcome complete, he marched off to what had to be the base operations building. There was not a lot to the base. At the moment, they were the only squadron assigned. They were shown the hut that was indeed their Dispersal building and the Officer's Mess for their sustenance and billeting. The base orientation took less than 30 minutes.

'Stack' gathered them back up. Their ground crew advance contingent had not yet arrived.

"OK, lads. Looks like we're done for the day," 'Stack' said. "We'll need to close up our aircraft before we secure. Assuming we can get our aircraft checked and refueled, we will fly an area orientation flight tomorrow morning. Once we debrief that sortie, we will set out our training program for the replacement pilots. I expect we will be here for several months before we return to 'B' and 'A' status. If we have combat patrols assigned, they will likely be convoy escort, as they clearly do not get much business up here these days. I do not expect any combat assignments for at least a month, unless there is an emergency of some nature. Any questions, so far?"

"No, sir," they answered in unison.

"I would like to get the orientation flight done and at least one training flight tomorrow before we go back to pick up our remaining aircraft. 'Harness,' you, 'Boxer' and 'Fog' will be flown back down to Middle Wallop. I am told a Hudson will arrive early Wednesday morning to take the three of you back. I'd like you back up here as soon as possible, so we can get on with our training program."

"Skipper," interjected Jonathan, "I'd like to take a day or so to go visit 'Hunter' before I return."

"Sure. That should be no problem. Take the rest of the day. I'd like you back up here Thursday morning."

"Thank you, sir."

"Say hello for us all," said 'Sparky.' "Wish him speedy recovery."

"Will do."

"OK. That's the immediate plan. Now, I would suggest you stow your flying kit, so we can close up Dispersal, and then close up your aircraft. We will head to the Mess to settle things in there, and then I would propose we head into the village to reconnoiter the place."

"Sounds like a plan, Skipper," responded 'Sparky' for all of them.

So far so good, Jonathan told himself.

—

Wednesday, 9.October.1940
Standing Oak Farm
Winchester, Hampshire, England

The early morning flight back down south had been slow and long, but comfortable, or at least as comfortable as any flight could be without the controls in his hands. Dodging rain clouds in transit added time to an already slow flight in the Lockheed Hudson Mark I configured for transport. The pilots descended through the solid cloud deck, broke out at about 500 feet, and entered a nice, crosswind approach and smooth landing at RAF Middle Wallop. The No.609 Squadron pilots thanked the crew for the ride, went to their waiting Spitfires, and checked their aircraft were ready to go. While Jonathan arranged for an automobile, 'Boxer' and 'Fog' mounted up and took off for the return flight to RAF Catterick.

The drive south had been equally uneventful. Two of the four checkpoints in place since last spring had been removed, making the drive a little faster. Other than the steady rain, the drive had been easy with virtually no traffic.

As he crested the last hill before reaching the main compound of Standing Oak Farm, Jonathan stopped to marvel at the idyllic scene, like some famous landscape painting. The lake that nearly claimed Brian's life spread off to the right, stretching beyond the shoulder of the adjacent hill. The beautiful main cottage was not as big as Carlingon Castle, but it certainly was quite ample, plenty for a widowed woman and her guest. The larger barn and associated holding pens dominated the building of the compound. Several other smaller buildings complemented the scene.

Charlotte Palmer stood outside under the front awning as Jonathan drove up and parked his government automobile off the main, gravel drive. Her smile spread as Jonathan extricated himself and approached her. They embraced and touched both cheeks.

"Great to see you, again, Mrs. Palmer," said Jonathan.

"Quite so, Mister Kensington."

Jonathan chuckled. "May we dispense with the formality?"

"By all means, Jonathan."

"How is your invalid charge?"

"He is napping at the moment, and I wanted to talk to you without him interjecting. He is improving physically. In fact, from what the doctors told me, he is doing better than expected. It is his nightmares that are most worrisome. He had a really bad one three nights ago."

"We all have nightmares, Charlotte."

"With night sweats, thrashing and screaming out?"

"He has been through a lot, and he is younger than most of us."

"That is precisely my point, Jonathan. He's too young for this. I feel like it is tearing him up. What kind of a life will he have when all this is done?" She did not wait for an answer. "He does not have to do this. It is not his fight."

Jonathan searched Charlotte's face as if some appropriate answer would magically appear in her gorgeous blue-gray eyes. "Whose fight is it?" he asked softly.

"Yours. You are British. He is an American, and America is not in this fight," she said, perhaps more harshly and bluntly than she intended. "I am terribly sorry, Jonathan. I did not mean that the way it sounded." She paused to consider her words. "I love him. I shouldn't, but I do love him very much. He is a special man. And," she paused again, lowering her eyes and grasping her lower abdomen, "I am carrying his child . . . our child."

Jonathan smiled. "Charlotte, I know he loves you, and he is thrilled that you created a child together. He knows what his avocation does to you. He truly does. But," this time Jonathan paused to find the correct words he wanted to convey his thoughts without hurting her sensitivities, "he was born to fly. He is a natural, if there ever was one."

"There are many other ways to fly. He damn well nearly drowned," she said, pointing to the lake, "in front of me. The pieces of his broken and burning aircraft passed over my head, and could have easily killed me as well."

Tears descended her cheeks. Charlotte began to softly sob. Jonathan reached to soothe her, but she held up her hand to stop him. She wiped her tears away. "I'm sorry. I did not intend to be so sobby."

"I know this is not easy, Charlotte. You've lost so much to this war and the previous one. I wish I knew what to say to assuage your fears. Perhaps there is nothing that can be said. I suppose the best I can say is, he is one of the very best at what he does and we need him. We need many more like him. He is an extraordinary man and an exemplary fighter pilot."

"Yes, he is," she said softly.

The front door opened behind her. She turned. Brian was wearing a knee-length, for him, heavy wool robe and stood on his crutches.

"What on earth could be so important to be standing out here in the damp?"

"Ay mate. Great to see you. How are you doing?"

Brian ignored Jonathan's query. "Are you two conspiring against me?"

They all laughed.

Charlotte admitted, "I am for certain. But, your mate here is defending your honor."

"Well, thank you for that, brother."

"So, how are you doing?"

Brian looked to Charlotte. "Do you want to come inside, or continue your conspiring out here in the damp weather?"

Charlotte stepped toward him, kissed him on the lips and walked past him. Brian shrugged his shoulders and moved aside to allow Jonathan to enter.

"Why don't you both go to the Sun Room," Charlotte said loudly without turning around. "I will make us a pot of tea."

Jonathan followed Brian into a peripheral, south-facing, well-appointed room with large, not quite floor-to-ceiling windows overlooking the pond and the green, rolling hills beyond. Brian ambled to a large overstuffed, leather chair, motioned for Jonathan to take the adjacent couch, and then lowered himself slowly.

"Are you going to continue to ignore my query and raise my apprehension?"

"What is that?"

"How are you doing?"

"Oh that. Sorry. I've become numb to the questions. I'm doing fine."

"No need to be cheeky, Brian. A large leg cast, head bandages and crutches do not qualify as fine."

Brian laughed and winced. "Well, right you are. I am rather busted up, but compared to my state a week ago, I would say I'm doing pretty good."

"Excellent."

Charlotte entered with a large tray holding the teapot, three cups and a plate of cookies. She poured tea for each of them. "Milk or sugar?" she asked Jonathan.

"You have sugar?" he said incredulously. "White, with, if you please."

"White, with, it is then. That is how Brian likes his tea as well."

Jonathan took the proffered cup and selected a cookie. Brian did the same.

As soon as Charlotte sat, Brian said, "So, I suppose my caretaker has told you about my nightmares."

"Yes, she has."

Brian looked to Charlotte. "They are worrisome. I am sorry for that, my dear."

She waved her hand dismissively. "I am just worried about you, my darling."

"Was it the same one?" interjected Jonathan.

"Yes."

"You've known about these?" she barked at Jonathan.

"Yes," he answered boldly. "We all have them, I suspect."

Charlotte kept her disgusted gaze on Jonathan, withheld the words of anger she felt, and then turned her stern eyes to Brian. "It is nearly lunch time," she finally said. "I need to get my mind off of this tragedy." She rose and departed the room for the kitchen.

"She is not happy," Jonathan observed.

"No, she is not. She has been angry with me for three days now. My nightmare must have really scared her, and that breaks my heart, actually. Hell, it scared me. She has been so kind and generous with me."

"Do you think she is correct? Perhaps, it is time for you to leave the fighter cockpit."

Brian hesitated and locked on Jonathan's eyes. "Really? Are you going to join with her, now?" he said, with a twinge of anger in his voice.

"Easy, mate. I am just thinking out loud here."

Brian wrapped his knuckles against his leg cast with a thud. "I've got to get this damn thing off my leg and get back into the cockpit, before I drive myself crazy."

"You must heal, first – body, mind and spirit. Otherwise, you will be a danger to yourself and the rest of us."

"I know! I know!" Brian exclaimed in frustration. "I'll be fine once I get back in the air."

"I hope so. We miss you. By the way, news from the squadron you may not be aware of just yet. The squadron moved to RAF Catterick in North Yorkshire, two days ago. 'Boxer,' 'Fog' and I were flown back down this morning to retrieve the remainder of the aircraft we have without pilots, including your replacement machine."

"Oh great," Brian responded with more frustration. "Now, it will be even harder to return to the air."

"Not to worry, my eager wounded friend. You will have an aeroplane ready when you are. Also, this morning, while I was waiting for an automobile to drive down to see you, the BBC broadcast an announcement by the Air Ministry that Seven One Squadron has formed at RAF Church Fenton, not far from Catterick, with all American volunteer pilots."

"Closer to Carlingon Castle."

"Much."

"And, farther from the action."

"That is true as well."

"Have you been home, yet?"

Jonathan laughed. "We just arrived, mate. I only convinced the Skipper to give me an extra day in retrieving our spare aircraft, so I could check up on you." Jonathan paused for a moment. "Have you reconsidered transferring to Seven One Squadron with your countrymen?"

Brian shook his head. "You are my countryman as long as I am in this fight, wearing an RAF uniform, and flying for 'Stuffy' and Fighter Command."

"Your loyalty is laudable and I cherish your friendship."

Charlotte appeared and announced, "Lunch is ready, gentlemen."

Jonathan and Brian made their way to the kitchen table. Elderly farm hands Lionel Bridges and Horace Morgan were already seated. Charlotte did not speak and served up the simple lunch of sandwiches, farm cheese and self-canned green beans with fresh tea and milk.

Lionel asked Jonathan, "How is the war going?"

Charlotte's angry glare ended any answer and further conversation. They finished their meal in tense silence. The two farm hands excused themselves, took their plates and utensils to the kitchen sink, and left to return to their chores. Charlotte rose, cleared the remainder of the dishes, stored the uneaten food, quickly washed the dishes, put them away, and joined Bridges and Morgan without words. She left Brian and Jonathan with fresh cups, tea and a plate of cookies. The two men sat at the table lost in their respective thoughts for countless minutes.

Jonathan was the first to break the silence. "If you want any kind of a relationship with Charlotte, you need to fix this."

"And, just how do you propose I do that?"

"Help her understand. Find some compromise."

"She wants me to stop flying fighters."

"Did you figure that out all by yourself?" Jonathan responded, dripping with sarcasm.

"No need to be nasty, Flying Officer Kensington."

Jonathan laughed, and then Brian joined him in a good laugh, which eased the tension they both felt.

"She is pretty hot at the moment, Brian. I would recommend you avoid the topic for a few days."

Brian considered his friend's advice and chose to let the topic rest there.

"My sister called me the other day to ask if you or Charlotte might object, if she could visit you here?"

"Do you really think Rosemary should be travelling through the invasion zone?"

"She has been down here before."

"True."

"What do you want me to tell her?"

Brian considered his answer to Jonathan's query. "I am her guest. This is her farm. Rosemary should ask Charlotte if she could visit. It is really up to Charlotte. If she is OK with it, I am OK with it."

"Does she know about you and Rose?"

At first, she did not want to know, and then she changed her mind. "Yes."

"Wait, really?"

"Yes."

"Does she know about your relationship with Anne and more significantly the impregnated Mary?"

"Yes."

"I will be damned. You are either extraordinarily foolish or she is incredibly accepting . . . or both."

"Both, I should think."

The two veteran fighter pilots continued their discussions about flying, the new pilots assigned to the squadron, news from other squadrons, and of course the war as they knew it.

As the milking time neared, Brian suggested they both assist Charlotte, Lionel and Horace with the afternoon's session. It was still a bit awkward for Brian to move around the cows. Regardless, collectively they made quick work of the task. Charlotte was still not talking to the two pilots, but Brian suspected she had confided in the two older helpers, as they were now wordless as well.

When their afternoon chores were complete with the barn cleaned and their tools stowed, Charlotte told them all to get cleaned up and ready for supper. It would be ready shortly. She had apparently prepared a large dish of Shepherd's Pie during the lunch preparation and she simply had to bake it for 15 minutes.

The five of them ate the delicious meal, again in silence. This time Lionel cleared the table and washed the dishes. The two older men promptly said good night and departed. Jonathan felt the need as well. He was not going to be a third wheel, when they clearly had a lot to discuss without interference from him. Jonathan offered his gratitude and farewell, and drove back to RAF Middle Wallop, so he could depart at first light tomorrow.

—

Saturday, 12.October.1940
Chequers Court
Ellesborough, Buckinghamshire, England

The early autumnal, damp, chilly weather made the pleasant warmth of the afternoon fire, all the more comforting in the spacious study within the classic 16th Century, Gothic Tudor mansion at the foot of the Chiltern Hills that had been the prime minister's country retreat since the estate was deeded to the nation in 1917. The elegant mansion offered far more space with which to entertain guests as well as magnificent and immaculately kept grounds for a casual walkabout and quiet contemplation. There was peace at this place, even in wartime and with the threat of impending invasion.

Assistant Private Secretary 'Jock' Colville continuously fed sorted and appropriately filtered documents to Prime Minister Churchill from the black dispatch box utilized to carry the paperwork of the premiership wherever Winston happened to be on a given day. They made quick work of the day's traffic. They went through section from 'Top of the Box' through 'Parliamentary Questions,' and 'General Ismay' to the final 'R Week-end' section – the most routine, catch as catch can, message traffic within Churchill's filtration instructions.

A revolving pool of half a dozen duty stenographers feverishly took down short hand notes of instruction, comment or query from the Prime Minister. The hour or so after naptime was usually a good productive time to keep the grease of ministerial action running smoothly, and to keep the Prime Minister informed. Colville had the duty this weekend. He usually tended the prime minister's door and communications, but on the weekends the duty private secretaries tried to give as much of the administrative staff time away to be with family. Mister Churchill proved himself to be a perpetual dynamo in his premiership, employing a small army of stenographers, typists, file clerks and 'go-for' research assistants, who worked long hours and multiple shifts day and night, seven days a week, to keep up with Churchill's demands and quests for information. This particular afternoon, Mister Churchill's valet, Frank Sawyers, guarded the door. A rather timid and faint knock on the closed door stopped the beehive of activity.

Sawyers opened the door and stood partially through the opening. "Mister Colville, if you please."

"Let us take a break," said Winston.

Colville went to the door.

Sawyers whispered, "Colonel Menzies and Commander Denniston are here to see the Prime Minister. They say it is urgent."

"Please show them in. I know the Prime Minister will want to talk to them."

Sawyers disappeared. Colville turned to face Churchill. "Sir, Colonel Menzies and Commander Denniston have arrived with the 'Buff Box,' I presume."

Churchill waved his hand to clear the room. Colville shepherded the small weekend staff out of the room and waited for the Director-General, Secret Intelligence Service, and the Chief, Government Code and Cypher School, to enter. They both nodded to Colville. Denniston carried the 'Buff Box' manacled to his left wrist. Colville left the room and closed the door to ensure their privacy. He knew that box carried special controlled intelligence information.

The 'Buff Box,' as they called it, was a newly created, hardened, secure, watertight, fire-resistant, beige-colored container. Only a few men on the planet actually knew what the 'Buff Box' contained. Once it was locked at Broadway House or at Bletchley Park, it could only be opened by a secure key upon return, or by a single key held on the person of Prime Minister Winston Churchill.

The three men dispensed with the salutations and cordiality of familiar colleagues. Denniston placed the 'Buff Box' on Churchill's desk with the lock facing the Prime Minister. Winston unzipped the light blue overalls he commonly wore on casual occasions for comfort. Even Winston had taken to the term 'siren suit' coined by someone on his staff, since he had begun donning the overall garment to cover his sometimes-naked body at night when the air raid sirens blared. Churchill extracted his key and unlocked the box.

"What have we today?" the Prime Minister asked. He paused to hear what the two senior intelligence leaders had to say.

"Not much today, I am afraid," answered 'C.' "However, considering our present state, we believe these were sufficiently important to make the journey to Chequers for your review." Churchill nodded, somewhat impatiently, for Menzies to continue. "As you will note from the information we have collected as of this morning, we do not have a clear view, but we are just now beginning to see signs the invasion training preparations are easing off somewhat."

"They have not informed the *Luftwaffe* just yet," Churchill snickered, referring to the nightly heavy bombing of London and other major cities.

Menzies ignored Churchill's quip. "We have deployed and tickled all of our available assets on the Continent to watch for changes in transportation and conduct."

"Tickled?"

"Sorry, sir . . . a term we use when sending instructions to field agents.

"Yes, right, then, now I know. Let's see what you have." Churchill opened the first section reserved for the deemed most significant decryptions of German message traffic. He read the first sheet.

MOST SECRET - ULTRA

```
SECRET
DATE: 9 OCTOBER 1940
TO: COMMANDER, AIR FLEET 2
FROM: AIR FORCE HIGH COMMAND
BEGIN
ORDERS ISSUED THIS DAY TO MAINTAIN FUEL RESERVE
AT FULL CAPACITY AND CURTAIL OPERATIONS NOT TO
EXCEED DAILY RESUPPLY BREAK UNITS AT ROSE AND
ANTLER MUST REMAIN AT READINESS UNTIL S PLUS 3
DAY BREAK ANY HARD OBSTACLES MUST BE REFERRED
TO HIS HEADQUARTERS IMMEDIATELY BREAK
END
SECRET
```

MOST SECRET - ULTRA

"Interesting," Churchill pronounced. "It could mean many things."

"Yes, sir," answered Denniston. "We have deemed it a positive for continued invasion preparations, largely so they may continue operations against this country and maintain their readiness to support an invasion."

"What are 'Rose' and 'Antler,' and what is 'S'?"

"Rose is Rotterdam and Antler is Antwerp, which we have confirmed by other means. 'S' we do not know. Our best guess may be their execution date for their Operation SEALION."

"Again, it could mean many things," Churchill observed.

"Quite so, I'm afraid," answered Menzies.

"This was three days ago?"

"We try to keep up with their setup changes to achieve near real-time decryption," answered Denniston, "but, they do surprise us from time to time. As you noted, sir, the message was intercepted three days ago, but we did not successfully breakdown the Enigma wheel settings until this morning."

"As you say, then. What is the next one?" The Prime Minister returned the first sheet to the box, and then read the second and last Enigma decryption for the day.

MOST SECRET - ULTRA

```
SECRET
DATE: 12 OCTOBER 1940
TO: COMMANDER, AIR FLEET 2
FROM: AIR FORCE HIGH COMMAND
BEGIN
47 ANTI AIRCRAFT REGIMENT DIRECTED TO CONTINUE
EMBARKATION DEBARKATION EXERCISES UNTIL
DESIRED PERFORMANCE ACHIEVED BREAK INFORM THIS
HEADQUARTERS ONCE ACCOMPLISHED BREAK
END
SECRET
```

MOST SECRET - ULTRA

"We deemed this one a positive as well," Denniston said, as Churchill absorbed the words.

"Understandable, I should say, and it was today."

"Yes, sir," responded Denniston. "Today is a good day."

Churchill offered a nervous, odd sort of chuckle at the perfidy of happenstance.

"On the negative side of the ledger," interjected Menzies, "we have a routine message from an agent."

Churchill returned the ULTRA second sheet to its place, checked the second section, other Enigma decrypts, nothing; the third section, other decoded enemy message traffic, nothing; and the fourth section, field intelligence, where he found a single sheet of paper.

MOST SECRET - CHICKEN

```
AT SITE 1762 AND 3017 TRAINS MOVING WEST NEARLY
EMPTY TRAINS MOVING EAST FULL ENEMY SOLDIERS
YESTERDAY ONE TRAIN MOVING EAST PERHAPS 50
TANKS
1940OCT12/0330Z
```

MOST SECRET - CHICKEN

"Chicken?"

"He has been an asset of ours for many years – a Dutch chicken farmer with a long family history of Anglophilia," continued Menzies.

"You know these sites he refers to? You can point them out on a map for me?"

"Yes, sir, if you have a map."

"Yes, I have many maps, but that is not necessary. I just wanted to ascertain the resolution of your information."

"Yes, sir. Both sites are on either side of his farm, a rather large farm, I must say. He grows several important crops and raises more than chickens. His products are important to the Dutch economy. Each agent has a unique location identification system known only to the specific agent and his field service handler. None of us," 'C's said, nodding to Denniston, "know the unique system in our effort to maintain the utmost operational security. You will note the date-time stamp on 'Chicken's message. It was his scheduled transmission time – another security measure."

"Are you sure this is not a repeat or extension of last year's debacle?" asked Churchill.

The Venlo Incident involved a successful clandestine operation by the German *Sicherheitsdienst* (Security Service or SD) in Venlo, Netherlands, on the border with Germany, just west of Düsseldorf. On 9.November.1939, SD agents captured and abducted SIS field agents Captain Sigismund Payne-Best and Major Richard Henry Stevens. They were imprisoned and extensively interrogated by an SD team led by *SS-Sturmbannführer* Walther Friedrich Schellenberg – a trusted and personally assigned operative of SS Chief *Reichsführer-SS* Heinrich Luitpold Himmler and SD Chief *SS-Gruppenführer* Reinhard Tristan Eugen Heydrich. The German operation compromised nearly two-dozen British agents in the Low Countries and Northeast France at a critical stage of the war in Europe, and generated considerable disinformation that took six months to unravel and correct. The Germans nearly blinded British intelligence during that crucial period.

Menzies lowered his eyes and head. "We may never know for sure," he answered softly.

"That is not reassuring, 'C.'"

"No," Menzies, raising his head and reconnecting with Churchill's stern gaze. "The security procedures I mentioned earlier and many others, of course, are a direct results of our eventual determination of what happened. There should be no doubt that we will lose agents in the future. We will continuously strive to refine and improve our security procedures to recognize the compromise of any field agent as soon as possible and implementation of aggressive actions to minimize the damage done by such compromise." Churchill just stared at Menzies without blinking. "I might add here that MI5 has been quite successful in turning more than a few captured German agents throughout the Empire."

"Yes, quite so," said Churchill, "but, do you think 'K' would claim he has them all?"

"No, of course not. No rational intelligence specialist would ever make such a claim. Our business is far too precarious and fragile."

Churchill decided to shift the focus. "As you may know, the military chiefs are my guests this weekend. Do you think they should see these latest intercepts?"

Menzies considered the question before he answered. "I do not consider these messages important or definitive enough to expand dissemination."

"If the Germans are moving their forces east, what do you think the meaning is?"

"It could be simple, routine, rest and recuperation for front line units in preparation for a spring invasion attempt. We doubt the scale of these moves is any kind of a feint or deflection effort. We still believe Hitler's primary objective is the Soviet Union. While these moves may mean nothing, we strongly suspect we will see progressively more signs the Germans have postponed or abandoned any cross-Channel invasion attempt, and they are turning their attention eastward."

"I certainly pray your suspicions are correct."

"We will most definitely keep our collective eyes open to develop a firmer assessment."

"It should be our primary intelligence objective at this stage of the war. If you are correct, it will be a major mistake and blunder, to fail to subdue us on our beautiful island. We will never again be as weak and vulnerable as we were this past summer, and we will not give him another such opportunity."

"I should think so."

"You can bet your life on it, 'C.' I know I have."

"Thank you, sir. If there are no further questions, Prime Minister, I will drop off Denniston at Site X with the 'Buff Box,' and I shall return to London."

"Thank you both for making the journey. Things are beginning to look upward, if we can only find a way to stop their damnable, incessant bombing of our cities and citizens."

Both men nodded, stood and departed.

—

Chapter 10

Ruin seize thee, ruthless King!
Confusion on thy banners wait,
Tho' fanned by Conquest's crimson wing
They mock the air with idle state.

-- Thomas Gray

Sunday, 13.October.1940
RAF Catterick
Catterick, North Yorkshire, England
18:30 hours

Week 15

Morning fog had kept No.609 Squadron from their planned training flight. After the lunch, the fog lifted, and they managed to complete the planned training flight – 'A' Flight versus 'B' Flight. The resultant melee gave their replacement pilots a good taste of how intense and crazy a dogfight quickly became. There were some close calls, as there invariably were in actual combat engagements, but no one was injured and no aircraft were damaged. After an extended debriefing and with the consent of the flight and section leaders, Squadron Leader Long-Roberts declared the squadron operational. Within 15 minutes of notifying Group Headquarters of the satisfactory completion of training, the squadron had been brought to Readiness, which had occurred two hours ago. It seemed quite appropriate to the veterans that the eager replacement pilots should have to wait for their first combat mission. They finally had been ordered to Standby status ten minutes ago. Their assigned mission now was a dusk patrol over a northbound convoy off the east coast of England, which meant even with no combat, they would be landing at night, or at best in late twilight.

Flying Officer Jonathan 'Harness' Kensington's Green Section was the only section still one pilot short, as Pilot Officer 'Hunter' Drummond remained on recuperation leave. He would be a short section for another few weeks and perhaps more. Jonathan sat strapped into the cockpit of his comparatively new, 'PR-K,' Supermarine Spitfire Mark IA fighter airplane with his pre-start and start checklist items complete to the point of pushing the starter button and bringing his big, powerful, Merlin engine to life. Landing at night in war darkened England was never a task to be taken lightly, and they often became dependent upon the radar operators to get them close to home plate as well as watch for trailers, who might use the landing area lights as their aim point for bombs and bullets while the fighters were landing. *I just hope we get this done before we need the lights. We are nearly a month past the autumnal equinox – one of two days each year when daytime and nighttime are equal – so, the end of evening twilight came sooner each day in season. Under British Double Summer Time in effect year around for the duration of the war, sunset would come at 19:19 today; end of evening nautical twilight at 20:40. The waxing, nearly full Moon would rise at 18:22, which would make a night landing a little less problematic. Well, at least the clouds have moved off to the southeast, and the skies to the west are nice and clear. With a near full Moon, perhaps a night landing would not need the lights.*

The distant bell ringing brought Jonathan to focused attention. He glanced toward 'Stack's 'PR-A' Spitfire. The Skipper's right hand scribed a circle above his head – the signal to start and launch. Jonathan gave the same signal to his crew chief. His Spitfire responded promptly and fully. They were airborne in just under two minutes. 'Stack' complied with the guidance instructions from Sector Control, heading southeast, climbing to 12,000 feet altitude.

Convoy PILE was not difficult to find before the squadron leveled off, and throttled back to loiter and conserve fuel. The wakes of 15 merchant ships along three, slender, escorting destroyers on the seaward flank of the convoy. As their Standard Operating Procedures dictated, 'Stack' settled the squadron into a long racetrack patrol pattern.

"Serpent Four Seven, Sorbo Leader calling."

"Sorbo, Serpent Four Seven, go ahead," the lead destroyer answered.

"Sorbo is on station."

"Roger Sorbo. Six requests a low pass to let the convoy know you are above us."

"No problem, Serpent. Break. Sorbo Green, Sorbo Leader, take your section down for a full length low pass, a good show, and get back up here."

"Sorbo Leader, Sorbo Green, wilco," Jonathan responded to acknowledge acceptance of 'Stack's order.

'Harness' nodded his head to the left, signaling right wingman they were rolling left and descending. Jonathan kept his throttle back, set up a steep, high-speed descent, and headed south. He judged his turn so that both Spitfires leveled off just above the wave tops, throttled up to full power to maintain their speed, and aligned their flight path to bisect the convoy. Jonathan glanced off his right wing and saw the 'PR-S' Spitfire in a level, close spread position. The few minutes at full throttle and low altitude would not overheat their engines. They were below main deck height with ships on the left and right. Several ships flashed their signal lamps that Jonathan noticed in his peripheral vision. The sailors appreciated the impromptu demonstration.

"Sorbo Green Two, we'll pull up, diverge and roll opposite."

"Green Two, wilco."

When they passed the lead ship, 'Harness' pulled up smoothly, separated by a few degrees, reduced his angle of attack, and then rolled twice. Jonathan looked out his right wing to see his right wingman make one additional roll before leveling his wings.

"Rejoin," 'Harness' commanded. The 'PR-S' Spitfire smartly returned to his proper position. 'Harness' adjusted his climb angle to maintain an optimal climb speed. As they reached 10,000 feet, he leveled off and began a slow left turn to scan for the squadron that should be just above them. They had not completed half a turn when he had the formation of the other fighters. Jonathan remained at lower altitude and adjusted his heading to rendezvous with the squadron. He looked behind his left wing. The sun was descending behind the mountains to the west. It was past sunset on the ground. *Why are we up here, taking this risk, with no threat around?* Green Section settled into their position in the squadron formation.

The sun had nearly disappeared behind the western mountains. Only a mere sliver remained, when the call came. "Sorbo, Echo calling."

"Echo, Sorbo, go ahead."

"Sorbo, Echo, return to base."

"Sorbo, wilco."

'Stack' did not broadcast any instructions. He simply led this squadron back to RAF Catterick. Twilight made the landing trickier. Fortunately, there was sufficient light to preclude any need for the landing area lights. They all landed safely, taxied to their parking positions and shutdown. Each of them reported the aircraft status to their respecitve crew chiefs. The pilots gathered in the Dispersal hut. Corporal Warren maintained her vigil and reported the squadron's return. The pilots stowed their flight gear and waited for their collective debriefing.

"What was that all about?" asked 'Sparky.'

"We needed a night landing and a routine mission, so Sector Control and I agreed on the mission."

"We didn't need that risk," interjected 'Harness.'

"Regardless, we completed the mission successfully. While we will likely have a month or so of low tempo operations here in the North Country before we head back to the line, we have now returned to being an operational squadron."

"As you say, Skipper," responded 'Sparky' for the pilots of No.609 Squadron.

The squadron was released for the night. The pilots departed for the Mess, a meal and a few beers. Corporal Warren secured the Dispersal hut for the night.

—

Monday, 14.October.1940
New Public Offices
Whitehall, London, England
22:55 hours

The blare of the air raid sirens punctuated the night over the darkened city. The sirens' wail dampened the pulsating drone of unsynchronized engines of the approaching gaggle of German bombers for their nightly insult of what had become known as The Blitz. The searchlights began to stab into the black night sky to the south. The lights eventually located and tracked a target aircraft. The 3.7-inch (94mm) anti-aircraft cannons of the Royal Artillery opened fire. The muzzle flashes followed by detonation of the shells at their programmed altitude. The thud of both events arrived many seconds later as barely perceptible reports. The sights and sounds moved progressively closer.

"Prime Minister," said Detective-Inspector Thompson, "I must protest this foolishness."

"Umph," grunted Churchill.

"Prime Minister!"

"Must we do this every time the shooting starts?"

"My mission is to ensure your safety, sir."

Churchill waved his hand dismissively. "Of course, and I am grateful for your service."

"But . . ."

"Some of our people are going to die tonight, Walter. The least I can do is bear witness to this attack."

Inspector Thompson knew further resistance was futile. If the Prime Minister was to die this night, then fate destined that he would die with his charge.

Prime Minister Winston Churchill stood on the elevated rooftop of the government building above the Cabinet War Rooms excavated and reinforced beneath the Whitehall administrative office building. His light brown siren suit, as they called his loose overalls, offered no reflection of his station in His Majesty's Government or in the free world's defiance of Hitler's fascism. His fists on his hips and the as yet unlit Havana *Romeo y Julieta* cigar – his favorite – clinched in his teeth offered a defiant image. The cool, night, autumn air did not bother or concern Churchill. *At least Clemmie and the children are safe*, he told himself. They stayed an extra couple of days at Chequers.

General 'Pug' Ismay joined the two men on the rooftop of the New Public Offices building. Ismay served in several positions – Principal Assistant to the Minister of Defense (Churchill), Secretary of the Imperial Defense Chiefs of Staff Committee, Deputy Secretary of the War Cabinet, and otherwise the principal conduit between the Minster of Defense, and the Chiefs of Staff and the military service branches. Ismay virtually repeated Inspector Thompson's admonition and received essentially the same response and result.

The sounds and light show moved closer to Whitehall. The flashes momentarily illuminated the barrage balloons tethered by cables at various heights and anchored throughout the city. High explosive bombs began exploding in Lewisham, Southwark, and Lambeth, south of the River Thames. Tonight, the distinctive white-hot blooms of white phosphorus explosions – incendiaries – were added to the high explosive bombs. The Germans intended to break-up the structures of the city and burn the pieces.

"Damn!" exclaimed Churchill. "The bastards are trying to burn the city." The Prime Minister began to pace, back and forth, without taking his eyes off the exploding German ordnance walking toward them.

The Royal Artillery guns did have some success. One bomber exploded in flight, raining fragments and debris down on the city. A half dozen other bombers burst into various levels of flame and dove or spiraled to the ground with their remaining fuel erupting into billowing fireballs.

Fires began promptly as the fragments of burning white phosphorus ignited everything they touched. The sirens and klaxons of the Fire Brigade were clearly audible amid the detonations. The explosions and fires had not quite reached the river when the sounds began to recede, and the bombs stopped exploding.

Churchill stopped his pacing and looked to the south. The fires grew quickly and spread. It was after midnight. The near full Moon and clear skies made the enemy bombardier's aiming task easier, but the fires simplified the nighttime bombing problem even further.

Churchill turned to Ismay. "'Pug,' pray tell me, why does it take so many shells from our guns to bring down just one lumbering bomber?"

General Ismay engaged the Prime Minister's waiting eyes in the moonlight. "Would you like me to query CIGS?" The Royal Artillery anti-aircraft batteries reported through the command structure to the Chief of the Imperial General Staff General Sir John Dill.

"No!" Churchill protested. "I don't want a ministry answer. I want your opinion."

"Very well, sir. The trajectory physics is not an easy problem to solve, I'm afraid. The gunners must set the proper setting on the shell before loading and firing . . . it all takes time and is part of the problem. Further, even when the shells detonate in proximity to an enemy aircraft, the shrapnel must strike something vital in the target aircraft, like the pilot.

"Then, we need bigger shells."

"If that was possible, I suspect larger shells would provide only marginal improvement. The research folks are making good progress on the radio detection equipment, they call it a proximity fuse, as well as the fire control predictor to aim the guns and properly place the shells within the influence sphere with the target. They believe they will be able to solve the trajectory problem and achieve direct hits, or close enough for direct effects. In that instance, a 3.7 inch shell will be more than adequate against an aluminium aircraft."

"Yes, but when, 'Pug'? When?"

"According to what I am told, perhaps early next year."

"That does not help our people tonight, does it?"

"No, sir. It does not. The scientists and engineers are quite aware of that fact. I might add that Sir Henry Tizard's exchange program with the Americans should help improve the time to obtain a working unit. They are farther down the lane in producing an effective predictor."

"I pray you are correct, 'Pug.'"

The sequence began again as the next wave of German attackers approached the burning city.

"What is the latest about the night-interceptor development?"

"About the same, I should say . . . good progress but not general success. The night-fighter lads have steadily improved their success rate. However, as we bear witness, they have not yet stemmed the tide."

"Quite so."

The bombs began exploding on the south bank buildings and walked across the river toward Whitehall. Churchill turned his stern gaze upon the imposing scene playing out before him. His fist returned to his hips and his cigar festooned chin appeared to jut out a little father, as the Prime Minister stood in defiance of the attackers. He showed no sign of flight or even flinching. Inspector Thompson and General Ismay were becoming noticeably more agitated.

"Prime Minister," shouted Thompson over the advancing cacophony of the German attack, "I must insist we get you to shelter."

"The bombs are walking toward us," Ismay added, as if none of them had detected the trend.

Prime Minister Churchill waved his left arm. "Go, go, both of you, go!" shouted Churchill in frustration. "I am not missing this travesty. I shall bear eternal witness to what these damnable Nawzees are doing to this great city. I shall have it no other way."

"Sir, this is foolhardy."

"You may well be correct, 'Pug,' but I am not budging. So, go, both of you. They shall not get me this night."

None of the three men moved. If tonight was to be the end, they would go together. Churchill walked to the edge of the roof and looked down on Horse Guard Road and King Charles Street. They were devoid of any life, no people, no taxis, no movement of any kind. Londoners had heeded the warning sirens and sought shelter. Winston nodded his head and smiled. Their warning system worked, even if their defensive systems against the bomber attacks were not yet up to the task. Churchill rejoined his two stranded colleagues. He looked to the west. The moon was well past its zenith in the night sky, but not quite to the western horizon. In another half an hour, the moon would disappear as the bombers' light source. The fires from the first wave continued to grow across the river. Two bombs exploded on Portcullis House, just across Bridge Street from Big Ben – the clock tower of Westminster Palace. Several more high explosive and incendiary bombs exploded in St. James Park just beyond their perch atop the New Public Offices building.

As the dust and debris settled, a bright incendiary bomb exploded near by. A fire started straight away. Bombs were exploding all through Whitehall. Churchill remained profoundly focused on the commotion around them. He watched as the Fire Brigade appeared and expertly leapt into their duties. The heavy, earthy smell of pulverized stone and concrete mixed with the acrid odor of burnt explosives overwhelmed them. The smoke of the attack dimmed the moon.

The second wave completed their horrific task and departed. Only the crackling sounds of the fires and the gallant work of fighting those fires could be heard on the roof. The 'all clear' siren had not sounded, which meant a third wave had been detected forming up over France and undoubtedly headed toward them to add insult to injury. Prime Minister Churchill had seen enough. He motioned with his right arm for them to make their way to the rooftop access stairs. While this night of The Blitz was not yet over, he had seen enough for this particular night.

—

Thursday, 17.October.1940
Air Ministry
Whitehall, London, England
14:00 hours

The pressure on the leaders of the Royal Air Force had been mounting for months, especially since the desperate days of the German aerial assault on Fighter Command began in earnest during mid-summer. The so-called Big Wing debate had consumed an inordinate amount of intellectual energy of Fighter Command and had actually reached the unwanted attention of the Prime Minister and the War Cabinet. The essence of the debate centered upon the time-proven principle of warfare – mass. No one argued against the notion of massing forces for battle; however, the rather thin, minimal depth of Fighter Command during the mortal aerial combat through the summer months demanded husbandry and critical, focused response. After months of heated discussion, it had all come down to this meeting.

The meeting had been called for and was chaired by Secretary of State for Air, 4th Baronet of Ulbster, Sir Archibald 'Archie' Sinclair – the principal minister for the air force of His Majesty's Government. With the minister at the head of the large conference table was Chief of the Air Staff Air Chief Marshal Sir Cyril Newall. The heir apparent as chief, Air Marshal Sir Charles Portal, was also present and would assume command in the next week, although it had not yet been announced. The leaders of Fighter Command had been called to attend and settle the debate as Minister Sinclair wanted the internal bickering to cease and especially for this particular issue to disappear from the Prime Minister's sphere of awareness and concern. The father of the Air Defense System of Great Britain, Air Officer Commanding-in-Chief, Fighter Command, Air Chief Marshal Sir Hugh Dowding, had been unable or unwilling to settle the argument within the military leadership of Fighter Command. Sir Cyril continued to maintain his defense of the Chief of Fighters. In addition to Sir Hugh, all four, fighter group commanders were present:

-- No.10 Group – Air Vice-Marshal Sir Christopher Joseph Quintin Brand KBE, DSO, MC, DFC, covering the southwest of England and Wales,

-- No.11 Group – Air Vice-Marshal Park, covering the critical southeast of England including London,

-- No.12 Group – Air Vice-Marshal Leigh-Mallory, covering the industrial Midlands of England, and

-- No.13 Group – Air Vice-Marshal Richard Ernest Saul, DFC, covering Scotland and the north of England.

Feeling very much the odd man out, Air Commodore John Spencer had been unilaterally invited by his immediate superior, Air Vice-Marshal Park, who wanted his Chief of Operations and Chief Controller in attendance for a very non-military reason. Everyone in the conference knew exactly who John's uncle was, and Park sought whatever leverage he could find. Park had opposed the Big Wing for a host of reasons, however his defenders were the aging and falling from favor Sir Hugh and Sir Cyril. John chose to sit in the otherwise vacant peripheral chairs behind Park. He did not like the feeling of being the junior officer in the room, especially on such an important and vital matter as the deployment of fighter resources, but he believed in and supported Park's position. This was clearly going to be a difficult meeting, as none of the staff were present . . . other than John. They did not even have staff secretaries present to take notes, and record the substance and conclusions of the meeting.

Distinctly missing from this particular meeting was the principal advocate and protagonist for the Big Wing, 'Tin Legs' Bader. The unabashed and vociferous Bader had managed to keep the pot at a roiling boil by his audacity and notoriety. He had lost both his lower legs as a consequence of crashing his Bristol Bulldog Mark IIA, during an impromptu airshow on 14.December.1931. He fought hard both physically and bureaucratically to return to full flight status, which he achieved in November 1939. He now commanded No.242 Squadron at RAF Duxford, flying Hawker Hurricanes.

"Gentlemen," announced the Air Minister, "we have a busy agenda today, and I must face the War Cabinet later today. Each of you knows precisely why we are here, so I do not need to summarize the purpose. I have tolerated this debate far too long and none of us needs this matter before the War Cabinet, or as an added burden for the Prime Minister. Thus, we shall be done with it today. What's more, once we have decided this, I want the position clearly, distinctly and definitively communicated to your commands to cease the corroding arguments among your pilots. Am I clear? Are there any questions before we jump into the deep end of the pool?"

Everyone remained silent. No one moved.

"Very well, then. Leigh-Mallory, if you would be so kind, please state your case."

"Yes, sir. We faced . . ."

"We have faced," interrupted Park.

Sinclair slammed his open hand on the table with a loud report that startled everyone. "Stop it! This is not some schoolyard. We are discussing the vital air defense of this country." The Minister stared at each officer to make sure his point was understood.

That was not smart, Keith, John Spencer told himself. John felt extraordinarily uncomfortable. Rivulets of sweat were already descending his back, and he had done nothing but sit there. *We have all faced this enemy more than once. You must resist the urge to defend your vanguard position.*

Leigh-Mallory began again, "Fighter Command has faced massed formations of enemy fighters during specific sweeps in their efforts to decisively engage our fighters, and as escorts for the large formations of bombers attacking this country. Two millennia ago, the Chinese philosopher Sun Tzu observed, 'The host thus forming a single united body, is it impossible either for the brave to advance alone, or for the cowardly to retreat alone. This is the art of handling large masses of men.' This has been an essential principle of warfare. Engaging our forces piecemeal simply depletes our resources. We need to mass our fighters to defeat the enemy decisively."

My dislike of this man mounts with each word, John thought. *Why has he chosen to be Bader's parrot?*

"Park," said Sinclair.

"I think we all appreciate the lesson in military philosophy. In the dark days of May, this year, Sir Hugh stood before the War Cabinet in opposition to the Prime Minister and argued that we could not send ten additional fighter squadrons to France. He also presented our considered and collective position that we need a minimum of 50 front-line fighter squadrons to defend the United Kingdom against the expected German air superiority campaign in preparation for their anticipated invasion attempt. We have yet to achieve that minimal fighter strength, and yet, we rose to defend our island with the resources we had. During the battle, we took the unprecedented step of stopping rest periods for our pilots, literally a fuse of finite length, and we bled down One Three Group and part of One Two Group to keep the strength of One One Group, which stood in the breach of the primary assault and expected invasion. In the particular reality of scant resources and the critical moments of time to optimize the limited endurance of our fighters, we engaged the massed enemy formations from multiple angles with separate squadrons and precise timing as best we could control it. By our experience, the critical element was time."

"Dowding."

"I appreciate the passion of my group commanders. I will add one additional, relevant point. Enemy fighter pilots captured and interrogated at Trent Park have given us valuable insight into the effects of our tactics."

The Combined Services Detailed Interrogation Centre moved to the renowned country mansion at Trent Park and began operations on 12.December.1939. All captured *Luftwaffe* pilots were processed, interrogated, and then held there with a complex system of surreptitious listening devices to record, transcribe and analyze their conversations in a more casual setting.

"They were demonstrably anxious about where and when our fighters would pounce on them – a distraction to say the least. I say this to complement Air Vice-Marshal Park's observations. The reports from Trent Park were quite clear. Our pilots have had little difficulty finding the large enemy formations. I dare say our enemy would find equal ease in locating Big Wing formations well before we could engage."

"The Big Wings might also add an intimidation element," interjected Leigh-Mallory.

"Yes, quite so. Although I must say, I have not seen much evidence of our adversary being intimidated by much of anything."

"Have you experimented with this Big Wing concept?" asked Sinclair of Dowding.

"To be blunt, Minister, no, we have not."

"Why not?" Sinclair did not wait for an answer. "Surely, a sortie or two to evaluate the effectiveness of the tactic would not have hampered our defense."

Dowding held the eyes of Sinclair as he considered his response. "Sir Archibald, since May of this year, we have been engaged with a numerically superior enemy, while we had marginally sufficient numbers of aircraft and inadequately trained pilots. We have depleted One Three Group below the threshold of viability, leaving the North vulnerable. Add to that the air defense demands of Scapa Flow further thinning our marginal capability in the North Country. We have been extraordinarily fortunate that the enemy has largely abandoned daylight, massed, bomber raids and turned his attention away from our aerodromes and factories."

"To London and the other cities . . . against our people," interjected Sinclair with some anger.

"Yes, tragic as it is, the change has allowed us to gather strength, replenish devastated squadrons, and more importantly to rest and train our pilots. We still must contend with these fighter-bomber raids during the day, intended to engage our fighters. And, of course, we now have these bomber raids at night."

"Yes, the night intercept problem, but that is not the subject here."

"Certainly. To your question, Minister, I am quite amenable to practical evaluation sorties after we have made the daylight skies safe."

"That may not be for months."

"Sir, our task is the air defense of the United Kingdom, not the placation of a dissenting faction."

"Damn you, Dowding," Sinclair exclaimed in frustration. "Why are you so damnably obstinate?"

"Minister Sinclair, again, my singular mission is the air defense of His Majesty's realm. Fighter Command has continually assessed our approach to achieve the mission before us."

The Air Minister ignored the statement by his chief of fighters. "We have not heard from Brand and Saul," said Sinclair, and looked to the two quiet group commanders.

Sir Quintin responded first. "Minister, we have debated the Big Wing for several months, now. Air Vice Marshal Saul can speak for himself, but as for me and our experience in the southwest sectors, we remain convinced the time required and difficulty maneuvering a Big Wing will simply allow the enemy to penetrate deeper unchallenged, or diminish the combat time of our fighters. We walk a fine balance between a wide variety of factors, not least of which are the limited endurance of our current fighter aircraft."

Sinclair looked to Saul. "I concur with Keith and Sir Quintin, and my group supports the Chief."

Leigh-Mallory cannot be feeling comfortable right now, John Spencer told himself.

"So, it is just Mallory as advocate," observed Sinclair.

That sanctimonious bastard will simply find another, less obvious manner to prove he is correct in throwing his weight behind Bader's Big Wing proposal. He does not give up that easily, not that it has been easy so far. Keith Park had been exceptionally tolerant of Leigh-Mallory incessant niggling on the topic.

"And, a few of his squadron commanders," added Park.

"Before we conclude, Sir Charles, would you care to add your perspective?" Sinclair asked.

"With respect, Minister, Bomber Command has nothing to contribute in this debate."

"Sir Cyril, you shall have the last word."

"Next to you," Newall responded with a smile and a nod to the Air Minister. "While internal discussion is essential to any command, at the end of the day, we have charged Sir Hugh with the aerial defense of the Kingdom. History shall record his fulfillment of the mission with which he was charged. History shall also record the wisdom of his decisions since he assumed command of the air defense system in 1936. I truly and thoroughly believe he remains correct with his decision regarding this topic. That said, we have breathing space from where we were six weeks ago, and thus, we have the time and capacity to perform a set or series of tactical evaluation of the so-called Big Wing technique." Newall paused and looked directly to the Air Minister. "If you would permit me, Minister," he waited for an affirmatory consent, "I would like to personally thank Sir Hugh and the group commanders for their exemplary defense of the Kingdom through the very tortuous summer."

Sinclair considered what he had just heard directly from the leaders of Fighter Command. "Very well, then, we shall stay on the horse we have in this race. I want this incessant backbiting to end today. I want this Air Ministry decision and soon to be instruction of His Majesty's Government to be clearly communicated to every pilot. Enough distractions I say. Lastly, I would respectfully ask His Majesty's chief of fighters to carry out an appropriate experiment and evaluation at his earliest convenience." Dowding nodded his head in consent for all of them. "Very well, then. Thank you for your time and patience. We are adjourned."

—

Thursday, 17.October.1940
RAF Catterick
Catterick, North Yorkshire, England
16:15 hours

"Scramble the squadron," announced Corporal Warren.

The pilots in the refuge of the Dispersal Hut did not hesitate. Those who did not have their flying kit attached or with them grabbed their equipment. They all dashed from the building to their aircraft. Engines began firing off as crew chiefs readied their charges for flight and combat.

Jonathan had only to strap on his parachute and seat harnesses, and connect himself to his aircraft. The Skipper had already begun to taxi. 'Harness' signaled his only wingman to advance. His aircraft swayed and bumped across the uneven grass. The entire squadron was airborne in respectable time, climbing to their pre-briefed loiter altitude of 20,000 feet. They would orbit above Cambridge, to fly as a covering force for most of No.12 Group that had been launched to participate in a large engagement with enemy fighter-bombers. The Germans had modified some of the Bf109 fighters to carry a single, 50-kilogram, high explosive bomb on the centerline of the aircraft beneath the pilot. They flew in large formations of 50 to more than 200 aircraft, generally half with bombs and the other half clean. They were not particularly accurate in dropping their bombs, but you did not have to be accurate when your target was a city the size of London. However, once they released their bombs, they became clean fighters. The fighter sweeps were formidable threats.

The mission of No.609 Squadron on this day was to keep any enemy aircraft from reaching the No.12 Group airfields, and to scrub off any trailers as their sister squadrons tried to disengage. How the German fighters were able to last longer or make it deeper into England was still a puzzlement.

'Stack' leveled them off at their assigned altitude and throttled back to conserve fuel. They could not hear the radio traffic for the engagement playing out south of London, but they could see the condensation trails, commonly called con-trails, spreading tangled streaks across the clear sky.

"Stay alert, lads," radioed Squadron Leader Long-Roberts. "We don't want to miss any Gerries squirting out of this furball."

'Harness' checked his right wingman. He was in position and his head moving. Jonathan scanned above and below the horizon to the south, with occasional scans back, over each shoulder. Their immediate sky remained clear. The process continued for another hour without further information from Sector Control. They ground along in the extended racetrack orbit to maintain their assigned position.

They remained on station until the new con-trails stopped and their fuel reached their bingo level – sufficient for their return to RAF Catterick.

Their landing back at their home base was uneventful. They debriefed the rather routine mission. The veterans appreciated the paucity of excitement, while the replacement pilots whined about the inaction . . . just as the veterans had done six months earlier. The veterans knew the fight was not over, and they would be back in it soon enough and for a lot longer. Now, as the veterans also knew, they waited, remaining at Available status. No.609 Squadron would not be released until halfway through evening twilight – a long day.

———

Thursday, 17.October.1940
The White House
Washington, District of Columbia
United States of America
11:00 hours

"**W**hat is next, Harry," President Roosevelt asked his principal confidant Harry Hopkins.

"The Attorney General is due any minute to discuss H.R. 10094 before you sign the legislation."

"The Voorhis Act?"

"Yes, precisely. 'Jerry' Voorhis did a fine job for us."

"Yes, he did. I will send him a personal note after I sign the bill. Also, Harry, my apologies, but my conversation with Bob Jackson will have to be private."

"Certainly, Franklin." Hopkins left Roosevelt alone in the Oval Office.

Roosevelt returned his attention to the papers before him. The never-ending stream of paper often became depressing and overwhelming, but Franklin knew it was the lifeblood of a democracy. Everyone needed the bureaucracy to be served, and often the President of the United States of America was the focal point of that necessary paperwork.

The knock at the door broke Roosevelt's concentration. His long-term, loyal, private secretary, Marguerite Alice 'Missy' LeHand announced, "The Attorney General has arrived for your scheduled meeting, Mister President."

Roosevelt wheeled his chair around his desk to his usual position between the two couches. "Please show him in," the President finally responded.

Attorney General of the United States Robert Houghwout Jackson entered the office and walked confidently to the far couch. "Good morning, Mister President."

"And, good morning to you, Bob."

"To get right to it, I wanted to discuss H.R. 10094 before you signed it."

"Yes, yes, I understand that, but first, we have not had the opportunity to re-address our wiretapping discussion of last May. How are you and Edgar getting along with that memorandum for the record I provided?

"I would be telling you a fib if I said it was a bust. J. Edgar has made good use of the authority. His surveillance program has confirmed the Sebold case is far deeper than we suspected. The FBI continues to develop the case, play out additional leads, and to keep track of all the suspects – 25 conspirators so far. We all believe there will most likely be more."

Shortly after arriving in the United States in early 1939, hopeful German immigrant Wilhelm Georg Debrowski legally changed his name to William George 'Bill' Sebold and fortuitously connected with other German immigrants. He quickly learned that several of the immigrants were collecting information for German military intelligence – *der Abwehr*. Sebold's concern for his intended future allegiance led him to contact the FBI with what he recognized as anti-American conduct. The information provided by Sebold had been quickly corroborated. The contact rapidly elevated from the New York Field Office to the Office of the Director. Coming on the heals of the Griebl-Lonkowski spy ring case in 1938, FBI Director J. Edgar Hoover aggressively sought and received an extraordinary secret presidential authorization the previous May, to conduct clandestine warrantless wiretapping to promptly develop the extent of the spy ring illuminated by Sebold.

"How badly have we been compromised?"

Jackson held the President's eyes as he considered his response. "It is quite serious, I'm afraid. Part of our problem is not just identifying the spies and gathering sufficient evidence to assure convictions, but we are earnestly attempting to precisely determine the extent of our compromise as well as limit any further exposure."

"Well, damn, this will not be good news to the British."

"The British?" Jackson responded with incredulity.

"Last month, we began an extensive technical exchange program with the British."

"I am aware of that effort, although not directly involved at this stage."

"Yes, well, one of Churchill's expressed concerns before we mutually agreed to the program was specifically the apprehension of His Majesty's Government regarding our ability to keep secrets, quite frankly."

"Well, between you, me and the fence post, they have reason to be concerned."

"We must plug these holes."

"Edgar and his special agents are working around the clock to do just that."

"I'm sure they are, and God only knows what else he is up to with this power of his," Roosevelt mused. "The devil we know . . ."

"Indeed, sir."

"Carry on, then. Now, you wished to discuss the Voorhis Act."

"Yes, as I am sure you are aware, the bill's language is liberal and quite open to interpretation, as we requested. I was surprised that the debate was comparatively brief. Yet, I am obligated by the law to advise you of my concerns – the same concerns I voiced in our discussion of the wiretap surveillance last May. This bill is so loose that it is ripe for abuse."

"I share your trepidation, Bob, but these are extraordinary times. We have come to this affair far too late, too ill prepared, and I dare say we are far too vulnerable. Extraordinary times demand extraordinary measures."

"Yes, sir. While it is my professional opinion that the law as passed by Congress will withstand judicial scrutiny, I am not so confident of the actions of the government's agents will survive comparable review."

"I am grateful for your wise counsel, Bob. Enforcement within the law, as we understand it, remains our responsibility and our burden. I have confidence in you, Bob, even if I retain my unease about Hoover. We must do our best to supervise this new power."

"Yes, sir."

"Is there anything else on this matter?"

"No, sir. I have spoken my peace."

"Then, what is your recommendation regarding H.R. 10094."

"I recommend you approve the bill, Mister President."

"Very well. I am prepared to sign it." The President wheeled himself back to his desk. Jackson stood and waited. Roosevelt opened the proper folder, picked up his fountain pen, and signed the Voorhis Act of 1940 into law. "There, it is done."

"Thank you, Mister President. I will add before I go that I have prepared instructions for Director Hoover. The German American Bund is at the top of our list for registration. We have several dozen on the list, including the Communist Party of the United States. We plan to execute the Voorhis Act within a month. We expect to have the initial registration completed by the end of this year."

"Excellent. Please keep me informed of your progress."

"Certainly, sir. Lastly, we are working with Congress to codify the wiretap surveillance authority you approved in May, at a minimum, against those organizations registered under the Voorhis Act or associated with them. I know Edgar has a much larger list. We will continue to vet the FBI's surveillance initiatives."

"Very well, Bob. Thank you. One final question for you, if I may . . . what is your opinion of Bill Donovan's strategic intelligence and counter-espionage proposal?"

"Edgar is not impressed."

Roosevelt laughed. "Of that, there is little doubt."

"I think most of us see merit in how the British intelligence services have evolved. They certainly have more experience than we do. I can only imagine your frustration with the fragmented, and I dare say parochial, intelligence products you are given. In that sense, for your purposes and the Justice Department's as well, I believe Bill's concept for a master strategic intelligence agency has merit worth developing."

"Would you be able to manage Edgar, if we pursue Donovan's line?"

It was Jackson's turn to laugh. "You give me too much credit, Mister President. I'm not sure Edgar is manageable anymore. The best I can say is, I will do my best to guide Director Hoover in the direction you wish and remain square with the law."

"Thank you, Bob. Anything else?"

"No, sir.

"Fine. If you would be so kind, please send Harry in, will you?"

"Of course. Good day, Mister President."

——

Saturday, 19.October.1940
RAF Catterick
Catterick, North Yorkshire, England
10:10 hours

As it was becoming more common with the autumn weather, a steady, moderate rain descending from rather low clouds masked most of the aerodrome. No.609 Squadron had been at Available status for two and a half hours so far. Most of the pilots dozed in the various interior chairs. They waited, and they tried to avoid thinking about what might happen or what could happen. They lived in the minute, and minute by minute, to avoid being overwhelmed by what might happen.

The telephone rang. A month ago, that ring would have jolted all of them to instant alertness and in some cases sent a few bolting for the door to vomit or to get a head start on what inevitably became a scramble command.

Corporal Warren answered and listened. It was a short, one-way conversation. "The squadron is released."

Half the pilots stood and began to secure their flight kit.

Squadron Leader Long-Roberts appeared at his office door. "Not so fast, gentlemen. I remind you this weather is expected to clear this evening, and we must be back to Available status by zero six thirty tomorrow morning. 'Harness,' you are next up for a two-day pass, and as luck would have it, you get a few extra hours tacked on at the outset."

"Thank you, sir."

"You are due back at here, ready to fly, by noon on Monday."

"Yes, sir."

"Are you going to see your dolly-bird?" asked 'Waggle' Davies.

"No, not this time. I'm going home."

"Enjoy."

The pilots loaded onto their assigned truck, to be returned to the Mess and the rest of their day. Flying Officer Jonathan Kensington did not waste time.

He did not disembark at the Officer's Mess, having arranged for the driver to drop him off at the Catterick Bridge Station on the North Eastern Railway to Newcastle. His father's driver would be waiting for him to take him the rest of the way home.

—

Saturday, 19.October.1940
Carlingon Castle
Newcastle-upon-Tyne, Tyne & Wear, England
12:30 hours

Jonathan's Mother and Father, Theona and George, stood together outside the main front entrance to the only family home he had ever known as the car drove up the driveway and stopped before them. Jonathan extricated himself from the limousine. He shook his father's hand and hugged his mother, kissing both cheeks.

"It's great to see you, Son."

"Thank you, Father."

"We held lunch for you," Theona announced, and led the two men into the mansion and the dining room. As they made their way, Mother Kensington asked, "Have you been well?"

"Yes, Mum. We moved to Catterick, to rest and refit the squadron."

"Just down the road," interjected George.

"Yes, sir, and a most welcome respite from the war for us."

The large dining room table was set for three at the end closest to the kitchen. Jonathan sat opposite his mother, to his father's right. Lunch was served.

"So, the war has eased off a bit?" asked George.

Jonathan finished his bite. "In a manner of speaking, I suppose. The huge formations of bombers and covering fighters we saw in August and September have disappeared. One One Group in the southeast still faces fighter sweeps, but even those seem to be tailing off as well."

"Then, you have won the battle?"

"I would not make such a claim, Dad. We remain under invasion alert. However, the simple fact that Fighter Command has pulled us, and other hard-pressed squadrons off the line for rest and refit is the most positive sign we have seen since the middle of July."

"How is your friend Brian?"

"You heard?"

"Yes, Rosemary told us."

"He was banged up quite badly in this last one, I'm afraid. He is on convalescent leave now, until the end of this month or early next month. He is recovering nicely, I must say."

"The last one?" asked Theona.

"Do you really want to talk about such things, Mum?"

"He is important to you. He is important to Rosemary. That makes him important to us."

Jonathan searched his mother's eyes to ascertain her sincerity. "Yes, he is. Very well, then. Brian is an aggressive and exceptional fighter pilot. This latest event was the fourth time he has been shot down since the war began."

"Fourth!" exclaimed George. "I thought the object was to shoot down the other fellow."

"Quite so, Dad. He has shot down 19, so far. He is an ace several times over. In this last one several weeks ago, combat damage to his aircraft prevented him from bailing out of his stricken fighter aeroplane. The crash nearly killed him. If it was not for some farmers close at hand, he might not have survived."

"Oh dear," Theona said.

"This is a nasty business, but no where near as nasty as being under the Nazi boot."

"Amen," George added.

At that moment, Jonathan recalled his conversation with Brian from 10 days ago. Given the content and context of his conversation, he had to know. "Where is Rose, Mum?"

"She said she was going to visit Brian . . . near Winchester, she said."

"Standing Oak Farm?"

"Yes, that is what she said." Theona must have seen something in his expression. "Why do you ask?"

"Mum, this is between Rose and Brian."

"Yes, it certainly is. She loves him."

"I know, Mum."

"What is this Standing Oak Farm?"

"It is the place where Brian is recuperating."

"Well, what is wrong with that?

"Mum, Standing Oak Farm is a comparatively large farm owned by Mrs. Charlotte Palmer."

"And?"

"How should I say this," Jonathan thought aloud, as he considered his choice of words.

"Just say it, Son."

"Well, Mum, to be direct, Brian is in love with Charlotte. He is staying at her farm for several reasons, not least of which is she saved his life from drowning in her pond."

"He doesn't love Rosemary?"

"Brian loves Rose in his own way, but Charlotte is pregnant with their child."

"Oh dear." Theona relapsed into contemplation, as they finished their lunch. "Rosemary said she talked to the owner."

"That is Mrs. Palmer."

"You've used her title several times, now. Is Brian involved with a married woman?" asked Theona.

Jonathan shook his head. "Widow. She lost her husband on *Glorious*, early last June. She lost her father and uncle in France during the first war."

"Poor child," said Theona.

"I don't know what is going to happen between Charlotte and Brian. I also do not want Rose hurt in all this." Jonathan thought of his sister and the situation. "*Iacta alea est.*"," Jonathan said with some solemnity.

"The die is cast," George muttered.

"Quite so, Dad. Now, if you will excuse me, I would like to take a ride into the mountains. I will return in a few hours, I should think. I would like to take a nap before dinner."

"How long will you be able to stay this time? Theona asked.

"I have to be back in the saddle by noon Monday."

"Good. I am glad you are home, Son. Enjoy your ride. We will see you later this afternoon."

Jonathan went to the stable, found his favorite horse, saddled him, and headed west to the mountains. Solo rides in the mountain air with its unique scents, sounds and peace always cleared his mind and enabled him to think. Jonathan considered calling Brian, but at the end of the day, as he mentioned earlier, this was between Rosemary and Brian . . . with whatever part Charlotte was to play. He knew he would eventually hear of the results and consequences.

—

Saturday, 19.October.1940
Chequers Court
Ellesborough, Buckinghamshire, England
14:10 hours

"**P**rime Minister," announced 'Jock' Colville. He waited for Churchill's attention. "Commander Denniston has arrived with today's 'Buff Box.'"

"Yes, yes, please show him in."

The chief of the government's signals intelligence group entered the Prime Minister's spacious office. The 'Buff Box' was manacled to his left wrist.

"Good afternoon to you, Prime Minister."

"And the same to you, Alastair. Let us see what the 'Buff Box' has for me today."

Prime Minister Churchill retrieved the only key able to open the carrier box outside Bletchley Park or SIS Headquarters at Broadway House. He looked through the various folders to find the single sheet of paper.

MOST SECRET - ULTRA

```
SECRET
DATE: 18 OCTOBER 1940
TO: HIGH COMMAND WEST
COMMANDER ARMY GROUP D
COMMANDER AIR FLEET 2
FROM: ARMED FORCES HIGH COMMAND
BEGIN
THIS COMMAND CONFIRMS PROPOSED SEALION
EMBARKATION POINTS ROTTERDAM ANTWERP OSTEND
DUNKIRK AND CALAIS BREAK DIEPPE LEHAVRE
CONSIDERED NO LONGER SUPPORTABLE BREAK
EMBARKATION FORCES MUST BE PREPARED TO EXECUTE
OPERATIONS PLAN WITHIN 72 HOURS BREAK EXECUTION
ORDER RESERVED TO THE LEADER
END
SECRET
```

MOST SECRET - ULTRA

"This sounds like they are close to launching their invasion of our precious isle, despite all the negative signs over the last few weeks. What happened?"

"It could be taken that way. This is perhaps the most positive message we have intercepted to date. It was sent from Berlin yesterday. I have discussed this one with 'C' and his German desk chief. We collectively agree this appears to be an off-timing message, in that it was overcome by other events."

"How is that?"

"We still gather daily photographic and field evidence their transport vessels are moving east. To be specific, we can detect no transport vessels in

Dieppe or Le Havre. While a few remain in Calais and Dunkirk, the number is a mere fraction of the quantity in those ports just one month ago. 'C' felt it important for you to be aware of this decrypt.

"Rightly so. What about parachute, airborne invasion? If they are not showing signs of waterborne invasion, what about airborne invasion?"

"We see only incidental transport aircraft, and not the substantial numbers of transport aircraft necessary for a parachute invasion. Further, with the changing fortunes of Fighter Command and the diminishing German fighter force, transport aircraft would be extraordinarily vulnerable."

Churchill stared at Denniston, and then nodded his head. "Is there anything else I need to be aware of in today's traffic?"

"No, sir. This is all we have."

"Very well." Churchill returned the classified message back to the 'Buff Box' and locked the lid. "Thank you for making the journey."

"My honor, sir."

"Have you had lunch, Commander? Would you like something to eat before you go?"

"No, thank you, sir. I downed a bite before the drive over."

"Very well, then, Commander. Have a safe journey back to Site X."

"Thank you, sir, by your leave."

Churchill nodded his head, and Commander Denniston departed smartly. The Prime Minister returned to the shipping report he was studying. Embarkation points sounded ominous, but they were part of any water crossing operations planning, and the message had clearly stated the execution order would be issued by Hitler himself, which in turn meant it had not yet been issued.

—

Saturday, 19.October.1940
Standing Oak Farm
Winchester, Hampshire, England
15:00 hours

Charlotte stood at the window, watching the gravel approach over the far hill. Brian sat in the large leather chair by the large, stone fireplace with only a modest fire burning. The warmth felt good to his bones. From Rosemary's telephone call a week ago, Charlotte's emotions fluctuated. Brian never knew what to expect. He had misgivings about the wisdom of Rosemary's visit. Charlotte had agreed to Rosemary's visit without consulting him, and truth be told, there was no reason for her to seek his consent. Then, like a lightning bolt from the heavens, Brian thought, *Surely she is not coming to tell me she's pregnant. Surely, she would have chosen a more appropriate venue, if that was her purpose.* Brian's level of apprehension jumped up an order of magnitude with those thoughts. He then realized Charlotte did not mention how long Rosemary intended to stay. *What is this visit all about?*

"A taxi crested the hill," Charlotte announced. "I suspect this is Miss Kensington."

"Forgive me, Charlotte. I need to remain here by the fire."

"Certainly, Brian. I think Rosemary will understand. Excuse me, I will go outside to greet our guest."

"Thank you, Charlotte."

Brian heard the crunch of gravel stop and the engine remained running. He heard the car door close, and then the boot close. The automobile drove away. Brian waited for the two women to appear. He could not hear them talking. *What is taking them so long? What is going on out there?* Brian closed his eyes and leaned his head back against the chair. Brian tried and failed to shake off the image of the last time Rosemary visited him in the hospital in August. She had done such a magnificent job and the nurses allowed her to do it. A smile briefly washed across his face.

The sound of the front door opening jerked him from his doze. He did not know how long he had drifted off into sleep. Rosemary Kensington walked swiftly toward him.

"You simply must stop this nonsense, Mister Brian Drummond," she pronounced.

"Not my choice," Brian responded, as he tried to stand up.

"Don't!" Rosemary commanded. "You are injured."

"Thank you for informing me. I was wondering what all these attachments, straps and bandages were for."

They all laughed.

"Would you care for some tea, Rosemary?" asked Charlotte.

"That would be lovely. Thank you." Rosemary pulled a simple wooden chair up next to Brian. "So, tell me about your injuries this time."

"What's to tell?"

Rosemary laughed. "He is so bloody predictable. Isn't he, Charlotte?"

"Yes, he most certainly is," Charlotte responded without turning around.

"Fortunately, for my curiosity and caring, Charlotte informed me of your injuries."

"So, you two had a little girly chat, now did you?"

"Yes, we did. Charlotte was so kind to show me where she pulled you out of her lake as well as some of the wreckage of your aeroplane from that day."

"Do tell. I've not seen those bits, as yet," Brian said, looking over his shoulder at Charlotte's backside.

"Yes, well, you never asked," she mumbled.

"Thankfully, Charlotte was most kind in allowing me to check up on you, since I seem to be making a habit of visiting your bashed and broken body from this damnable war." Rosemary paused and stared into Brian's eyes. She was not smiling. "And, for the record, I am with Charlotte on this one."

"What one?"

"With not going back to flying and those terrible fighter aeroplanes."

Brian looked down at nothing in particular. He had to break her penetrating gaze. Brian could only shake his head in meager response. He knew better than to react to the combined position of both strong-willed women. The urge to extricate himself pushed him to say, "I'm tired. I need to lay down before dinner."

Pilot Officer Drummond struggled to rise from his chair and stand. Rosemary Kensington stood to help him. He grabbed his crutches and hobbled off to the bedroom.

"You don't have to run away from a fight, Brian."

He stopped. This time his better judgment did not prevail. He spun around, as best he could. "I've never run from a fight, and I am not starting now. I need to rest, so I can return to my Spitfire cockpit as soon as possible, and that is the end of this topic. Go ahead and eat without me, if I do not wake in time." Brian did not wait for a response or a reaction. He entered the bedroom and closed the door. He pulled the heavy comforter, blanket and top sheet back, and sat down on the edge of the bed. "Damn these women," he muttered aloud to himself. "They just do not understand." Brian kicked off the house slippers and removed the heavy robe Charlotte had given him. He moaned and groaned as he lay down on his side, lifted his legs up and covered himself with the sheet, blanket and comforter. He tried not to think about Rosemary's words, but mercifully, fatigue took him quickly.

———

It was dark outside by the time Brian regained consciousness. He could not see the clock, so he had no idea how long he had been asleep. Brian felt for Charlotte, but she was not in bed, and the dim light under the door suggested the two women were still awake and talking – conspiring actually, as his mind perceived things. Brian considered just remaining in bed. After all, he was warm and comfortable laying down. However, he knew he had to face the music, better to get it over with.

Brian donned and tied off his robe, got his slippers and crutches, and wobbled back to the living room. His eyes hurt as they adjusted to the light. All the blackout curtains had been lowered and fastened to the wall, including the large heavy black curtain over the entryway and front door. Neither Charlotte nor Rosemary could be seen or heard in the house. A plate of food sat alone on the kitchen table. The food was cold and the glass of milk was warm, which meant they had been there for a while. He sat heavily, ate quickly, and managed to get his empty plate and glass to the sink without breaking anything.

His curiosity got the better of him. *Fresh air will feel good.* Moving past the blackout curtain on crutches was not an easy maneuver. He stopped to allow his eyes to adjust to the dark, and to make sure there were no light leaks past the curtain. He opened the door to be greeted by bright moonlight. Brian moved outside into the brisk, night air, closed the door behind him, and then looked around. The moon was a few days past full. The sky was clear with no detectable clouds – *a good bombing night for the Germans,* he thought. Again, he could not see or hear any clues as to the whereabouts of the women. The cool air actually felt good, although that sense of refreshment would not last. The chill would eventually work its way to his broken bones. He decided to enjoy the fresh air while he could. Brian moved to and sat on the sturdy, wooden, garden bench positioned between the front door to the house and the barn.

Brian heard the two women before they appeared from behind the house. They noticed his white robe immediately.

"Welcome back to the land of the living," said Rosemary.

"Did you eat?" asked Charlotte.

"Thank you and yes. A delicious meal, as always."

"But, cold."

"Delicious hot or cold."

"You are such a cad," Charlotte said with an air of humor.

"You decided to take another walk, this time in the moonlight," observed Brian.

"Yes," answered Charlotte. "Rosemary helped me with the evening milking, and I showed her more of the farm . . . well . . . at least as much as can be seen by moonlight."

"It is a glorious night."

"Yes, it is."

"We need to get you inside by the fire," said Charlotte. "You are not dressed for the night air."

"It felt good . . . at least for a little while, but I am cooling off; so, yes, it is time to go inside."

Charlotte opened the front door. Both women insisted he enter first. There was barely enough room for all three of them, and then to shut the door. Once the door latched, Rosemary pulled back the blackout curtain. Brian moved to his chair by the fire. Charlotte added three split logs. *She is planning to stay up for a while. I hope they are not going to gang up on me about flying.*

"Rosemary is going to spend the night," Charlotte announced, as she prepared tea.

Brian looked to Rosemary, who simply smiled back at him, and then looked at the clock – 20:30 – the night was still young. Charlotte served tea and offered a plate of freshly baked cookies.

"Dare I ask what did you two co-conspirators talk about?"

"We talked about the farm."

"And, Rosemary's studies at Oxford. She is quite an accomplished woman already."

"Thank you, Charlotte."

"You are most welcome, my dear, and thank you for coming to visit our wounded warrior."

"And, to visit with you."

"Yes, certainly. You have been most gracious in filling in some blanks for me."

I wonder what she means by blanks?

"Are you going to beat me up about flying some more?"

"Life is not about you, Brian," answered Charlotte with a calm, even voice, "or, flying for that matter."

"OK."

"Now, I am afraid I must claim fatigue," Charlotte professed. "I must be up early tomorrow to tend the cows. I know you both have plenty to talk about, so I shall respectfully ask to be excused. Rosemary, the upstairs bed to the right and then left should offer the most warmth from the fire. I have laid out towels for you and a good country robe for your use. If you should need any additional blankets or linens, they are in the closet at the head of the stairs."

"Thank you so much for your generosity, Charlotte."

"You are most welcome, my dear. Have a good night. Now, if you will excuse me."

Charlotte left them without kissing Brian. *I am not sure if that is annoyance or just courtesy.* They both sipped their tea and watched the fire for several minutes.

"Charlotte told me you are mending well," said Rosemary finally.

"Yes, at my latest check-up with the doctor earlier this week, he was quite pleased with my recovery progress."

"I am sorry you feel we have been abusing you, Brian, but we love you."

"Did you say that to Charlotte?"

"Yes, as a matter of fact, I did."

"We both freely acknowledge our paucity of appreciation regarding your passion for flying; however, I can assure you we both accept it as a fact of life."

"There is much more to this than just the love of flying."

"Yes, certainly . . . the ol' patriotic bravado."

"Make fun of what we do all you wish, Rose, but it is your brother and our brethren who are keeping the Hun had bay. You should not forget that reality."

"I can assure you, Brian, none of us do. We are all genuinely grateful for the enormous sacrifices made by our armed forces. However, it is our caring for our loved ones that spurs our efforts to secure the safety of those we love."

"If everyone stood back from the line in deference to those who loved them, who would guard the parapets and defend the realm?"

She laughed. "You are sounding quite British, my amazing, big American."

"Yes, well, I believe in what we are doing."

"Thankfully. I shutter to think of how many more of these . . . events . . . you will survive. Each of these has been worse than the previous crash. I worry about you. I worry about my brother. I do not want to lose either one of you to this war." Neither of them spoke for what seemed like minutes. "Charlotte has lost so much already, more than any person should ever be asked to endure. She is carrying your child, Brian. She loves you very much. She fears losing another person she loves."

"I know. She had made that point very clear. All I can say is, I have no intention of losing."

"I'm sure no one does. Yet, it happens. Nonetheless, you are a fortunate man to have women who love you and care for you."

"Plural?

Rosemary smiled, and the laughed. "Surely you are not that unaware." Brian shrugged his shoulders, not wanting to admit his imagination or naïveté. "I love you, Brian. I have for some months, now."

"Rose," Brian said and held up a hand.

Rosemary Kensington shook her head. "Don't worry, my dear American. I am not going to interfere with your relationship with Charlotte. In fact, I confessed my sins to Charlotte, and . . ."

"Rose!"

"Stop, Brian. Listen to me. Hear me out before you protest." Brian stared at her, and then reluctantly nodded his head in consent. "She asked me directly if I loved you. She has been so generous in tending your recovery, and extraordinarily gracious in allowing me to be her guest. I was not going to lie to her. She did not ask and I did not offer the intimate details, but I did tell her how we met and the relationship we have had."

"Rose!"

"Brian, I am not some child with a schoolyard crush. I know and fully acknowledge your love for Charlotte, and not for me, and more importantly, her love for you. I will not and cannot interfere with your relationship with Charlotte. I have assured Charlotte as well. She is the one who suggested we chat tonight."

"Really?"

"Don't be so surprised. She is an amazing woman."

"I know that."

"Why did you come here, Rose?"

"I am hurt that you would even ask."

"Don't be."

"I have visited you each time you ended up in the hospital after one of your silly crashes. This is just another time. You are not in a hospital. You are here. And, to be honest and forthright, I wanted to meet Charlotte, and hopefully talk to her. She has been honest, frank and open with me. I felt obligated to return her honesty and generosity. If I cannot be your lover, I want to be your friend and Charlotte's friend."

"Rosemary, I have known that you are a very special woman since the very first moment I met you last Christmas. I care for you a great deal. I share your brother's concern; I do not want to see you hurt. I want to be with Charlotte for many reasons, some of which I understand, most I do not. If Charlotte can accept you as her friend, then I know I can."

"I cannot ask for more." Rosemary stood, stepped toward Brian, leaned forward and kissed him on the forehead. Brian winced and sucked in a breath. "I am so sorry. I did not mean . . ."

Brian laughed a good, hearty laugh that actually hurt his ribs. "Just kidding."

"You are incorrigible," Rosemary protested. She started to slap him but pulled up short. "You are lucky you have not fully healed yet. Don't do that, again!"

"I won't. I promise."

"The hour is getting late. Your future wife is most likely fast asleep, and you need your rest. Let me help you to bed."

"Thank you, Rose. I can manage the to-bed part, if you would be kind to place the spark screen and extinguish the lamps. There is a torch on the kitchen counter, so you don't stumble on anything. Sleep well. We will see you in the morning."

This time, Rosemary Kensington rose on her toes and kissed Brian on the lips. It was not a passionate kiss, but it was more than a sisterly peck.

—

Chapter 11

Nothing in the World can take the place of persistence.
Persistence and determination are omnipotent.
The slogan 'press on' has solved and always will solve
the problems of the human race.

-- Calvin Coolidge

Sunday, 20.October.1940
Standing Oak Farm
Winchester, Hampshire, England

Week 16

"That was so nice . . . Rosemary helping the lads and me with the morning milking before she departed," said Charlotte.

"Yes, it was, but I should be helping as well."

"No, Brian, not yet . . . soon, but not yet."

Brian knew there was no point in arguing with Charlotte. She joined him on the front bench. The cool, crisp, morning autumn air felt good. Birds sang and darted about adding a peaceful tone to the moment.

"Do you have any plans for the day?" asked Brian.

"Just the usual afternoon chores, I should think. Lionel and Horace are off tending to the boundary fence near the northwest corner."

"I'm feeling much better these days," Brian offered in a subtle attempt to work into the topic he recognized as unavoidable.

"Yes, I'm sure you are, and I am happy for that. So, with the pleasantries dispatched with, let us get to the point."

Brian turned his head to his right shoulder. Charlotte kept her eyes straight ahead without blinking or even twitching toward his look. *This is not going to be easy, but it has to be done*, he told himself. "You know what I am going to say."

"Yes, I do, and you know how I feel about what you are going to say."

"Charlotte, the last thing I want in my life, or yours, is to cause you any pain, apprehension or worry."

"Then don't."

Brian thought about her words and her position, yet again. In his heart, he knew the struggle within her had to be very deep and quite strong – her father, her uncle, her husband, and now the potential of her impregnator. Being with child – her first child no less – with a man younger than her, who is preoccupied with flying fighter aircraft in yet another deadly war had to be mind-numbing and bone-chilling. Yet, as deep as her fears were, his elation in flight was comparably great.

"Charlotte, I cannot fly forever. This damnable war will end."

"I'm sure my father and my husband said the same thing, but neither of them is with us in the land of the living."

"You know I cannot promise to survive this war. I can only do my best."

"Rosemary was knowledgeable enough and kind enough to share with me the two previous times you had to jump . . . into the water, no less. There are so many things that can go wrong up there. She has also experienced your nightmares, Brian. I need a father for our child. Running this farm is hard enough. Being a solo mother with an absentee or deceased father might well be a burden too great."

"I cannot desert my brothers."

Charlotte sprang to her feet and squared herself before Brian. "You can't desert your flyboy brothers," she shouted, "but you can sure bloody well desert me!"

Her outburst stunned Brian for a moment. "That is not fair," he responded meekly.

"Life is not bloody fair, Brian," she continued to shout, and then lowered her voice. "Life is what happens to us everyday. The choices we make can affect those outcomes. You have deposited your seed, and now you want to fly away like all male birds do, leaving me with the consequences." Charlotte turned away from Brian, walked several paces, and then stopped, to stare into the distance. After what seemed an eternity to Brian, she turned back to face him. "Get up," she commanded. Brian rose on his crutches. "This is not going to be easy, but you are coming with me."

Brian moved a couple of steps toward Charlotte. She walked off to the east. He turned awkwardly and followed her. She made no effort to look at him or slow her pace. Only the sound of his crutches and the footfall of his one good leg offered any sign that he was behind her. She stopped at the spot of his rescue, turned to face the pond, and waited for him to catch up. Brian arrived beside her and took a couple of deep breaths to clear his lungs.

"I have told you this story before," Charlotte said with considerable solemnity. "You have no way to remember what happened here, but I remember every horrifying detail every single day. I often sit out here, like I did that day, and I cannot escape the persistent nightmare of how close I came to never knowing you . . . and . . . ," she lowered her voice even further to almost a mumble, "losing my life as well." Charlotte placed both her hands on her not yet distended abdomen. "I am grateful I will finally be a mother. I have and still do want children. I do not regret what we have done. I am only afraid of the future I do not know." She turned toward him and kissed him passionately. "Now comes the hard part. Follow me."

Charlotte walked up the small hillock behind the pond. Again, she did not slow her pace or offer to help him.

Using crutches on flat ground was one thing. Trying to hobble upslope through tall grass was altogether more difficult. Brian had to stop several times to catch his breath, but he eventually joined her at the crest.

Charlotte waited for him to catch his breath and stand up straight. She stared off to the distance. It took Brian several minutes to finally see what she was looking at . . . several burned spots and bits of wreckage. He saw a curved piece that had to be a Spitfire vertical stabilizer.

"My aircraft?"

"Yes."

Brian moved first. He wanted to see what was left of his machine.

"Brian, stop!"

The recovering RAF fighter pilot halted off the crest, but not quite to the primary slope.

Charlotte came up beside him. "You don't need to do this. I know it is not easy moving on these slopes. I just wanted you to see the pieces. These were the fiery bits that flew over my head that day. This is what I remember every day."

"Charlotte," he said with sympathy in his tone. She held up her hand to interrupt him. Brian waited for her to gather herself.

"It is a long way down there. I shouldn't have brought you up here. You need to recover."

Brian continued to look at her face in profile. "For what?"

Charlotte turned to engage his eyes. "Brian, you do not have to be cruel . . . for life."

I had better not press this point. "I need to see what is left," he changed the subject. She looked away and nodded her head.

"Do you want me to go with you?"

"I do not want to cause you more pain than I already have, Charlotte. You do not have to do this for me. I can manage. It will take me a while. I did not realize the wreckage was comparatively this close. I need to see what is left."

"It is what it is . . . can't hurt worse than it already does."

"Then, I would welcome your company in case something happens out here." She nodded her head. Brian leaned toward her and kissed her.

Brian began working his way down the slope of the hill. This time, Charlotte stayed beside him. Brian struggled more against the grass than he did with the uneven ground. The first clump of wreckage they reached was the crumpled and broken tail section. The vertical stabilizer had surprisingly little damage, but half the rudder and the left horizontal stabilizer were missing, clearly broken off in a violent event. Enough of the empennage was recognizable, including the last four digits of an aircraft serial number – 4319 with part of a preceding 'X' – this was indeed his aircraft that had been shot out from around him at the end of July. They moved to the next set of pieces – more of the tail – and then, the back end of cockpit. Both wings were missing, ripped off the fuselage. Brian looked around but saw nothing that even remotely resembled either wing. The structure was quite distorted. He could see a portion of the twisted canopy rail and the bulkhead that would have been directly behind his seat. The seat fittings had all been ripped away. There were also signs of fire now. The grass beyond had been burned, but new grass sprouts appeared throughout the area. Charlotte remained behind Brian, out of his way, and holding his crutches when he needed to bend over or examine a part more closely. Twenty yards further on, Brian saw what was left of the seriously mangled forward portion of the cockpit and what was left of the instrument panel, gunsight and forward windscreen. Fire had clearly burnt away wires and melted some of the aluminum parts. The forward cockpit bulkhead had been torn, twisted and melted. Another 30 yards further lay the burnt and abused remnants of the Merlin engine the twisted and curled up blades of the Rotol propeller, or airscrew as the Brits called it. Most of the engine cowling was missing, torn away at some time during the break-up sequence. *With what is left of my aircraft, it is absolutely amazing anyone survived. I cannot imagine how on earth I survived this? How I just happened to be thrown clear, separated from my seat, and somehow my parachute ripcord pulled? No wonder Charlotte has had her own nightmares after seeing this and knowing that I had been in this aircraft, just moments earlier.*

Brian turned on his crutches to look in Charlotte's eyes. Tears streamed down her cheeks. "I am truly sorry you had to see this," Brian said softly.

"Yes, well, as you say, this is life."

"I know, but . . ."

"No, buts, Brian. I see this wreckage and I understand your nightmares."

"They started before this crash."

"And, that is supposed to make them more tolerable?"

"No."

"When I think of you returning to these machines, this is what I see. This is pastureland for the cows. The grass will grow back."

"Have you asked the Emergency Services or the Air Force to haul this stuff away?"

"They offered. I said no. I want to preserve this stuff, as you say, as it is as a memorial to how bloody lucky you were that day."

"OK."

"Have you seen enough?"

"Is there anything else to see?

"No, this is it on my land."

"I have seen enough."

"Very well, then. I want you to find a spot off the burnt area. I want you to lay down and take a little nap. I'm going to fetch Lionel or Horace. They will bring a cart to bring you back. You have had enough punishment for one day."

Brian laughed and Charlotte joined him. "Punishment, indeed," Brian said.

Charlotte's laughter and smile evaporated. "Go ahead," she commanded. "Find your spot, so I know where to tell them to look."

Brian kissed her, again. "Thank you." Brian scanned the ground and pointed to a nice dry area with plenty of grass about a yard beyond the edge of the burned area. Charlotte nodded her acknowledgment and turned to go. "Before you leave me here for retrieval, this is probably the appropriate place to ask you a sensitive question."

Charlotte stopped, turned back to Brian, and shrugged her shoulders to gesture her non-plus'ed acceptance.

"I would like to visit the site of my last crash, and if possible, to thank the men who pulled me out of my fighter."

"I have no idea where or how to find that place."

"All I was told was a field south of Romsey."

"That is not far from here."

"We can probably get some help from Jonathan."

"You would make me go to see another crash that nearly took your life," Charlotte responded with a particularly stern expression.

"No, Charlotte. I was just informing you that I would like to visit that site and thank those men . . . before I return to duty."

Charlotte stared at Brian. She shrugged her shoulders, again. "I know some farmers near Romsey. I buy feed from them, when I need it. I'll see if I can find the information, before we bother Jonathan. He's still on duty."

"Are you sure?"

"Yes. I should be able to find it, and against my better judgment, I will drive you. I will confess to a modicum of morbid curiosity."

"Charlotte, you do not have to do any of this. I know I have caused you pain, concern and considerable apprehension. I do not want to add to your burden."

"I am an independent, grown woman. I make my own decisions. Thank you for the acknowledgment of the bumps you bring to my life. This is who we are." Charlotte held Brian's eyes. A smile slowly bloomed across her magnificent face. "Now, get your arse on the ground."

The relief felt good to finally be off his one good leg and those irritating but necessary crutches. Brian watched Charlotte's backside as she walked away. The distinct odor of scorched earth remained quite pronounced. Brian closed his eyes. Fatigue and the warmth of the sun took him to slumber in short order.

—

Monday, 21.October.1940
RAF Catterick
Catterick, North Yorkshire, England

Despite the low, dark, heavy clouds and intermittent spells of heavy rain, No.609 Squadron had been at Available status since the time the clock said it was dawn, except for the step down to Available-in-30-Minutes status to allow pilots and ground crews time to have a proper hot lunch at their respective mess halls.

The movement north two weeks ago and the resumption of off-duty rest breaks after three hard months of combat had done wonders for their alertness, energy, concentration and spirit. The dreariness of autumn weather had not helped their efforts to bring the replacement pilots up on step with the remaining veterans. Today was another one of those lost days.

Nice, Jonathan told himself as he looked around the Dispersal Hut. The newest and last replacement pilot had arrived on Saturday, fresh out of the advanced training unit – Pilot Officer Victor Hanson Clegg from Lancashire – bringing the squadron to full strength. *Well, minus 'Hunter' who was still recovering from his combat wounds.* No.609 Squadron had not seen a full complement of fighter pilots since early July, and they needed 'Hunter' back to finally achieve that level.

Within a week of 'Hunter's crash, four fresh, eager pilots arrived on the same day – Pilot Officer Ian Stanley Noring from Suffolk, assigned to 'Waggle's Yellow Section; Pilot Officer Angus Bevan from the Scottish Highlands of Argyllshire, assigned to 'Sparky's Red Section; Flying Sergeant George Carrwood from Hampshire, assigned to 'Stack's Blue Section; and, Flying Sergeant Adam Joseph Lowry from the mean streets of Belfast, County Antrim, Northern Ireland, also assigned to 'Sparky's Red Section.

The idle, non-specific chitchat of the morning turned to more professional topics in the gloomy afternoon of waiting.

"There were rumors going around at OTU," opened Clegg, "that the Germans have figured out how to bomb through clouds."

"It's actually quite easy, just hit the release button," responded 'Sparky' Morrow. Everyone laughed hard, even Corporal Warren joined the humor. "Those heavy bombs fall right through the clouds."

"I mean and hit something."

"Yes, well, that takes a little more sophistication, I'm afraid," added 'Waggle' Davies.

"How do they do it?" persisted Clegg.

Squadron Leader 'Stack' Long-Roberts appeared at his door to join this particular part of the conversation. "They have some kind of crossing beam system, so I am told by Group."

"And, we sit on the ground," said Carrwood.

"You will get your turn soon enough, my eager mate," 'Waggle' said.

"We can still go after them, if it is clear above a cloud deck," said 'Stack,' "like we practiced last week. Today, this stuff goes up to Angels Two Five and higher in places, so the Met lads tell us. Again, I am told, the night-interceptor lads can still find them even in the clouds with their RDF kit."

"How on earth does that work?" asked Clegg.

"Seems we have another curious one," interjected 'Sparky,' "like 'Hunter.'"

"Who is 'Hunter'?"

"You will meet him in a few weeks, I should think," 'Harness' said, jumping for his friend. "He is the squadron's highest scoring ace – 19 confirmed victories, so far – third or fifth highest in all of Fighter Command. He was seriously wounded in combat just before you newbies checked in."

"Hear, hear," said several of the veterans in unison.

"Quite so. Back to your query, Pilot Officer Clegg," 'Stack' said. "RDF stands for Radio Direction Finding. The night-interceptor lads are experimenting with radio beams that detect other aircraft, and determine the direction and range to their target. They are using it at night mostly, since the Gerry bombers have pretty much given up on daylight bombing. The night-fighter fellows with their RDF kit can see through clouds as well, so what we cannot handle, they can."

"Amazing."

"Indeed."

"I can't imagine groping around in the black of night for another aircraft," added Clegg.

Several of the pilots laughed. The day-fighter pilots shared Clegg's apprehension.

"We still have to deal with these damnable fighter-bomber sweeps, but they are not much use in this weather."

"They don't have beam equipment?"

"No . . . not that we know of."

"The One One Group lads went after a gaggle of those Gerries earlier this morning. They apparently pickled their bombs somewhere in the countryside of Sussex, got into a bit of a scrape with our lads, and then high-tailed it back to France," explained 'Stack.'

"I can't imagine them being very serious about hitting anything," observed 'Boxer.'

"No, indeed," said 'Waggle.' "I suspect the bombs are to make the Fat Man Göring happy, while they try to lure our mates into a fight."

"Yes," said 'Stack,' "I suppose it did."

"Will we get to have a go at those fighter-bomber fellows?" asked Carrwood.

"Not likely up here," 'Stack' answered. "The fighters can't make it this far north. "Do not fret, however, I am certain they will move us south once Command is satisfied with our progress, so they can relieve another squadron for rest and refit, as we are doing."

"There is plenty of shooting left to be done in this war," added 'Sparky.' "Enjoy the respite while it lasts. There is much more ahead of us, and all of you newbies will have plenty . . . ample . . . opportunity to show your mettle, so as you have been told before, patience my junior darlings."

The telephone rang. All heads, eyes and ears turned to Corporal Warren. "The Squadron is released."

"There you have it. Corporal Warren, if you would be so kind, call us a lorry for transport to the Mess. Let's secure our kit for the day. I expect everyone back here at dawn tomorrow. 'Sparky,' you are up for a rest break tomorrow. We will have to see to the weather tomorrow morning, but if the forecast remains valid, I should be able to release you early."

"Thanks, Skipper. I'm good with whatever should play out."

"OK, lads, we are done for the day. Don't get into any trouble that I might hear about tonight. See you at Mess, or in the morning."

The pilots who still had their flight kit on doffed their paraphernalia, hung their flight gear on their respective hooks, and waited inside out of the rain for their transport – another day done.

—

Tuesday, 22.October.1940
Bridgewater House
Burma Road & Spaniard's Lane
Romsey, Hampshire, England
13:30 hours

During the 40-minute drive from Standing Oak Farm, Brian learned from Charlotte and her inquiries that Peter Carrwood was the farm foreman who saved him, along with his two, young, hired hands – Darren Eagleman and John Duckworth. Peter was the oldest child and son of Elizabeth and Justin Carrwood. Peter's younger brother George was a serving flying sergeant, who had just joined No.609 Squadron. The senior Carrwood purchased the Bridgewater farm in 1914, before the Great War began. All of the Carrwood children – Peter, George and two younger sisters – had been born on the farm. George entered military service before being conscripted. Peter was of prime conscription age, but had so far not been called up.

Charlotte expertly navigated the country roads without any assistance from Brian. She passed through the village of Romsey, and then drove south on a narrow gravel country lane that was Burma Road. As the country lane curved east, a small wooden sign – Bridgewater Farm – stood on the left side of the lane, just beyond a modest, virtual tunnel into a thicket of hedgerow shrubs and evergreen trees. Charlotte turned into the tunnel that opened onto a looping drive around a well kept, grass field to a rather spacious and elegant manner house. As Charlotte drove around the driveway and stopped in front of the large portico, the Carrwood family, including the two younger daughters, appeared from the house to welcome Charlotte and Brian. Introductions ensued. The Carrwoods had done their homework regarding their visitors. They began almost immediately regaling their visitors with the noted accomplishments. Neither Brian nor Charlotte felt comfortable being treated like famous celebrities. Peter was the first to break the moment.

"I know you came to see Mister Drummond's crash site," Peter said.

"Yes, if it is no imposition," responded Brian.

"The field is still a little too wet for an automobile. I arranged for Johnny and Darren to harness up two of our Shire horses to our largest wagon. You should be comfortable."

Brian looked to Charlotte. She nodded her agreement and said, "You are the cripple, darling."

Brian sneered and responded, "Well, there is that. I should be OK, as long as I can get up there."

"We will help you, Mister Drummond," Peter added, "if necessary."

Peter Carrwood is older than me. It seems quite odd for him to be calling me Mister Drummond. That's my father, not me. "I should be fine."

Peter signaled to a large outbuilding north of the main house. Brian saw two men in a large four wheel wagon drawn by two, enormous, magnificent, draft horses come from the outbuilding toward them.

"If you will excuse us, Miss Palmer and Mister Drummond," said Justin Carrwood, nodding to both of them, "we will leave you to your tour."

"Thank you very much for allowing us to visit your farm," Brian added.

"It is an honor. It is not everyday we see an air force ace in rural Hampshire." The older Carrwood couple and their two teenage daughters returned to the house.

Charlotte had wisely worn light brown, wool trousers and her field boots, along with a heavy wook sweater and leather overcoat. She easily ascended to the wagon bed and sat on a bench seat behind the two field hands. Fortunately, there was a makeshift ladder built into the side of the wagon. Brian awkwardly climbed onto the wagon, while Peter stood behind him, presumably in case Brian missed a rung or had a stumble. Brian sat beside Charlotte. Peter virtually lept onto the wagon and sat behind his two guests, and then completed the introductions. Duckworth held the reins, gently commanded the horses, and they were off heading south.

Peter offered a running commentary on their property and the family's agricultural pursuits. They passed through several hedgerows and eventually opened onto a large, freshly plowed field. The rich, aromatic smell of the earth reminded Brian of home. Duckworth halted the team of horses about halfway down the edge of the field. Peter jumped down off the wagon, went forward to pat the heads and necks of both horses, and then jumped up onto the rigging just in front of Duckworth and Eagleman, and between the two massive Shire horses.

Brian saw no evidence of any disturbance to the idyllic scene and certainly not an aircraft crash site.

In fine pilot fashion, Peter Carrwood waved his hands to explain the sequence of events they witnessed that early afternoon three and a half weeks ago. He described in surprising detail the final moments of flight for a flaming and smoking Spitfire fighter. He distinctly remembered seeing the tail letters 'PR-F' just before the first impact and pointed to the spot about 20 yards into the field from their position. Brian remembered nothing after the first impact. Peter indicated the aircraft bounced three times on its belly before pitching forward and burying its nose in the freshly plowed soil.

Brian repeatedly checked Charlotte's facial expression during Carrwood's detailed explanation. She remained stoic, expressionless and surprisingly attentive. He was impressed by her emotional control. Charlotte even asked a few questions during Peter's recounting of their rescue actions and what they found when they got the unconscious and striken pilot out of and away from his destroyed aircraft.

The air force had sent a recovery team of a half dozen men within a week of that day, with a crane truck and several flatbed trucks to cut up and cart off the wreckage. They had even cleaned up the soil that had been saturated with fuel, oil and other fluids, and secured the unexpended ordnance. Peter, Johnny and Darren had helped the recovery crew fill in the impact craters and final lodgement hole.

"Thank you so very much for everything you did for me that day," Brian said.

"I imagine that was not a good experience for you," answered Peter, "but, we are very proud that we could do our part for the war effort." Eagleman and Duckworth nodded their heads in agreement. "And then, when Dad found out who we had rescued," Peter nodded to Brian, "we were doubly proud."

"I must interject here," Charlotte said, "I am also immensely grateful for your decisive and swift response. Our young American here was lucky beyond measure to have landed in your field. I am also thankful the air force did well by your property."

"She is saving hers," Brian added and immediately regretted his words.

Charlotte did not take the bait or display any response. None of the men chose to pick up on that observation either. Brian was quietly appreciative of their discretion.

They returned to the house. Charlotte and Brian paid their respects to the Carrwood family for their hospitality, before departing on the return trip via the same route. Charlotte did not speak and Brian was wise enough to let the silence remain. She would talk, if or when she was ready. He was grateful for her efforts to fill in the blanks of that day three and a half weeks ago. Brian closed that chapter of his life.

—

Tuesday, 22.October.1940
The White House
Washington, District of Columbia
United States of America
10:15 hours

"Good morning, Mister President," said Cordell Hull, and then nodded to Hopkins, "Harry."

"Good morning to you, Cordell. How are the affairs of state this fine fall morning?"

Secretary of State Cordell Hull moved to the closest couch and to a position closest to the President's desk. He remained standing. Harry Hopkins pushed the President's low-profile wheelchair to the open space between the two, facing, long couches, and then pulled up a simple, straight back, wooden chair behind and to the left of the President, and then sat.

"First off, Mister President, I received just this morning Joe Kennedy's letter of resignation, to be effective at your discretion."

U.S. Ambassador Joseph Patrick 'Joe' Kennedy, Sr., of Massachusetts – actually, Ambassador Extraordinary and Plenipotentiary to the Court of St. James's, the principal representative and senior American diplomat in London – had been in the position since January 1938, through the tumult of the German absorption of Austria, the dissection of Czechoslovakia, the Munich Accord, the Night of Broken Glass, and the invasions of Poland, Denmark, Norway, Holland, Belgium and France. He consistently, persistently and in some aspects defiantly had been publicly and vocally supportive of Irish Republican efforts to rest control of the predominately Protestant province of Northern Ireland. Perhaps as a consequence of that support, Kennedy had also remained demonstrably vociferous in his support for Adolf Hitler and the Germans, even through the Battles of France and Britain. He continued to disregard guidance from Washington and publicly espoused capitulation of the British, and worse collaboration of the United States with Germany.

"Long overdue, I must say, but thank you. He liked all the trappings of being Ambassador to the Court of St. James's all too much, and he clearly resisted my hints for him to move on with his ambitions. Please convey my appreciation for his service at your earliest convenience . . . nothing more."

"Yes, sir."

"You did not make the journey from Foggy Bottom about your Inbox."

"No, sir. I assume you would prefer he not remain in London until his replacement can be confirmed."

"Correct."

"Obviously, we need to find an appropriate replacement as soon as possible."

"Yes. Isn't Herschel Johnson the *Charge d'Affaires* in London?"

"Yes, sir. Good man."

"Can he handle our affairs until we get a vetted and confirmed replacement for Kennedy?"

"In my opinion, yes, sir."

"Then, make it so. I am long past fed-up with Joe's pro-German, negative advice and influence. Who do you have in mind?"

"We have an evolving list of just over a dozen candidates."

"Is Winant, Governor John Gilbert Winant of New Hampshire, on your list?"

"The Republican?" Roosevelt did not respond and only held Hull's eyes. "He is now."

"Please give him a good look-see. How long will your process take before we can submit our nominee to the Senate for confirmation?"

"We have anticipated this day, so I suspect we can make a recommendation to you by the end of the month."

"Please proceed with a sense of urgency, Cordell. I'm sure Johnson is quite capable, but I am particularly concerned about the image, the impression, this change leaves with Winston and His Majesty's Government at this critical time in history."

"Yes, sir. I think everyone involved in the process fully appreciates your concern, Mister President."

"Excellent, then proceed."

"Yes, sir. Next, did you hear Churchill's broadcast to the French people, yesterday?"

"I did not listen to the whole speech, but I certainly heard enough. His French could use some polish." All three men laughed. "No one can say the man does not have balls. You have to admire his backbone, given the circumstances since June. It is most unfortunate the Anglo-French Union proposal could not be completed in time."

"And, the French collapse was so quick."

"Certainly. Where are we on getting an ambassador on the ground in Vichy? We cannot do much, but we must do something to keep Pétain and his so-called government from leaning too far toward Germany, and especially enabling French Army and Navy assets to fall under operational control of the Germans. My efforts to convince 'Jack' Pershing to take the post came to naught."

'Jack' Pershing was none other than General of the Armies John Joseph 'Black Jack' Pershing [USMA 1886], Commander of the American Expeditionary Force during the War to End All Wars. He had picked up his nickname as a consequence of his service with the 10th Cavalry.

"He would have been perfect. His relationship with Pétain, and many of the senior French generals and admirals may have made neutralization of the Navy easier."

"Quite so. We need a senior, respected, retired, military officer for that role. Pershing said he was flattered, honored to be asked, and tempted to take the assignment, if for no other reason than to be reunited with his old colleagues, but he said his health was failing and would be unable to serve properly. I have to respect his candor . . . too bad but understandable. Last night, another name came to me. What do you think of 'Bill' Leahy?"

Admiral William Daniel 'Bill' Leahy [USNA 1897] had served a full career in the Navy, with his last assignment as the previous Chief of Naval Operations (CNO), serving in that post until the month before the shooting war in Europe began.

"Governor of Puerto Rico?"

"One and the same," responded Roosevelt. "I first met him during my tenure as Secretary of the Navy and of course he was CNO early in my second administration. Good man, stern character and very well respected."

"Do you want us to vet him for an ambassadorship?"

"Yes, unless you have a better candidate."

"No, sir. We'll get that completed promptly."

"As soon as you do that, please let me know. I would like to call him myself. If he agrees, we need to get him to Vichy as soon as possible, after I have a long chat with him about the assignment."

"As you command, Mister President."

"Excellent."

"I have one last item on my list."

"Yes."

"The fifth group of surplus destroyers is in Halifax Harbor and is also expected to transfer tomorrow morning. That leaves one last group in the original plan of 50 warships in the bases-for-destroyers trade deal, and that group is on schedule to transfer later this month, according to Frank Knox. We have given the British what Churchill asked, but let if suffice to say, our cousins across the Atlantic were not particularly impressed with the serviceability of those ships."

"What did they expect for surplus ships that are nearly 25 years old and have been in mothballs for 20 of those years."

Roosevelt chuckled. "We did the best we could with what we had in our grasp. At least, they are under their own power and are fully provisioned with fuel, ammunition and stores. The rearmament program will enable us to build many more new, modern destroyers."

"I doubt Congress will look the other way, if we try to pawn off new destroyers as surplus."

"You are most likely correct, Mister Secretary, but we will cross that bridge when we get to it. We need to build those new destroyers first. Thank you for your assistance in making the transfer happen."

"I must confess, more than a few of my staff and the diplomatic corps were quite apprehensive when the program was originally proposed last summer."

"They were not alone. Justice was not particularly enamored with the proposal either, but we managed to do what had to be done. The British need much more from us, and we need the British to remain in the fight, to keep the Germans at bay, if nothing else. I am now convinced war with Germany, and probably Japan, is inevitable. It is only a matter of time."

"Not everyone at State shares that assessment, I'm afraid."

Roosevelt chuckled, again. "Of that I am certain, Cordell. They do not see what I see."

"Yes, sir. Eventually, you will have to help them and the American people see what you see."

"Quite so . . . unless the Germans, or the Japanese, do that for us."

"We hope not."

"One question, while I have you, Cordell, I noted with interest from London that the British have officially formed Seven One Squadron with all American volunteers and a British commander. Where does that leave our citizens, who are standing the line over there?"

"That is a bit of a sticky wicket, as our British brethren would say. Each one of them, to our knowledge, has now fought for a foreign belligerent in direct violation of the Neutrality Act of 1939, and prior I must say. As I am sure you recall, you issued a full pardon to the Drummond lad, who by our records was the first to go over."

"Yes, I remember. Winston has been and remains very keen on the young man . . . a fighter ace several times over . . . already awarded a Distinguished Flying Cross and soon a Military Cross, as I understand it. Impressive young man."

"Yes, sir. However, the others are not so covered. Drummond's passport has been reinstated in accordance with your pardon, but the others are at various stages of disenfranchisement. Several have sworn allegiance to a foreign sovereign or government, which additionally places them in violation of the Citizenship Act of 1907."

"I would like you to find a way not to punish any of these young men. Whether Congress knows it, recognizes the reality, or acknowledges their sacrifice, they are in service of freedom, which is in our interests."

"A pardon for each of them would resolve any ambiguity."

Roosevelt considered Hull's words for a few moments. "Yes, certainly, I would prefer not to take that path. I supported Winston's request in the Drummond case, but there is a steady stream of young men going over and a mounting number that will soon make pardons an impractical method to remedy their situation."

"What does Justice say?"

"I've not discussed this topic with Bob Jackson, yet, as I think we have thwarted the legal violations element for the time being, but it is their citizenship and thus their passport situation that concerns me the most. To my knowledge, none of them have made an attempt to return to our territory, but it will happen, and I do not want the embarrassment to us or the trauma to them of being in a limbo state."

"Yes, sir. I will have our passport and immigration folks look into how we might reconcile their status in accordance with the law."

"Thank you, Cordell. Anything else for us to discuss?'"

"No, sir. That does it for me."

"Anything from you, Harry?" Roosevelt asked, glancing over his left shoulder.

"No, sir, Mister President."

"Very well, then. Thank you for stopping by, Cordell. Have a splendid rest of the day."

Secretary Hull stood. "Thank you, Mister President," he answered, "good day," and then he departed the Oval Office.

———

Wednesday, 23. October. 1940
Special Operations Executive
No. 64 Baker Street
Marylebone, London, England

The rather ordinary, five-story, building had a ground floor entrance with a simple, black, solid oak door and brass numbers to identify the address, a haberdasher store and a run of the mill restaurant on the other side of the shop from the entrance door. The building bore no signs or other descriptors.

Former naval intelligence field operative and now SOE agent Trevor Thomas 'Diamond' Andersen, alias Robert Henry Stone Johnston, knew the building and the routine security procedures for the offices above the ground floor. He made his way through the various checkrooms to the office for his appointment. Trevor did not have to wait long for the anticipated meeting.

"The director will see you now," announced the young man in a simple, light grey, ill-fitting suit with a white shirt and narrow black necktie.

The director he was scheduled to see was Director of Operations & Training Colonel Colin McVean Gubbins, DSO, MC – the well-known special operations officer, now assigned to SOE. His leadership of the Independent Companies during the Norway Campaign, the previous spring, had pumped up his reputation further, and he was awarded the Distinguished Service Order for that work.

Trevor entered the modest office. The two men shook hands, exchanged pleasantries and took seats in facing leather-upholstered chairs.

"Let us get down to it," said Gubbins. Trevor nodded his agreement. "I understand you and 'Blocker' performed well in our inaugural training course."

"Thank you, Colonel. It was fairly straight up stuff, I should say. I did learn a few new tricks and tools, so the training was worthwhile."

"Excellent. With your field experience, I invite you to help us improve the training program. We need a small army of field operatives to raise hell on the Continent. Minister Dalton has a most ambitious charge from the War Cabinet."

Under the mandate of the War Cabinet last July and the personal instructions of Prime Minister Churchill, Minister of Economic Warfare the Right Honorable Edward Hugh John Neale 'Hugh' Dalton, PC, Member of Parliament for Bishop Auckland and a member of the Labour Party in good standing, formed the Special Operations Executive (SOE) to attack German infrastructure and logistics facilities, and encourage and support indigenous resistance movements in the occupied territories.

"What do you have in store for me?" asked Trevor.

Gubbins stared into Andersen's eyes looking for something. "How would you feel about going back into Germany?"

Agent 'Diamond' smiled. "A rather sporty proposition, I should think."

"Yes, certainly. You have the appearance and German language mastery with a nice Upper German accent." Andersen nodded his head in recognition. "As you may know, MI6 maintained contact with known resistance groups inside Germany through the Phony War and up to Hitler's move west."

"The Gestapo is a very effective suppression force."

"Indeed."

"They are not just intolerant of dissent, they follow the slightest whiff like a bad dog on a good bone, even just a idle comment, and from my perspective, that scent is sufficient to send an individual to a *konzentrationslager*, like Dachau, Buchenwald or Sachsenhausen, where he will either be executed swiftly, or slowly by forced labor and starvation. Beyond that reality, I have not been inside Germany since late last year."

"We understand all that. 'C' and his experts believe they have kept up with security and travel documents to cover your movements. Your previous boss, Admiral Pike, speaks very highly of your ability to blend in and move within Germany."

"Yes, well, that was before the shooting started."

"Then, my good fellow, are you trying to tell me you do not feel capable of such a mission."

It was Andersen's turn to study Gubbin's face. "No. I am just saying we need a measure of realism in assessing the risk equation."

"On the other side of that equation, our task is to diminish the ability of the Nazi regime to wage war. We know there are dissenters inside Germany. None of us know what they can do, but in the case of the Germans, His Majesty's Government seeks to keep the dissenters alive, out of the camps, and a potential resource when the time comes, as it most assuredly will come. As I understand your history, you have trustworthy contacts in the Ministries of Transport, and in Food and Agriculture."

"Yes, I do . . . if the Gestapo has not liquidated them, yet."

"That would be your first objective, re-establish contact."

"They are in Berlin."

"Yes, we know."

"My normal methods of entry are now closed."

"Quite so. Unless you have a better approach, we think entering via Switzerland under the guise of a logistics analyst for Friedrich Krupp AG, working on improvements to rail component production for rapid repair and increased speed."

"That is plausible, but I have some learning to do, if we use that cover. I do not know much about the steel business or railway components, and I do not know anyone inside Krupp to validate my cover."

"We have the experts to bring you up to speed. As far as a cover contact, we think we can help there, and that will be part of your mission training."

"Then, I am good to go.

"Excellent. However, that mission is a little farther out. We have another short term mission for which we can apply your language and operational skills." Andersen raised his left eyebrow to gesture 'and?' "We lost contact with a former French Army intelligence officer, who successfully went undercover after the surrender. We have reason to believe he is alive and well, and more importantly not compromised. We would like to clandestinely insert you into Northeast France, south of Abbeville, make contact, re-establish communications, collect what intelligence you can from incidental presence, and assist as required to develop a local network."

"Are we certain he has not been compromised?"

"We can never be sure." Gubbins paused, perhaps to consider how much he wanted to discuss before the mission was agreed upon. Trevor waited. "We believe his code sequence expired and he failed to renew. There were no signs of compromise."

"I knew Major Stevens. It took us six months to figure out the Gestapo had him and Captain Payne-Best. We lost more than a few agents as a consequence."

"True. To be more specific, then, we received a clear text message that meant he was still active but has no code sheet."

"I can do it. I presume you have the skeleton of the mission."

"Excellent. Yes. The plan is to parachute from a night bomber raid passing overhead. We would expect you to be on the ground for three or four months. Major Jacques Merton is your objective. He goes by the name of Yves Moulin and is working as a hired hand on the farm of Madame Marie du Clerc, east of the village of Drucat. You can work out the details with your handler. I'd like to insert you before winter weather sets in and get you out by early spring."

"That should be no problem, I should think. What about 'Blocker,' if I may ask?"

"He is in mission preparation training. He has agreed to reenter Poland with essentially the same mission as we just discussed."

"Is that wise? Certainly the intelligence, security and police functions know who he is and exactly what he looks like. He has to be high on their target list. Even the Soviets in the east will be hunting for him. 'Blocker' is not a fan of the NKVD or KGB, and I'm sure the feeling is quite mutual."

Colonel Stanislaus 'Blocker' Pordonski had been chief of the Polish secret security service until the fall of Poland at the end of September 1939. The Red Army had invaded Eastern Poland, once the German attack had begun, to settle old scores. As the government, military and security services disintegrated in the days after the fall, 'Blocker' began a tortuous exfiltration with both German and Soviet forces searching for him, while his combat wounds healed and he made is way to Egypt. He finally arrived in England last July.

"Understood. We are working on altering his appearance. We are also refining his credentials and cover story."

"He is a very good man, Colonel. He is far more valuable for his knowledge and experience than he could ever be in the field."

"We will be the judge of that," Gubbins responded with a new sharpness to his tone.

Trevor considered whether to press his point. "He is not just a colleague, Colonel. He has been a valuable and loyal friend of mine for better than 10 years."

"Friendship in this business is tenuous and fleeting."

"Yes, sir. All I ask is, do not press him past his threshold of caution."

"Duly noted, Mister Andersen."

Trevor nodded his head. "Do you have anything else for me?"

"Yes, actually, I have one more point . . . more of curiosity than specific operational necessity." Trevor nodded his head in consent. "I read your field report from April 1939. It is rather thin and does not say what the mission or purpose of that operation was, and perhaps, it is not one of those questions I should ask."

"I am unable to discuss the purpose of that mission, sir," Trevor responded, knowing there were only a handful of people who actually knew the product of that mission.

"Understood and I shall not press. My question was actually about 'Blocker's colleagues with you in that operation."

"You should ask Stan. He may have more accurate or current information than me. When I was reunited with Stan last August, I asked. He told me then that his agent, Carmin Kraczhuk, had been reported killed during the defense of Warsaw. Of the other three operatives with us that night – I knew them as Lech, Georgi and Milo – Georgi was killed near the border at the beginning of the German assault. The whereabouts of Milo and Lech are unknown. We do not know their status, and they may very well turn up eventually, like Stan did."

"Is there anything else you can recall?"

"That is the extent of my knowledge, sir."

"I have not had the opportunity to discuss my curiosity with Pordonski. I will do so at the first window."

"Why do you ask?"

"Beyond my curiosity, they would appear to be worthy candidates for the SOE mission, especially in Eastern Europe."

"Yes, sir, they would. All of them were accomplished, capable men. Stan has a keen eye for those skills."

"Poland has a long history of being a battleground between east and west. None of us knows how this fracas is going to play out. Best to be pre-pared."

"No one could argue with that perspective. I will listen for any clues. Any other questions?"

"No, I think we are done."

"Who do I see for my additional training?"

"Take a good rest to visit with your family and friends. One of our supervisors will contact you on Monday, with your instructions."

"Very well. Thank you, Colonel Gubbins."

"Thank you for your service, Mister Andersen."

Trevor left SOE HQ, but stopped a few paces from the front door. He instinctively scanned the city street in both directions for anything out of place or anyone who might be unduly interested in him. His experience made him instinctively attentive to his surroundings and cautious with his movements. Satisfied things were as they should be, Trevor walked north on Baker Street to the Baker Street Underground Station, took the Metropolitan Line to King's Cross Station, and the next railway coach to Ipswich, his parent's home. He knew he should take advantage of the opportunity, since there was no way to know when the next time would come.

—

Friday, 25.October.1940
Cabinet War Rooms
New Public Offices
Whitehall, London, England
14:30 hours

"Gentlemen, let us convene," announced Secretary to the War Cabinet Sir Edward Bridges. He waited for everyone to be seated and quiet. The underground conference room was demonstrably less spacious than the equivalent Cabinet Room at No.10. The full War Cabinet was present and fortunately this session was without the Cabinet Secretariat staff, chiefs of staff, or other ministers. "First agenda item . . . the Prime Minister's report."

"As my colleagues recall," began Churchill, "I took a bit of a break from the war for an inspection tour to Scotland. Day before yesterday, I spent the day with General Sikorski's Polish forces in exile."

After the fall of Poland in the fall of 1939, General Władysław Eugeniusz Sikorski evaded capture through Romania to Paris, and then from France after the collapse of the French republic the following summer. General Sikorski was appointed Commander-in-Chief, Polish Armed Forces in exile with his headquarters in Eastend House, near Thankerton, Lanarkshire, Scotland, southeast of Glascow. General Sikorski collected up elements of the surviving Polish Army, Navy and Air Force that were now garrisoned throughout the United Kingdom and preparing to join His Majesty's Forces in the field.

"My assessment in one word . . . enthusiastic. The Poles have upgraded their weapons with our assistance, and they have significantly improved their training, again with our assistance. We are still collecting up stragglers from the Continent, a year after the fall of Poland. General Sikorski already has roughly four divisions, so far. I think we all agreed that another couple of battalions worth of exiles or recruits will enable him to form another full division. Fighter Command has two full squadrons of Polish pilots and may well form another one in the coming months. I had with me on this tour Generals Brooke and Ismay."

Commander-in-Chief Home Forces Lieutenant General Sir Alan Brooke currently held operational control of General Sikorski's Polish Army units.

"I believe it safe to report to the War Cabinet, General Brooke was impressed by the significant improvement in the Polish forces and has already directed integration of the Polish forces in the defense plan of the Home Islands. After thorough briefing by General Sikorski's staff at his headquarters, we witnessed a number of exhibition demonstrations, including armor maneuvers and a parachute jump by elements of their parachute brigade.

"The following day, we boarded HMS *King George V*, the first of her class and the newest of the Royal Navy modern battleships. She just completed her sea trials, and is due to be commissioned and enter service in the Home Fleet month after next. She was at anchor in the Clyde Estuary, off Greenock, and a most impressive ship of the line, I must say. The crew is rightly proud of her and is quite eager to assume their place in the line. Captain Wilfrid Patterson offered me a thorough briefing on their sea trials, which I am happy to report, have gone exceptionally well. As the first of her class, she has additional tests to complete prior to commissioning. Her sister ships *Prince of Wales* and *Duke of York* are shortly behind *King George V*. We have every reason to be proud.

"With your consent, gentlemen, I would like to send our congratulations and appreciation to Captain Patterson and his crew, as well as General Sikorski and his Polish forces, for their exemplary service and extraordinary efforts for their contributions in the defense of the realm." Churchill looked to each member of the War Cabinet. He received a head nod or verbal affirmation from each member. "Sir Edward, if you would, please prepare the appropriate letters."

"Yes, sir. Next up, Prime Minister, we wanted the report of the Air Minister regarding the Big Wing issue."

"Is Archie here?"

"Yes, sir. He is waiting in the hallway."

"Very well, let us have his report."

Sir Edward opened the closed door and called for the Air Minister. Sir 'Archie' Sinclair entered and took the nearest open seat at the end of the closest leg of the 'U' shaped table. Sir Edward returned to his desk in the center of the 'U' table, facing the Prime Minister.

"The Cabinet is ready for your findings on the Big Wing matter, Sir Archibald."

"Thank you, Sir Edward. A week ago, Sir Cyril and I met with Sir Hugh and the fighter group commanders to settle the controversy with respect to the tactics utilized by Fighter Command to counter daylight operations by the Germans. There are good and valid points on both sides of the debate. We have agreed to maintain our current procedures as the best approach to dealing with daylight raids, which I must add have been tapering off this month. The Germans are concentrating their bomber force on night raids. Sir Hugh and the group commanders have agreed to conduct several experimental operations to evaluate the so-called Big Wing tactic in realistic conditions. I'm no longer convinced we will face massed daylight raids, thus the applicability of Big Wing procedures has significantly diminished. I would like to report that we can take the Big Wing issue off the table."

"Archie also has some personnel changes to make amid all this debate," said Churchill.

"Yes. As the War Cabinet has already approved, this afternoon I will meet with Sir Cyril. He has elected to retire, so his retirement will take effect today."

"He will be awarded the Order of Merit, and we have submitted Sir Cyril to his Majesty for elevation to Knight Grand Cross. We have discussed other appropriate assignments for Sir Cyril."

"Who do you propose replace him?" asked Arthur Greenwood.

"Air Marshal Sir Charles Portal, Air Officer Commanding-in-Chief Bomber Command," answered Sinclair. "He will also be promoted to air chief marshal when he assumes his new position of Chief of the Air Staff at the simple office ceremony in my office this afternoon. At that time, Sir Charles will become a member of the Chiefs of Staff Committee and the Defense Committee."

"Archie, please inform the War Cabinet of your other proposed leadership changes," suggested Churchill.

"Yes, well, the Prime Minister and I have discussed other changes as a consequence of this whole Big Wing debate, and my belief we need new blood to lead Fighter Command into the next phase of the war."

"Surely you are not proposing to remove Sir Hugh," Attlee said with disdain.

"Yes, actually, I am."

"The man . . . singularly man," continued Clement, "led our fighter forces since their inception, and as history will mark, he successfully defended this country at our most vulnerable and threatened moment. You propose to sack the man at his moment of triumph."

"He has refused to embrace change," Sinclair responded with more timidity than he probably intended.

"Dear God, man, he is the literal father of the Air Defense of Great Britain schema, including the incorporation of radio direction finding, and our evolving and vital night intercept capability. You simply cannot be serious." Attlee turned to Churchill. "Do you support this change?" he asked Winston directly, eye-to-eye.

"Yes. I support Archie. It is his department to run, as he deems best. I must say I am an unabashed admirer of Sir Hugh, and for the record, I advised against removing Sir Hugh. However and conversely, I believe Archie is doing what must be done, sooner or later. We have survived the immediate onslaught."

"Survived," responded Attlee with anger now in his tone, "our cities are being bombed every night. Citizens are being killed every night. I dare say it may be a bit premature to claim we have survived the German abuse."

"I appreciate your anger, Clement, but the day will come in the not too distant future, I should think, when we will transition from the defense to the offense. Fighter aeroplanes are not just defensive tools. Sir Hugh has done an exemplary job defending us, keeping the Hun at bay, but we need to change our focus."

"Are you suggesting the threat of invasion has passed?" asked Arthur Greenwood.

"No. While there are positive signs, the threat of invasion still exists, and we must continue our preparations to thwart such an invasion attempt, if it should come when the weather improves in the spring."

"The King elevated him to Knight Grand Cross upon our recommendation," interjected Lord Halifax.

"He already had one . . . the Victorian Order, as I recall . . . so, he is supposed to feel better about his service, adding the Order of Bath? He deserves a peerage, and a high one at that, for what he has done in defense of the realm."

"He has become an impediment to change," Sinclair said with solemnity. "Sir Hugh has simply not put sufficient effort and attention into solving the night intercept problem. It is time for this change. As the Prime Minister noted, we need men in position who will help us take the fight to the enemy until victory is won."

"Did I hear incorrectly," interjected the Right Honorable Sir Howard Kingsley Wood, Kt, PC, Member of Parliament for Woolwich West, member of the Conservative Party, a former air minister himself and now Chancellor of the Exchequer, "aren't the night-interceptors achieving something like a 75 *per centum* success rate in recent weeks?"

"Yes, well," stammered Sir Archie, "that is true but only recently, and we have grossly insufficient numbers of night-fighters to deal with these damnable massed night raids."

"All very noble, Archie," Clement responded with a more subdued tone. "Who do you propose to replace him?"

"Deputy Chief of the Air Staff Air Vice-Marshal Douglas. Once approved by the War Cabinet, I propose he be promoted to air marshal. I also proposed moving Leigh-Mallory from One Two Group to replace Air Vice-Marshal Park at One One Group." Although it was not stated to the War Cabinet, Churchill accepted Sinclair's recommendation that Park would be ignominiously shuffled off to form a new training group in Gloucestershire, and eventually to supervise fighter defenses of Egypt and Malta.

The changes Air Minister Sinclair proposed were not popular among the political leaders of the War Cabinet. Arguments contained flare-ups of anger and disgust. But at the end of the day, the War Cabinet approved, even though notably and reluctantly so, the changes in leadership within the Royal Air Force and specifically Fighter Command.

"If you will indulge me one last question," said Attlee, "what if anything will you do with Sir Hugh, or is he to be retired to pasture like some used up stallion?"

Churchill did not wait for Sinclair to answer. "We all appreciate and share your loyalty to Sir Hugh, Clement. I have suggested and Archie has accepted that Sir Hugh take up the most urgent and useful Air Mission to the United States. He can use his experience to guide the Americans and perhaps even influence their air forces build-up and preparations for war. We shall do our best to ensure Sir Hugh appreciates the full accolades of His Majesty's Government."

Attlee snorted and waved his hand dismissively. He remained clearly unhappy with the changes.

The War Cabinet concluded their afternoon meeting and dispersed to their other tasks. Prime Minister Churchill departed earlier than usual for the prime ministerial country estate at Chequers. He knew his wife Clementine and most of his children, their spouses and children, would be waiting for him. He needed the embrace of his family after yet another difficult week of war.

—

Friday, 25. October. 1940
Air Ministry
Whitehall, London, England
16:00 hours

Secretary of State for Air Sir Archibald 'Archie' Sinclair waited for his next appointment, trying to address as much of the perpetual paperwork on his desk as he could. The subtle hourly chime from the shelf clock on the bookcase next to his desk marked the proper time.

The knock turned his attention to the door. "Yes."

His assistant cracked the door and announced, "Air Officer Commanding-in-Chief Air Marshal Sir Charles Portal has arrived as you requested."

"Please show the marshal in."

After the cordialities were dispensed with, Sinclair motioned toward the overstuffed leather chairs facing each other and with a small table between them. "I shall get directly to the point. This afternoon, the War Cabinet approved your appointment to be Chief of the Air Staff effective now."

"Thank you, sir."

"You are also promoted to air chief marshal also effective today. I apologize for the lack of ceremony, but these are volatile and troubled times. As we have previously discussed, Bomber Command must expand rapidly and punish the enemy in a relentless bombardment campaign. Fighter Command must become more effective in dealing with these damnable night raids of the Blitz, and transition from the defense to offensive operations against enemy forces on the Continent."

"Yes, sir. I understand precisely. I am ready."

Sinclair stood. Portal followed his lead. "Then, I shall leave you to your task. Good luck. I look forward to working with you to make His Majesty's Air Force the best in the world." Sinclair shook Portal's hand. "If you would be so kind to ask Sir Cyril to join me."

"Yes, sir," Sir Charles responded and departed.

A knock on the door jam preceded the appearance of Air Chief Marshal Sir Cyril Newall. "Good afternoon, Minister."

"If you say so, Sir Cyril. Please come in." Sinclair took two steps toward Newall. He did not motion to the chairs."

"So, this is it, then."

"I'm afraid so, Sir Cyril. Sir Charles has assumed the responsibilities of Chief of the Air Staff. You are officially relieved of those responsibilities. If I may be so bold to ask, what are your intentions?"

"I have not been given any options other than the obvious . . . retirement, I suppose."

"Very well, Sir Cyril. The administrator will process your retirement promptly. I apologize for the paucity of ceremony here, but this is wartime after all. On behalf of His Majesty's Government, we sincerely and deeply express our gratitude for your long and venerable service to the Crown. I will say, the Prime Minister has discussed further assignments outside and in uniform."

"I remain at His Majesty's service." An awkward pause swallowed any sound beyond the ticking of the shelf clock. "I would like to say I did my best to serve the air force and the government. I regret my counsel was not more useful. If I may be so bold, what is to happen to Sir Hugh and Keith Park?"

"We must allow Sir Charles to settle into his new assignment."

"Quite so."

"The final decision has not been taken by the War Cabinet or myself. We await the recommendation of the new Chief of the Air Staff. I will say I expect Sir Hugh to be relieved of his Command probably next month, and I suspect Air Vice-Marshal Park will be relieved the following month, but we shall see how this plays out."

"They are very good men, Sir Archibald. I pray they will be treated properly and with all due respect."

Sinclair nodded his head. "Thank you for that, Sir Cyril . . . loyal to the end. Please keep the Ministry informed of your whereabouts, in case further opportunity presents."

"Of course . . . as it should be. Now, I know you are a busy man. Unless you have further need of me, I shall request your leave."

"Certainly. Thank you, again, Sir Cyril." Without further adieu, Newall departed.

Sinclair nodded his head to himself with satisfaction, and then returned to his desk. The first of many changes in leadership had been completed without rancor.

—

Saturday, 26.October.1940
Shepherds Tavern
No.50 Hertford Street
Mayfair, London, England
19:30 hours

Flying Officer Jonathan Kensington arrived in London at King's Cross Station on the Great Northern Rail Line. He took the Underground Victoria Line train to Green Park Station, where he emerged past the black out curtains into the darkened, nighttime streets of London. The not-so-fresh air felt good, actually invigorating in the cool, dry, night air of the city. The particular odors of explosives, burnt wood and the nearly ever-present dust of brick, concrete and churned dirt unmistakably tinged the night air with the tokens of war.

The waning half moon offered sufficient illumination for street navigation despite the scattered clouds. Bomb damage to buildings and the rapid, temporary repairs to the roadways displayed the brutality of The Blitz. While not entirely clear, the rubble of destruction littering the Mayfair streets was surprisingly greater than he expected. Jonathan knew the air raid sirens might begin their blare at any moment. He quickened his pace to reach his girlfriend before the confusion of people seeking shelter made the process far more difficult, if not impossible.

His walk down Piccadilly, up Shepherd Street, to the fighter pilots' preferred pub in London was harder than the same transit prior to the beginning of The Blitz. Yet, people still moved with stoic purpose.

Jonathan spotted the golden, shoulder length hair that had to be Linda Mason waiting patiently outside the pub. She turned her head, recognized Jonathan, and then leapt into his arms. They kissed passionately. Linda was tall for a woman and nearly Jonathan's height.

"Why are you out here in the chill?" Jonathan asked.

"Too much smoke and too many hands."

Jonathan smiled at the audacity of fighter pilots. "Do you want to go in, or go somewhere else?"

"It's OK, Jonathan. I know these men are important to you. The crowd is not as great as before The Blitz began. I think we can find a small table. We have no way to know when the sirens will announce the approach of nightly bombs, so I would suggest we get some pub food, a drink or two, and you can see some of your friends."

Jonathan smiled at Linda's acceptance. "That should work quite well, since we may well spend the night in an Underground station."

"Before we go into the noise," said Linda, "I haven't seen you in several months. Anne Booth and Virginia North have disappeared. I was questioned a month ago by some Scotland Yard detective-inspectors and some others, who I suspect were with the Security Service. They asked me some very intimate questions about you, about us, and about Anne, Virginia and Brian. What is going on?"

"I learned from Brian that Anne and Virginia were arrested and tried for spying for Germany. According to Brian, both of them were executed in August."

Linda inhaled sharply, covered her lower face with her hands, and said, "Dear God above! I had no idea. They were such nice women."

"Apparently not."

"Jonathan," she protested.

"They were spying for Germany, Linda. Brian doesn't know more, so we don't know why or exactly what they did, but it had to be very serious for the government to hang two young women."

"Speaking of Brian, I saw in the casualty lists that he was injured in combat."

"Yes, fairly serious I must say, but he appears to be recovering well. He is recuperating with the woman who saved his life last July."

"So, he is OK?"

"Yes, he remains in Hampshire, but we expect him to rejoin the squadron next month."

"Good. I worry so much about you boys up there against the Germans."

"The situation has improved substantially from a month ago. Daylight raids have nearly ceased. Fighter Command moved our squadron and others north to rest and refit. None of that happened a month ago. These nightly bombing events are troubling, but I am confident we shall gain the upper hand soon."

"As you say then," Linda responded. "Let's see if we can get something to eat and drink. I'm famished and thirsty."

Jonathan laughed heartily. "That's my girl."

The interior of the pub was as Linda described it, although the sea of Royal Air Force blue uniforms was not as thick as he had seen it before the Germans began attacking Great Britain. Virtually all of the pilots were standing around the bar. They found an empty table.

"What are your wishes, my lovely Linda?"

"How about a pint of bitter?"

"Two, then. I shall return with our libation."

Jonathan greeted several fighter pilots he knew on his way to the bar. He waited his turn, ordered the beers, and returned to the table. Linda was reading the table menu.

"Here we go." He handed the tall glass to Linda, sat next to her on the bench seat, and then held his glass up to her. "To the most beautiful woman in London."

"Not England?"

"In all of England."

"You are just smitten, young man. You will say anything to get in my knickers."

They both laughed. "Is it working?"

"You will find out soon enough, my darling Jonathan."

Flying Officer Kensington smiled broadly. They clinked their glasses and took their first drink.

"'Harness,' you wily bastard."

Jonathan recognized the voice. He looked up to see Squadron Leader Lord Jeremy Robert Kenneth 'Mud' Morrison, Esq., approaching. "Good evening to you, 'Mud.' Congratulations on your promotion," Jonathan said, motioning to his new sleeve stripes. Morrison had a half empty pint in his hand.

Morrison ignored Jonathan's salutation. "I've not had the pleasure," 'Mud' said, staring intently at Linda. He extended his right hand to Linda, who in turn placed her right hand in his. He leaned over and kissed the back of her right hand. Jeremy did not release her hand.

"Miss Mason, may I introduce Squadron Leader Lord Jeremy 'Mud' Morrison. 'Mud,' this is my lady friend, Miss Linda Mason."

Without breaking his gaze, 'Mud' said, "You know better than to do that, 'Harness.'"

"My apologies. I do like to be proper." Jonathan turned to Linda. "'Mud' does not like us to refer to the fact that he is the younger brother of the 8th Duke of Cottingstone."

"A honor to meet you . . . ," Linda paused, as Jeremy kissed her hand delicately, again, "how am I to refer to you, your Lordship?"

Morrison looked up, but did not move far from her hand. "Everyone is entitled to one *faux pas*, Miss Mason. I prefer to ignore my birthright, if you would be so kind. Jeremy, or 'Mud' if you prefer, is quite adequate." He kissed her hand again, before he finally released her. "May I join you?"

Jonathan quickly looked to Linda, who nodded her assent. "By all means, please do."

Morrison sat across the table from them. "So," he said, "if my information is correct, I believe you were friends with my precious 'Virgin.'"

Jonathan noticed Linda's expression of puzzlement. "His familiar reference to Virginia North."

Linda smiled and nodded her head. "Yes, I met her a number of times. I was school chums with Anne Booth, so I knew Virginia through my friendship with Anne."

"Not intending to be indiscreet . . . are you in the profession, as well?"

"'Mud,' damn, that is grossly inappropriate," protested Jonathan.

"No offense intended, my dear," Morrison said to Jonathan, and then to Linda.

"None taken," answered Linda. "No, I am not and never have been in the profession, as you call it. It is terrible what happened to them."

"Yes, my dear, quite terrible. But, what they did was far worse, if I may be so frank."

"What did they do?" asked Linda, innocently, intending to engage Jeremy.

"Please allow me to be blunt," 'Mud' said, and waited for Linda's consenting head nod. "They were accomplished prostitutes. They preferred the French term *courtesan*. They apparently used their womanly wiles to engage members of Parliament and the House of Lords, as well as generals of all three branches of the armed services, to gain information they passed along to their German handlers."

"Why?"

"Who knows? I never got to talk to either 'Virgin' or Anne, either before or after their arrest. Truly, I wish I knew."

"They paid such a dreadful price."

"Indeed, they did," Jeremy responded. He lapsed into contemplation. No one spoke. Morrison took a long draw from his beer. "Changing the subject not so discreetly, how is our young magician 'Hunter'?"

"I saw him two weeks ago. Rosemary saw him after I did. He is under the loving care of Charlotte and recovering quite well, as I understand it."

"Will he come back?"

"Charlotte is not making it easy for him. I am told the doctors will likely return him to flight status early next month. It is unclear whether he will return."

"He is a volunteer, after all."

"Indeed . . . as we all are."

"But, he is an American volunteer. We are English. There is a dif- ference."

"Not to Brian."

"Quite so. That is why we love him so much."

The silence between them amid the cacophony of a raucous public house not quite filled with devil-may-care fighter pilots left the two men consumed by their thoughts.

Linda broke the poignant lull. "I am now verging on ravenous."

"Where are your manners, lad? Let's get the lady fed."

'Mud' took their choices and went to the bar to place their orders. Jeremy returned. "Our meals will be out in 10 minutes."

"Congratulations on your assumption of command."

"Thank you, 'Harness.' I am gobsmacked the Air Ministry would entrust such an awesome responsibility to a malcontent and tainted officer like me."

"Three Two Squadron."

"Precisely. Hurricanes rather than Spitfires . . . and they moved us to Acklington for recovery and refit."

"Like us. Six Oh Nine moved to Catterick for the same reason."

The waiter arrived with their requested meals. They ate with social chitchat that filled the space between mastication of their bites. Fortunately, for their sustenance, the nightly air raid sirens waited for them to finish their meals. They swallowed the remainder of their beers and followed the sea of blue to the Hyde Park Corner Underground Station rather than the Green Park Station. The flashes in the night sky announced this night's rendition of The Blitz before the sounds of detonating explosive came to them. The nightly fireworks moved closer, as they descended the first set to steps into the protection of the Underground caverns. Two levels below the street, they found a spot to sit, leaning against the tiled wall. Not exactly the night Jonathan had in mind, but fate was like that. Soon the thumps of aerial bombs exploding not far above them, punctuated by clouds of dust stirred up by the German attack. Jonathan wondered, *where are the night-interceptors to stop those bastards?* He leaned over and kissed Linda, and winked at her. *This will not last forever.*

—

Chapter 12

Virtue is harder to be got than a knowledge of the world;
And, if lost in a young man, is seldom recovered.

--John Locke

Sunday, 27. October. 1940
RAF Catterick
Catterick, North Yorkshire, England
07:45 hours

Week 17

The lightening sky to the east marked the beginning of another day. The pilots of No.609 Squadron, less Pilot Officer Brian Drummond, silently gathered in the Dispersal Hut for their duty day. At the direction of Squadron Leader Long-Roberts, Corporal Warren notified Group the squadron was at Available status, even before the sun broke the horizon. An overcast cloud layer covered the sky with just a narrow band of clearing at the eastern horizon. If the weather guessers were correct, they would only get a brief glimpse of the sun in the gap, and then it would be another dreary day. The forecast indicated just cloud cover, no rough weather or rain. They would not get their first glimpse of the sun for another hour, if the gap in the cloud deck lasted that long.

"I don't think I will ever get used to this Double Summer Time non-sense," announced Pilot Officer Noring.

"Well, used to it or not, it is what it is," rebutted 'Waggle' Davies. "Whining about it does not make it better or make it go away."

"The days are getting shorter as it is. We start before the sun rises and we finish after the sun sets."

"So, what's your point?"

"No point . . . just an observation."

"I think we can all note the time as well as the astronomical and me-teorological conditions, Noring. So, unless you have something constructive to say, why don't you put a cork in it."

Silence returned . . . at least for a few minutes, until the telephone rang.

"Group for you, Skipper," announced Corporal Warren. Once she heard 'Stack' on the line, she returned the handset to the cradle. Muffled words from the squadron leader's office were unintelligible. Beyond that fact, the conversation was predominately one-way. Ten minutes later, 'Stack' appeared, seeking attention without a word.

"Seems Gerry decided to make an early morning of it. Fighter Com-mand has a raid forming over northeast France. Headquarters expects to deploy a Big Wing."

"So, 'Tin Legs' won," 'Sparky' Morrow proclaimed.

"I suggest we drop the politics. We will have a mission to perform in short order. We will be one of four Spitfire squadrons in the second wave. The first wave will be six squadrons – nearly 70 Spitfires."

"Well, that should be impressive enough," added 'Waggle' Davies.

"Indeed. We can expect to go directly to Standby in 15 to 30 minutes, and launch soon after that. Our wing will form north of London. Our station will be north of the Thames."

"We're going to use a lot of petrol just to get there."

"Yes, and we have been temporarily consigned to Twelve Group. We can expect to fly a full duration sortie and to land at Northolt or Duxford to refuel . . . and rearm, if necessary. Group expects this to be the first of several fighter-bomber sweeps today. We may be down there most of the day, if this plays out the way Group expects. For the last week or so, Gerry has been confining their daylight operations to high altitude fighter-bomber sweeps to lure us into a fight where their supercharger performance and fuel injection give the One Oh Nine an advantage. Today, we are going to give them a fight. So, once we go to Standby, I want each of you . . ."

The telephone rang. Corporal Warren listened, and then announced, "Squadron to Standby."

"I want each of you to check your masks and oxygen equipment carefully. If you do not have a full bottle, you do not go. OK, lads, this is what we have been preparing for. Let's do it right. Now, mount up."

Jonathan gathered up his flight kit, as his brethren did, and walked to their machines. There were no words of bravado. In fact, there were no words whatsoever passed among the pilots. Morning twilight brightened, but the sun had still not cracked the horizon. This is going to be an interesting . . . and long day, Jonathan told himself.

Each pilot strapped into and connected with his airplane, like he was putting on and zipping up a flight jacket. Jonathan made sure all his switches were properly positioned for start. They would wait until the launch command arrived, to save as much fuel as possible. Jonathan attached his oxygen mask, checked to make sure it was fully functional and his oxygen bottle was full, and then switched it off and disconnected the right side of his mask from his leather helmet. Jonathan stared off to the east and the approaching dawn. Minutes passed. As the first sliver of dawn arrived, the launch bell rang and 'Stack' signaled for start.

The 11 fighters of the nearly full squadron took to the air, headed south, and entered a slow climb at reduced power, again to save fuel. They reached their initial assigned altitude of 20,000 feet. 'Stack' kept his throttle back and allowed their speed to build slowly. The other three squadrons of their wing orbited where they were expected and were readily located. They could see the mixed and twisting condensation trails to the south of them, which had to be the first wing engaging the Germans. The join-up was simple and direct. No.609 Squadron adjusted their formation with each section in trail and step-down, and then took their assigned position at the bottom on a diamond formation of the four squadrons. Jonathan liked the formation. He could see most the wing aircraft ahead and slightly above him. The mass of fighter airplanes was certainly impressive, if nothing else.

The radio communications remained confusing. The first wave appeared to be having some success, but enough of the enemy aircraft got through. The second wing wheeled toward the south and began another climb to 30,000 feet. Until the large formation stabilized on their closure heading, the maintenance of formation position proved more consuming and difficult than expected. Their condensation trails . . . they called them con-trails . . . began at 24,500 feet on this day. The water vapor in the engine exhaust condensed in the cold air of high altitude and clearly marked the location, direction and speed of the source, in this case, 44 Supermarine Spitfire Mark IA fighters in a broad, staggered, diamond formation.

From the twisted ball of white streaks south of the city emerged perhaps a couple of dozen straight con-trails . . . the remaining Germans pressed their attack. They had to know the second wing of fresh Spitfires was headed toward them. The two groups closed to the point that Jonathan began to consider how this massed engagement was going to play out, and then, on command, every German fighter rolled inverted and dove for speed and separation. No.609 Squadron watched the Germans quickly dive through their lower altitude on the formation and ahead of them. Instinct gave Jonathan the urge to push the stick forward slightly, lower the nose and give chase. Yet, not one British fighter altered its flight path. They droned on, as if the Germans might miraculously reappear ahead of them. Through the now broken clouds below them, Jonathan could see they were over South London.

Jonathan glanced down and noted his fuel gauge was less than a quarter remaining in his tank. They were going to need to break it off soon and find a place to land for fuel . . . and oxygen replenishment. As he waited for the bingo signal to terminate this phase of their mission, Jonathan noticed the cold at these altitudes for the first time. He never got used to it. The cold was just part of the mission. Fortunately, the condensate from his expelled breath did not produce much ice on the inside of his canopy. A few shivers punctuated his idle thoughts.

"Wing Two, this is Wing Two Leader. Break up. Return to base."

"Sorbo, roger. Sorbo flight, switch to channel five."

Jonathan changed his radio frequency from channel two to channel five. Each member of the squadron checked in."

"Sorbo, throttle back." Jonathan immediately began to slowly pull his throttle back while he watched his spacing on 'Stack's Blue Section and checked to make sure Clegg was holding his position off his right wing. "Sorbo, left turn." They followed their leader. "We will recover at Northolt."

'Stack' had throttled way back to save what fuel he could for the squadron. They steadied up on a northerly heading and a nearly gliding descent. Squadron Leader Long-Roberts picked his way through the broken clouds. They were below the clouds at 2,000 feet. Jonathan picked up the features that marked the location of RAF Northolt airfield. They were lined up well for a straight in approach. 'Stack' made the radio calls and received clearance to land for the squadron.

After they were all safely on the ground, they were directed to a transit area for aircraft, where they parked and shutdown. Ground crewmen with fuel tanker-trailers and ammunition carts waited for them to begin their work of preparing the fighters as quickly as possible for flight, if necessary. An aircraftman directed the pilots to a large field tent nearby. With the aid of the local aircraftman, 'Stack' telephoned No.12 Group to check in and receive further instructions.

"Well, that hardly seemed worth the effort," said 'Waggle.'

"Quite so," 'Sparky' added.

"Perhaps so, but we accomplished our mission," 'Harness' said.

"If our mission was chasing the bad chaps away, then yes, we did; but, we did not fire a shot and we did not kill any Germans," 'Waggle' responded.

"It was impressive, though," noted Clegg."

"Aye, that it was," said 'Waggle.'

'Stack' returned to the tent. "We have been put back to 30 Minutes, so we will go the Mess for an early lunch, and back here at Available by half eleven. Group expects another fighter sweep this afternoon, so we will stage here. Since Wing One engaged this morning, we will most likely take the lead as part of Wing Two, if another raid comes. Wing One will then be the second echelon, as we were this morning. Any questions?" There were none. "A lorry should be waiting outside to take us to the Mess for lunch."

Each of the pilots found a peg to hang their flight equipment on, and then silently filed outside and into the covered back end of the waiting truck. Lunch was simple sandwiches and tea. Their banter during lunch found other topics than the impressive but uneventful morning mission.

When they returned to their assigned Dispersal tent, 'Stack' reported the squadron at Available, and received no additional information or instructions. They waited. Hours passed. They napped, joked and tried to not think about what might happen at any moment. The afternoon German raid did not materialize as expected. The squadron was released at 17:30 hours for their return to RAF Catterick and No.13 Group. They took off by sections and joined up during their climb out.

Just north of London, the clouds joined to a solid overcast layer. The cloud deck remained solidly overcast but thinner than experience in the morning. Sector Control gave them vectors for their cloud penetration. They broke out at about 1,500 feet over the North York Moors, west of Scarborourgh. 'Stack' turned them west. They landed before sunset at RAF Catterick.

The pilots quickly debriefed the day's mission with Flying Officer Royster and his intelligence aircraftmen. Another day's work done.

—

Monday, 28.October.1940
RAF Middle Wallop
Middle Wallop, Hampshire, England
10:00 hours

Charlotte was a careful driver. The journey north had been uneventful. The half dozen checkpoints had become as much a part of their routine of life as the sunrise . . . well, other than armed guards with serious expressions, and attitudes of distrust and suspicion. With all the stops, the trip took a little longer than the usual wartime slowness.

Brian had not seen the familiar environs of RAF Middle Wallop in several weeks. He liked his memories of flying from the air base. A Hurricane squadron occupied the southwest corner, where his squadron had staged from, and they drove past their old staging area.

As the hedgerow shrubbery thinned on the east edge of the roadway, he noticed a section of three Spitfires taxiing for take off. As they approached an opening in the hedgerow, Brian said, "Stop the car and switch off the motor."

Charlotte stopped. "We are blocking the carriageway."

"There is no traffic. Anyone who comes along can go around us. I want you to see and hear this."

"What?"

"Just switch off the motor." Charlotte did as she was requested. She pulled over as far as she could to the shoulder of the road before switching off the engine. "See those three Spitfires in front of the hangars."

"Yes."

"They will take off shortly. I want you to watch and listen as they takeoff. Let's get out of the car." Brian did not wait for Charlotte's answer.

He had to contend with the foliage close to his door. Even with his crutches, Brian made it to her side, before she stepped out and closed the door behind her.

"This does not seem like a good idea," she said.

"This won't take long. Now, they're just about ready. Watch and listen."

Brian stood next to Charlotte. He leaned on his crutches and held her hand as they both watched. They heard the engines open up, as all three aircraft began their takeoff roll in front of them. Halfway through their ground roll, the engines advanced to full takeoff power. All three aircraft were airborne before they reached Charlotte's and Brian's position. The undercarriages retracted into the wings. They stayed low to build speed rapidly. They disappeared behind the hedgerow shrubbery, but the throaty, melodic sounds of the three Merlin engines at full power remained with them until they saw the three Spitfires begin their climb and fade into the distance to the south.

Brian leaned toward Charlotte and kissed her. "Now, wasn't that the best sound ever?"

"Yes, Brian, very impressive. They are gorgeous machines."

"Yes, they are . . . the best."

"We need to move along. You are already late for your appointment."

"Yes, yes, quite so." Brian opened the driver's door, waited for her to enter and situate herself, closed the door, and then made his way through the foliage and eventually into his seat next to Charlotte.

Charlotte drove up to the armed sentries at the main gate. Brian presented his credentials to the Corporal of the Guard.

The corporal smiled, returned his identification card and saluted smartly. "Great to see you again, Mister Drummond."

"Thank you, corporal. Likewise," Brian responded, returned the salute, and nodded his head for Charlotte to proceed.

Brian directed Charlotte to the Medical Services building. They found a parking spot and went into the dispensary together. Brian did not have to wait and was ushered to an examination room. Charlotte waited in the lobby. A female nurse joined Brian to take down his vital signs and took them to the doctor. An RAF flight lieutenant doctor entered.

"Well, Mister Drummond, how are you going?"

"Fairly well, I would say. I'm ready to get this impediment removed and get back to flight status."

"I understand you are quite an accomplished Spitfire pilot."

"I like to fly and I really want to get back to flying."

"Let us have a look at you, then."

The doctor carried out a thorough, complete, medical examination from top to bottom, including x-ray images of his ribs and leg. The whole process took just over an hour.

"We have examined everything we need. Your recovery is nearly complete. Your ribs are healed. Your burns are out of the vulnerable phase and will take several more months to achieve full recovery, but they look pretty good."

"What about this?" Brian said, rapping his knuckles on his leg cast.

"The best medical counsel says you have another two weeks to go."

"What if I ask you to remove it now, so I can get back to flying?"

The doctor chuckled, more to himself. "You are not the first pilot to display impatience with their medical recuperation. That said, first, I cannot justify that action. Second, I will not be able to certify your medical readiness for flight duty. Third, a tibia fracture like yours generally takes six weeks to heal sufficiently to be satisfactorily healed to be load bearing. Letting you go too early risks re-injury and actually risks more serious injury."

"That is very disappointing, sir."

"Certainly. Our tasks as flight surgeons are to ensure your readiness for flight and to protect you medically."

"Thank you, sir."

The doctor consulted the papers attached in his folder. "I'll tell you what . . . you broke your leg on the 27th of September. It has been four weeks. You are a young, healthy man. Since you have been a good sport about this, I will compromise with you. I will see you back here in one week – the five-week mark. We will remove the cast, check your leg, and if you have sufficiently healed, I should at least be able to return you to limited duty, if not full flight status."

"Very well, thank you, sir. So it shall be."

Brian left the examination room and joined Charlotte in the lobby. Their eyes connected immediately. Brian shook his head that he had not been successful, but that reality was obvious with his crutches and leg cast still in place.

"I'm sorry, Brian," she said. "I know that was important to you. Let's get you back to the farm. You can take a nap to speed your recovery."

Brian nodded his head in resignation. He considered asking Charlotte to drive the flight line, so he could get closer to the aircraft, but he decided that would not be fair to Charlotte.

During the drive back, Brian recounted his examination and the flight surgeon's findings. Brian sensed that Charlotte was gradually warming to the approaching clearance for flight and return to duty. Yet, he did not feel the time had arrived to broach the topic, again. The journey back to the farm was virtually the same as their inbound trip, except in reverse. By the time they saw the familiar farmhouse, Brian acknowledged his readiness for a nap. *Maybe my recovery is not as advanced as I thought.*

—

Monday, 28.October.1940
Standing Oak Farm
Winchester, Hampshire, England
16:10 hours

Brian awoke from a longer than usual nap. He sat on the edge of the bed for several minutes, allowing his temps and pressures to adjust. He rose onto his crutches, went into the kitchen where Charlotte was baking something.

"Did you enjoy a refreshing nap?" Charlotte asked.

"Yes. Thank you."

"Would you care for a snack of cheese and crackers to tide you over until supper?"

"Yes, please."

Charlotte prepared a plate of her homemade cheese and baked, unleavened crackers as well as a pot of tea for both of them. She placed the service items in the center of the kitchen table, served a portion for both of them, and sat down across the table from Brian.

"I received a telephone call while you were napping," said Charlotte.

"OK?"

"It was that journalist who talked to us late last August after the award ceremony."

"Murrow?"

"Yes, Edward Murrow."

"What did he want?"

"He asked to visit you . . . us . . . out here. He wants to do a more in-depth interview for a feature report . . . that's what he called it."

"What did you tell him?"

"I told him he would be most welcome."

"I wish you hadn't."

"Really?"

"I do not want anymore attention than I have already attracted." Her confused expression told him he needed to explain more. "Charlotte, I am an American, a young American, who defied my parents and federal law to come over here and fly for the Royal Air Force. I just cannot see anything good coming from making myself publicly more visible."

"I am sorry, Brian. I simply thought some recognition for your accomplishments would be appreciated. Should I call him and cancel?"

Brian considered her question. *He seemed like a nice enough fellow. Maybe it wouldn't hurt. After all, I do have that presidential pardon letter.* "When is he supposed to be here?"

"We agreed on day after tomorrow."

Brian shrugged his shoulders. "Well, what else do we have to do? No, let's let it stand."

"Are you sure? We have no obligation, Brian."

"Yes. It should be OK."

They sat in silence for several minutes, enjoying their cheese, crackers and tea. Charlotte was the first to break the silence.

"You are going to go back to your squadron," she said with solemnity.

"Charlotte, we don't need to talk about this now."

"Yes, Brian, yes, we do. If we are going to talk to this Murrow fellow, this needs to be settled. I do not want you appearing all wobbly when he inevitably asks the question."

"What question?"

"You know perfectly well what question," she said. Charlotte held Brian's eyes. "When do you return to flying? Or, what are you going to do next?"

Brian lowered his eyes and head. "I do not want to hurt you or cause you any pain, Charlotte."

"I know, Brian. I also know this issue is not going away between us. Further, I refuse to make this a me-or-that ultimatum." This time, Charlotte lowered her eyes and head. "I have no idea where our relationship is headed, but I do know I have had my heart broken too many times. I do not want to find out how much is too much for me to bear."

"I understand that as clearly as day."

"But . . ."

"I wish I could find a way to let you enjoy the thrill of flight."

"Brian, I am certain flying is a very enjoyable endeavor, but getting shot at and risking death cannot be even remotely enjoyable."

"The Prime Minister once said, 'There is nothing more exhilarating than to be shot at without result.'"

Anger filled her eyes. "Oh, so now you are going to quote Churchill to me."

Well, that was not a good decision. "I'm sorry for that."

Charlotte stared intently into Brian's eyes, and then shook her head. "You can make light of the serious risks you take every day. I shall respectfully ask you not to ever do that again with me, or where I might hear it. I do not need your bravado, or Churchill's for that matter, in my life. I have suffered enough loss. Against my better, sober judgment, I have allowed my love for you to grow far beyond where I should have allowed it, and I am carrying our child, my first child – a product of the love I think I share with you."

"Again, my apologies for my ill-chosen words, and I do agree."

"I have nursed you back, or at least nearly back, to full health. I shall now say, I want you to go . . . go back to your squadron and your flying. I will birth our child, and I will nurture and raise our child without you. I shall endeavor to retain my love for you, but I can make no promises, as neither of us knows how long this war shall go on or what challenges lay ahead of us. So, I want you to put my misgivings and fears aside, out of your mind, and concentrate on doing the best you can at flying your machines and staying alive."

"I can do that."

"Now, Brian, listen carefully . . . in case you have not been listening up until now." Brian concentrated on Charlotte's eyes and nodded his head slowly in agreement. "I must protect my emotional stability through the remainder of this war . . . however long that may be. While you are doing what you want to do, I must put you out of my mind, out of my thoughts, and pretend we never were, and you will not return. I will raise our child with love and the memory of you. If I need help, I have the means to gain that assistance, so you should have no worries. When this war is over, and I do believe it will be done some day, we can talk about making our relationship more permanent, if that is what we both wish when that time comes."

"Charlotte . . ." Brian said and stopped when she raised her hand and furrowed her brow.

"No, Brian. There is no debate, no discussion, no negotiation. You are going to do what you must do, and I must do what I must do to maintain my sanity and productivity as a consequence."

"Are you saying you do not want me to see you when I am able?"

"No, that is not what I am saying. I will appreciate and cherish every moment we are able to share through this tragedy around us. I am only telling you, I will not be writing to you, or pining over you, or what you meant to me. I must treat every segment of your absence as if you no longer exist."

"That sounds rather harsh. I will be thinking of you every moment I am with you . . . or without you."

"No!" she shouted uncharacteristically in protest. "You cannot, you must not, be thinking of me. You must focus every fiber of your being on victory and surviving this war, so that you can come home to me."

"This is really how you want it?"

"No, Brian, this is not how I want it, but it is how it must be. I truly do appreciate your passion for flight, even though I do not share that passion. I must respect that passion, that loyalty, that commitment to keeping us safe and defeating the oppressors. All I ask is that you recognize that same patriotism took my father, my uncle and my first husband. They had comparable commitments to the cause."

Brian smiled at Charlotte. She furrowed her brow again and cocked her head to the side. Brian pressed on. "First?"

"What?"

"You said your first husband."

"Brian, please do not make light of my grief."

"I am not, Charlotte. First implies there will be a second. Does that mean we will eventually get married?"

Charlotte stared at Brian, and a smile grew ever so slowly. "That is a rather obtuse manner of pressing my acceptance of a fortnight ago."

"You were very plain and direct, Charlotte. I intend to respect your position, your wishes, but rest assured, it is my intention to marry you as soon as you will stand before a magistrate or justice of the peace. I love you, Charlotte. I love you as I have loved no other person in my life. I will wait patiently until we are finally and officially joined, once I have satisfied your conditions."

"It is comforting to a degree that you feel that way."

"War time is not easy for any of us. I appreciate your candor, and thank you for not throwing me out."

Charlotte actually laughed. The laughter felt good for both of them. "You will be welcome here anytime you are able to be here. At least, I have a few more weeks. I will enjoy what you are able to give."

"You are an amazing and generous woman."

"I am glad you think so. Now, milking time is here."

They cleared the table and stored the storable items. Charlotte walked as Brian hobbled beside her to the barn. Lionel and Horace were already in the barn for the afternoon milking. Brian was actually getting to know the cows, their characteristics and traits, and enjoying the milking chore. Standing Oak Farm was also growing on him. Brian felt his relationship with Charlotte was a fully-grown tree with deep roots . . . in only the three months he had known her.

—

Tuesday, 29. October. 1940
Cabinet War Rooms
New Public Offices
Whitehall, London, England
10:15 hours

"I will not attempt to sleep another night on that damnable cot or in that oppressive room," declared Prime Minister Churchill to his on-duty Assistant Private Secretary 'Jock' Colville. "The concussions from last night's proximate bombs sent cascades of dust down on me. I want the Annex livable . . . *rapidement – plus rapidement.* That is my only concession to safety."

Construction of the Cabinet War Rooms had begun in the days when war clouds darkened the scene before the flames flared into full conflagration. The basement ceiling of the New Public Offices building had been heavily reinforced with steel. Additional construction began just the previous week to install a five foot thick, reinforced concrete slab in the overhead and expand the office space available to ministers, clerks, communications technicians and the other essential operatives of His Majesty's Government. They also recognized that No.10 Downing Street along with the Whitehall government buildings had become a target for the German aerial bombardment. His Majesty's Government directed the modification of the ground floor of the New Public Offices building as living quarters for the prime minister and his wife. The makeshift apartment had a living room, a dining room and kitchen, and a bedroom and bathroom. The facility would be known as No.10 Annex or just the Annex. After German bombs seriously damaged No.10, two weeks prior, Winston and Clementine began using No.10 Annex, although neither of them liked the accommodations.

"These facilities are intended for your protection and the security of the government."

"I know that," protested Churchill, "but, I refuse to allow the Germans to make me live like a rat cowering in these catacombs. This subterranean cavern with its incessant light is marginally rudimentary for the nobility of His Majesty's affairs. We can conduct the government's business down here, but I will not live here."

"I will see to it, sir."

"Where is the 'Buff Box'?"

"It has not arrived, as yet."

"Then, where are the Dispatch Box and the incessant flood of paperwork? We are late."

"Sir, Mister Martin is completing his review with the staff. Your first appointment is not until the cabinet meeting at thirteen hours."

"John had a difficult day yesterday."

John Miller Martin had been the private secretary to the prime minister since early in the year, before Churchill had even become prime minister. The private secretarial office handled all the administrative functions for the prime minister from travel arrangements and record keeping to the incessant daily paperwork that passed through the office. Five civil service men served as private secretaries, to Churchill. A small army of clerks, stenographers, typists, runners and support personnel supported the private secretariat office and had one mission – fulfill the administrative needs of the prime minister.

"Please allow me to go check on our progress."

"By all means, 'Jock.'"

"Is there anything I can get for you?"

"No," Winston said and waved his hand dismissively. Churchill turned his attention to yesterday's shipping losses report. Every daily report since June confirmed and validated what the *Kriegsmarine unterseebootmänner* called *Die Glückliche Zeit* – The Happy Time. The Germans continued to sink significantly more merchant shipping than the British were able to replace. The numbers scared him, yet he endeavored to remain optimistic with those around him or listening to his words. The American 'surplus' destroyers would help, once they had been refitted with the minimum acceptable level of anti-submarine equipment. The nation was a long way from out of a very dark forest. The knock on his door broke his concentration.

Martin and Colville entered, followed by the first wave of two stenographers.

"No, no," objected Churchill. "You must leave, now, John," he said to Martin. "You have been on duty too long, especially after yesterday."

"I am quite alright, Prime Minister."

"That may well be. I do not want you here. You will leave, go home, and get some rest. 'Jock' can handle your box."

"As you command, sir. Thank you for your understanding." Martin handed the sorted expand-a-file inside the carrying case to Colville.

"Nonsense, John. You've done yeoman's work. Now, shoo, shoo. Your day is done for now." Churchill turned to Colville without waiting for Martin's departure. "Let's get to it. Time and paper wait for no man."

They went through three relays of stenographers and comparatively quickly worked their way through the daily movement of paper in the form of telegrams, messages, letters, reports and even personal, social letters.

"That completes the daily file, sir."

"Excellent . . . with time to spare today," Churchill said with a more subdued tone. "Now, if you would be so kind, please see if the 'Buff Box' has arrived."

Colville ushered out the support staff and returned to announce, "Sir, Head of SIS Operations Acton has arrived with the 'Buff Box.'"

"Please show him in."

Acton entered the Prime Minister's underground room, carrying the 'Buff Box' manacled to his left wrist and containing the daily yield of sensitive intelligence material. "Good afternoon, Prime Minister. My apologies for our tardiness. 'C' asked me to attend you personally."

"Good afternoon to you, Mister Acton. Pray tell me, what have you?"

Acton placed the case on the Prime Minister's desk. Neither of them held the blue key to remove the manacled case. Winston retrieved his red key on a chain around his neck – the only such key outside SISHQ and GC&CS – and unlocked the case itself.

"Before we start into the box, Prime Minister, 'C' wanted me to inform you that we had a setback yesterday at Station 'X'" – the code name for the Government Code and Cypher School (GC&CS) at Bletchley Park, the code breaking service of His Majesty's Government. "The Germans apparently changed their Enigma wheels or the wheel sequence. We went blind at midnight yesterday. Thus, the box is rather light on material today, which is also why we were late getting the 'Buff Box' to you."

"That is not good news. I was reading yesterday's shipping loss report from the Admiralty before you arrived. It is not good, either. Pray tell, what is your estimate for cracking this latest hiccup?"

"We expected, or rather hoped I should say, to have the changes broken down this morning, but we were not successful. Based on past experience with changes like this, it could be a week or so before we start yielding results and that is under the assumption they continue to change their daily initialization settings at midnight Berlin time."

"That is definitely not good news. The Navy needs all the help they can get."

"We understand that reality quite well, sir."

"Please let me know immediately when you have been successful."

"Yes, sir. If 'C' or myself cannot deliver the news, we will have the Foreign Office pass one sentence to you through your secretariat office with the word 'lightning' in the simple sentence."

"Very well. Good luck."

"Thank you, sir. The lads at Station 'X' are working around the clock without much rest to resolve the problem."

"Certainly." Churchill opened the cover of the 'Buff Box' and removed one folder of aerial photographs.

"We have selected a series of photographs of specific harbors on the northeastern coast of France as well as Belgian and Dutch ports of significance. Each sheet is a comparative with yesterday's image on the right and the equivalent photograph from one month ago on the left for direct comparison."

"Most impressive, I must say."

"We have provided a representative sample of a dozen staging sites for the German Operation SEALION preparations. All of the available aerial photography is the same. Where we have been able to obtain field information, it has confirmed and validated the photography. The Germans are dispersing their invasion fleet back into the Rhine and ports east of Amsterdam. One of the last decrypts before ULTRA went dark was an OKW message to the services instructing them to continue SEALION training according to plan. Training would be normal, especially since the Army has never attempted a major amphibious operation, as a cross-Channel movement would be. With today's photographic evidence, we are lowering our threat assessment from imminent to serious. They have made no move that would suggest SEALION has been cancelled . . . at best for us, an invasion attempt is postponed until mid to late spring. The Combined Intelligence Committee met two hours ago and unanimously concurs with this estimate."

"Very well . . . some modicum of good news for a change." Churchill placed the photography folder back in the 'Buff Box,' and then looked through the other compartments and found nothing. "So that is it? Rather sparse today."

"Yes, sir, for physical items. There are two additional items I am to brief you." Churchill nodded his head, as he closed and locked the cover to the 'Buff Box.' "First, we have information from several Greek sources and confirmed by two of our field agents in the region that the Italians crossed the frontier at dawn this morning from Albania with a significant armor and infantry force. It appears Mussolini decided to invade Greece and to expand his holdings in the Balkans. The Greek Army is holding its own against a numerically superior force."

"They have the advantage of the home pitch."

"Quite so, and they are using the mountains to their favor."

"Has the information been passed to Wavell and Middle East Command?"

"Yes, sir."

"Very well. I shall raise the issue of expanding our support for Greece at the War Cabinet meeting in 40 minutes."

"We have issued instructions, redeploying some of our assets in the Balkans to improve our insight into both the Italian action and Greek defense."

"Make sure to keep General Wavell current as you develop the information."

"Certainly, sir."

"The last item comes from the United States. As you may recall, the American president signed into law their Selective Service Act last month. This morning, they began drawing lottery numbers for the draft under the new law. This is the surest sign to date that the United States has begun the process of mobilization. While there has been some grumbling from the isolationists, the protests have been surprisingly meek and timid, which in turn may be an indirect sign the mood among Americans is changing."

"Hopefully, the Americans shall have more time and commitment to rearmament and mobilization than we did before the shooting started. We need them in this fight."

"Yes, sir, which is why we thought the information important."

"Anything else?"

"No, sir. We shall remain vigilant."

"Very well, then, thank you."

Acton rose from his chair, checked to ensure the 'Buff Box' cover was securely locked and the case firmly fastened by the chained handcuff to his left wrist.

"By your leave, sir."

"Certainly. Please convey my appreciation to 'C.'"

"I will do so," Acton answered and departed, leaving the door open.

'Jock' Colville soon appeared. "Do you need anything before War Cabinet?" he asked.

"Yes, please ask Sir Edward to see me prior to the meeting. I am afraid we must amend the agenda."

"Yes, sir. I'll ask him to come immediately."

Churchill nodded his head, but his eyes had already returned to the shipping report in front of him. The situation in Greece just rose on the hierarchy of priorities. Too many things were vying for most important on the list, but he could not imagine anything exceeding the dreadful Battle of the Atlantic and the U-Boat menace, now that the looming invasion had stepped down from the number one position.

—

Tuesday, 29.October.1940
RAF Catterick
Catterick, North Yorkshire, England
15:10 hours

"**W**e have been at Available status all day," complained Noring.

"We've heard nothing since that fighter-bomber raid this morning, and we didn't even get to launch for that one," added Clegg.

"Are we going to be stuck up here in the quiet North Country for the rest of the war?" Carrwood asked, somewhat rhetorically.

"And, what is wrong with my native land?" asked Bevan, accentuating his Scottish brogue.

Everyone in the Dispersal Hut laughed to one degree or another.

"Nothing . . . it . . . it's just . . . ," stammered Carrwood.

Bevan held up his hand. "No need to apologize for the offense," he said. "None taken."

"Enjoy the quiet and rest while you can," answered 'Sparky' Morrow.

"We did not join the RAF and make it into Fighter Command to rest," said Noring with a tone of defiance. "We came to fight, to defend the Motherland."

"Ease up, you damn neophytes," shouted 'Harness' Kensington. "You are not the only ones in this squadron. The rest of us survived a very long, tortuous summer. Many of our brethren did not survive. You, each of you, should consider yourselves blessed and thank the Good Lord above, that you have this time to learn and improve your skills. Many others before you did not have that luxury. They were thrown into the meat grinder of those horrific, summer, aerial battles. Most did not live. Our best pilot is still trying to recover from his wounds and return to us. But, I suppose they were blessed in a form . . . they did not have to listen to your fucking drivel. Now, shut the fuck up about your bravado. We do not want to hear it. Those of us who have been in combat do not want to listen to another word from you fledgling fly-boys, unless it is too improve your knowledge of and skill in aerial combat."

'Sparky' and 'Waggle' Davies clapped. Even 'Stack' appeared at his door and clapped.

"Well done, 'Harness,'" said 'Stack' Long-Roberts. "Listen up, you newbies. There is a very real reason this squadron was moved so far north, into the relative quiet of One Three Group. We were bled in combat, bled nearly to death, as many other squadrons were bled. We were on the cusp of ineffectivity, nearly, no longer operationally viable. If we had remained in Ten Group, we would not have been able to impart our experience to you. 'Harness' is quite correct. You should be grateful you have this relative calm to learn, to hone your skills. This respite may well be the key to your surviving the war."

"My apologies," mumbled Carrwood. The other replacement pilots offered unintelligible concurring grunts and grumbles.

"This war is a long way from being over," added 'Stack.' "There will be more than enough fight remaining for each of you. The veterans need the rest and decompression from what they have just been through. Let them rest in peace. There may well come a day when you will be wishing for such an opportunity to recharge."

"Yes, sir," said Clegg.

The squadron remained at Available status until an hour before sunset and 90 minutes before the evening meal at the Officer's Mess. Not another word was spoken. The telephone rang only once and that was to release them from flight duty. Jonathan felt a little regret for being so harsh on the charged up replacement pilots, but at the end of the day, he knew he had done them a favor and found some relief for his colleagues. Fortunately for all of them, the off-duty rest rotations had been reinstated by Fighter Command, so each of them could look forward to a 24 or 48-hour pass away from waiting and flight operations. Jonathan knew as well as any of them how debilitating the long days of summer with the multiple combat sorties each day with not even a hint that they might get a break to rest. He had no desire to experience that degree of mortal intensity again; yet, he knew there would be periods of such intensity ahead; they were unavoidable . . . until the war was won.

———

Wednesday, 30.October.1940
Standing Oak Farm
Winchester, Hampshire, England

Brian sat in one of the living room, over-stuffed chairs with his leg cast raised and resting on an Ottoman. Charlotte stood at the window, as she often did when they were waiting for a visitor. The low cloud and mist, verging on fog kept everything outside damp and nearly obscured the gravel drive.

"Before he arrives, I want to thank you, again, Charlotte, for everything you have done for me and for your tolerance of what I do," Brian said.

Charlotte turned to connect with his eyes. "You are most welcome, Brian." She winked at Brian and turned back to the window. "A taxi is approaching. I suspect this is Mister Murrow."

"Thank you," he answered and began the process of standing on his crutches.

Charlotte walked with Brian to the front door. They made it onto the entry porch together, before the approaching vehicle came to a stop in front of them. The automobile was not a taxi. Rather it was a chauffer driven, black, Vauxhall Fourteen Series J sedan.

Murrow flung open the passenger door and leapt to the coverage of the porch. He wore a very British tweed jacket, white shirt, thin, solid black tie, and a long beige overcoat. He raised and tipped his Fedora to them both. Murrow first extended his hand to Charlotte. "It is a pleasure to see you again, Mrs. Palmer."

"Welcome to Standing Oak Farm, Mister Murrow. The pleasure is mine."

". . . and you, Pilot Officer Drummond," Murrow said, as he extended his hand to Brian.

"Mister Murrow."

"Please, would you call me Ed. Mister Murrow sounds like my father."

"Certainly, Ed, thank you."

"Shall we get out of the damp air," suggested Charlotte.

Murrow nodded his head. Charlotte motioned for him to enter the house, but Murrow motioned for Charlotte and Brian to lead him inside. Once inside, Charlotte took his overcoat and hat, and hung them on a hanger and the entryway coat rack.

"May I give you my jacket as well?" asked Murrow.

"As you wish, Ed."

Brian led them to the living room, motioned for Ed to sit on the adjacent couch, sat in his usual chair and again raised his leg cast on the Ottoman.

"Would your driver like to join us?" Charlotte asked. "It is warmer and dryer in here than out there."

"Thank you for the offer. He has another assignment in Southampton and expects to return around mid-afternoon. I need to be back in London this evening for my nightly broadcast."

"Very well. Would you care for some tea and homemade biscuits?" asked Charlotte.

"That would be most welcome and generous," Ed answered. Murrow removed a small, slip, spiral-bound notepad and a short pencil from his shirt pocket. He opened the pad to presumably the first open page and poised his pencil to take notes.

Charlotte went to the kitchen to prepare their refreshment tray.

"This is beautiful country," observed Ed.

"Yes, it most certainly is. Charlotte owns nearly a third of a section. It is a very productive farm that has been in her family for generations. It is very peaceful here."

"You are quite fortunate to have such a nice place for your recovery."

"Indeed, I am most fortunate."

Charlotte returned with the tray, served the tea for each of them, passed the plate of fresh cookies that she had baked earlier in the morning, and sat on a small chair across the low table from Murrow.

"Again," Murrow began, "thank you for agreeing to meet with me and allowing me to visit." Charlotte and Brian nodded their acknowledgment. "Let me start by saying that after talking to both of you at RAF Middle Wallop at the end of August and my broadcast of your story, I received several requests to expand my coverage of your story for both my company, CBS News, and a feature article for the New York Times newspaper. Thus, here I am. I will also say in the form of a preface, the mood in the United States is shifting dramatically, as we report on the Blitz in London. I am told subscribers are hungry for more of your story." Murrow looked to both Charlotte and Brian, and again received an acknowledgement head nod. "Our talk last August was very short, without a lot of detail. Plus, if you will allow me to be so bold, I sensed a relationship between the two of you last August, but your recovery on Mrs. Palmer's farm suggests a deeper relationship than incidental familiarity."

Brian looked to Charlotte, to answer for them both.

Charlotte nodded to Brian, and then looked back to Ed. "I do not know how intimately you wish to go," she answered. "Let it suffice to say, we are closer today than we were in August, and I think it safe to say . . . Brian can speak for himself . . . we will likely remain friends for life."

"Are you going to be more than friends?"

Charlotte chuckled softly. "Brian wants that."

"But . . ."

"I want that, too, but I have had far too much loss in my life. I am already emotionally attached to him, more so than is probably healthy for me."

"I did not intend to pry." Murrow quickly changed the subject. "How is your recovery progressing, Brian."

"Fairly well, I would say."

"I am informed that you are to receive your second Distinguished Flying Cross next week."

"So I am told. I have been ordered . . . well, actually, both of us have been asked to go to London."

"Are you getting another award?" Ed asked Charlotte.

"No, sir. Not that I know of. I was just asked to attend with Brian, and who am I to question such requests."

"Very generous of you, Charlotte," he said. She nodded her acknowledgment. Murrow turned back to Brian. "If it is not too difficult, would you be so kind to tell me the action for which you are to be recognized next week, and the event that caused your present injuries. I was informed that was combat as well."

"Yes, it was," Brian answered and looked searchingly at Charlotte.

She did not hesitate or need prompting. She stood and Murrow followed her. "Please excuse me, Ed. I have some chores to attend."

"By all means. I do not want to intrude."

Charlotte waved her hand dismissively and walked to the front door. "I shall leave you two men to your war talk."

After the front door closed behind Charlotte, Murrow said, "I did not intend any offense."

"No, no, it is quite alright. She is just very sensitive to what I do, to what has caused this," Brian said, waving his hand over the length of his body. "She knows you have questions you must ask. She just does not want to hear my answers, to be reminded of what happened."

"I am sorry for all this, Brian."

"It's nothing. So, to answer your questions, I have been told I am to be recognized for a specific action on the 13th of August. For reasons I know not, the Germans surprised us and began bombing our aerodrome. I took off as the bombs were falling. My aircraft took a lot of shrapnel damage. I was lucky to get off the ground. I climbed as fast and hard as I could, and plenty of targets. I shot down two Stukas before they dropped their bombs. Several lads behind me did not make it up. It was all over in a matter of minutes that seemed like hours. When it was over, I realized I had been in such a hurry to get airborne I forgot to strap into my parachute or seat." Brian laughed, as if it was some big joke. "It was a crazy day."

"Crazy, indeed. God bless you for risking your life to that degree. Many credit you with saving lives that day."

"I don't know about that. I simply did what I do."

"Simply," Murrow chuckled more to himself than to join in Brian's 'joke,' "is an understated adverb, I suspect." Brian shrugged his shoulders. "Those two Stukas made you an ace two times over, by then; and, you have gone on to chalk up 19 victories, third highest in all of Fighter Command." Again, Brian shrugged his shoulders. "What about this?" Murrow asked, motioning his hand toward Brian's injuries.

"We jumped into a tangle of One Oh Nines. I couldn't shake 'em. They got me . . . stopped my motor. One of the hits jammed my canopy, and I couldn't jump. I crashed in a farm field near Romsey, north of Southampton. Fortunately for me, dirt dowsed the fire, and farmers were nearby and pulled me out . . . saved my life . . . again."

"You have had more than your share of close calls."

"We all have. It is the nature of the fight. There are more them than there are of us."

"Are you going to return to your squadron, and if so, when do you expect to do that?"

"Yes . . . although I must admit, Charlotte is not particularly happy about that. It is the one major issue between us. I love her and I want to marry her."

"Pardon me, Brian, but you are quite young and in the middle of fighting this dreadful war. Marriage hardly seems like a wise move."

"Perhaps not, but those are my feelings, nonetheless."

"Can you describe what is it like in the tiny cockpit, flying at such high speeds, so high up in the atmosphere, against a determined and capable enemy with two, three, five times your numbers?" Murrow asked, swiftly and expertly changing the subject, again.

Brian laughed softly. "Well, since you put it that way, I suppose we should just give up."

Murrow joined Brian's laugh. "Perhaps I overstated my question. However, from the perspective of a humble journalist standing firmly on the ground, that is how it appears."

"Most of us, and here I include me, do not dwell on those aspects of what we do. We concentrate on getting the most out of our machines, and to be blunt and perhaps a bit crass, killing the other guy before he kills you. We depend upon each other. We know it is a dangerous business, but we simply cannot think about that danger. This country, and I will argue the world, is depending upon us to defeat anyone who enters our airspace."

"From everything I am hearing, the threat of invasion appears to have diminished."

"I do not know. I just fly Spitfires. You will have to ask the Prime Minister that question."

"You do not feel like you and your colleagues have won the battle?"

"I've not flown since the end of September. Heck, my squadron has moved north for rest and refit after losing half our number, so I am hardly current. From what I hear, the daylight raids have virtually ended, especially compared to what we had to deal with two months ago. You live in London, don't you?"

"Yes."

"Would you say we have won?"

"No, I suppose I would not. The Blitz, as the popular moniker has taken hold, is brutal and terrorizing."

"Yes, and as long as the Germans continue to bomb London and the other cities, we can make no claim to winning the battle. The fight has just shifted."

"I saw the pond as we drove up. I presume that is the pond from which Charlotte saved you and was awarded the George Cross."

"That is the one."

"May we take a look?"

"I would love to show you, however the medical folks tell me I must not get this cast wet."

"Oh, yes, certainly."

"I suspect Charlotte is right outside waiting for us to finish our war-talk, as she calls it. She told me the story of what happened that day. She can also show you the fragmentary remains of the Spitfire fighter I flew that day. What is left of it lies beyond the hill to the north. I have no idea how I survived, other than I did . . . and, my sorry ass was saved by the beautiful woman outside."

"She should not be outside in the chill and damp."

Brian rose from the chair, onto his crutches and led Ed Murrow to the front door. They both donned jackets and overcoats. Ed held the door open for Brian. The mist had stopped. The cloud lifted but remained overcast. Sure enough, Charlotte was sitting on the bench alone with her thoughts. She was dressed for the weather. It was also drier now than Brian expected. Charlotte turned her head toward them.

"Are you fellows finished?" she asked.

"For now," Brian answered. "Ed wanted to see the lake, hear your story of that day, and if it not too wet out there, he might like to see the wreckage."

"My little monument to your folly," she said and smiled.

"Yes, my darling, that one."

"We do not need to do this, if it is at all painful for you, Charlotte."

"No. It is quite all right. This is not my first time, and probably will not be my last either."

Brian watched them walk toward her spot, as Charlotte began telling her story. He went inside, doffed his jacket and overcoat, poured himself a fresh cup of tea, threw another small log on the fire, and settled into his big chair and thoughts. The same questions kept plaguing his thoughts. *Why was a famous journalist like Edward R. Murrow so interested in Charlotte and me? There are many more lads out there who are far better and more accomplished than me. Why me? I am not the only American serving with Fighter Command.*

Brian closed his eyes and dozed off in the warmth of the fireplace. He had no idea how long he had been asleep. The sound of the front door opening brought him back to consciousness.

"Thank you, Charlotte."

"You are most welcome, Ed."

"Did you see everything you wished?" Brian asked.

"Yes. Absolutely. Charlotte was most generous. Thank you, again." Charlotte nodded her head in acknowledgment. Murrow took a seat near Brian. "Now, I understand why you are baffled by your survival of that crash. There is almost nothing left of that airplane."

"Excuse me," Charlotte interrupted, "would you gentlemen care for lunch?"

"That would be most kind of you, Charlotte. Yes."

Brian nodded his head. Charlotte returned to the kitchen to prepare their meal.

"I should know from our last meeting, but I do not have the date. When did that happen?"

"Wednesday, the 31st of July."

"A date you will remember."

"Yes, you can say that. Life had been taken away from me and was returned by the gorgeous woman in the kitchen. Whether she tosses me out on my ass, I shall be forever indebted to her."

"Quite understandable, I should say. I must ask you, Brian, when you see that wreckage, doesn't it scare you or give you pause . . . that you came so close to death that day? My gosh, it gave me chills just seeing what remains and knowing you were sitting in that airplane just moments before it ended up in that field."

"Well, the honest answer is, yes, sure as hell, but if I kept those thoughts in my head, I would probably never fly again. Malcolm taught me years ago to get back on the horse, when you get bucked off, to hold the passion, and not lose what you love doing."

"Malcolm?"

"Malcolm Bainbridge, my flight instructor. He taught me how to fly and polished the skills I have today."

Murrow flipped through his notebook. "Ah, yes, you mentioned Malcolm that Thursday in late August last."

Charlotte appeared. "Lunch is ready, gentlemen, if you please."

They each took a healthy scoop of Shepherd's Pie, along with a freshly baked roll.

"I understand this farm has been in your family for some years, now," stated Murrow, avoiding anymore war-talk for the moment.

"Yes, for generations, now. I inherited the farm with my father's passing. I was only five years old."

"Oh my."

Please do not ask her about her father's service.

"How much land, if I may ask?"

Well done. Now, stick to this line.

"One hundred and ninety acres."

"That is quite a bit of land for England, I should think."

"I do not know what is normal. My father's great-grandfather received the land by Crown grant for service with the 37th (North Hampshire) Regiment of Foot and the East India Company during the Sepoy Mutiny of 1857. The paternal side of my family has successfully farmed this land since the mid-19th Century. Now, the farm is entrusted to me."

"Who helped you run the farm until you came of age?"

"Actually, Lionel and Horace began working with my father as young men. They still work for me, as they can. They are getting a little long in the tooth, so to speak, and there are no young men available."

"This war can't . . . won't last forever," Murrow said with some solemnity.

"Yes, well, until that day, this farm will be a struggle."

"I must say, Charlotte, you have a most impressive history and a beautiful farm."

"My family, perhaps . . . I am just a woman, trying desperately to keep things running properly during another dreadful war."

A knock at the door broke their conversation. Charlotte went to the front door. Murrow's driver arrived and announced that the local constabulary advised an early departure for London. Charlotte offered the driver some lunch, but he had already eaten. Murrow conveyed his gratitude for Charlotte's generosity and the tolerance of both of them to his questions. The article would take a few days to complete, he informed them, and he would send along a copy of the article once it was completed and published.

Brian and Charlotte did not talk. *The afternoon milking is just ahead. It has been quite a day.* They held to their thoughts. Brian did not want to stir things up, so he decided to wait for Charlotte to talk when she felt she was ready.

———

Saturday, 2.November.1940
Chequers Court
Ellesborough, Buckinghamshire, England
21:20 hours

Dinner had been unusually quiet for a dining room with Winston Churchill at the table. The Prime Minister seemed uncharacteristically withdrawn into contemplation. While he was certainly cordial and modestly engaging, he had not been particularly attentive.

The meal, meticulously prepared by Mrs. Georgina Landemare, the Churchills' chef, had been exquisite as was so often the case. The large table of guests had been amply supplied – a rarity among the people of Britain, but not the Prime Minister's wartime table, or rather more specifically, Winston Churchill's table. Winston justified and sloughed off the occasional criticism with a simple belief – he was conducting the Empire's business.

The ladies had excused themselves, retiring to the parlor for an evening of card games and gossip. The gentlemen moved to the large, manly library and study. Winston's favorite Hine cognac and Cuban *Romeo y Julieta* cigars were served to his male guests. This weekend and this night, the Prime Minister's male guests included: Air Minister Sir 'Archie' Sinclair, First Lord of the Admiralty 'A.V.' Alexander, War Minister Anthony Eden, Foreign Minister Lord Halifax, and the newly joined Home Secretary Herbert Morrison.

The Right Honorable Herbert Stanley Morrison, Member of Parliament for Hackney South, replaced Sir John Anderson as Home Secretary in the cabinet reshuffle of a month earlier. Morrison was another Labour Party leader and had been Minister of Supply from the outset of the Churchill government, until the assumption of his new post.

The Right Honorable Sir John Anderson GCB, GCSI, GCIE, PC, Member of Parliament for the Combined Scottish Universities, had been elevated to the War Cabinet as Lord President of the Council. He was not affiliated with a political party.

"The weather is simply dreadful out there tonight," observed Eden, to open the evening's conversation at a low-key level. "Thank goodness we are in this elegant library with an ample fire to keep the chill away."

"The weather may well help Fighter Command, I should think," Alexander added.

"The Germans have perfected their bombing technique through clouds and bad weather," interjected Churchill. "I suspect there will be no respite for London this night . . . while we sit here safe and cozy in this comfortable country estate."

"Is that what is bothering you, Winston?" asked Eden.

Churchill chose to ignore Eden's query and motioned for everyone to find a comfortable chair, of which there were plenty. He moved toward the fireplace, faced the fire for several minutes, and then turned and remained standing a couple of yards in front of the fire. "Where are we on the Wizard War, Archie?"

"We have made progress. Our success rate with the night-fighters has been steadily increasing. The lads are getting better at using the airborne radar units. The new Bristol Beaufighter is proving to be a very capable machine, and I understand the pilots like it much better than the Blenheim and Defiant versions we used earlier. The results support their assessment. We have had success with both the Blenheim and Defiant platform variants, but the pilots say the new Beaufighter is easier to handle for the crew."

"Is Max helping you get those new fighters?" asked Churchill.

Minister of Aircraft Production 'Max' Aitken, 1st Baron Beaverbrook, focused his aggressive, no-holds-barred, management style solely on ramping up, amplifying and protecting British aircraft production since Churchill created the ministerial position after his premiership ascendency in May.

"Yes, sir. Lord Beaverbrook has increased deliveries of the Beaufighter treble-fold in the last month and has informed me, the manufacturer should be able to sustain that production rate through the first quarter of next year."

"Let's talk to 'Max' to see that they maintain the necessary production rate, until you have sufficient night-fighters."

Sinclair knew that meant he was to talk to Beaverbrook to sustain production. It also meant his ministry must determine precisely how many of the advanced, radar-equipped aircraft the Royal Air Force needed to adequately defend the Home Islands against the German night-bomber attacks, now popularly known as The Blitz.

"Yes, sir. I shall tend to it posthaste." Churchill nodded his head in concurrence. "I must also report that Dr. Jones is still working with the Research

Establishment to institute a dependable means of reliably identifying and locating the broken leg beams promptly enough to configure our countermeasures."

Doctor 'R.V.' Jones had argued before Churchill and the War Cabinet against the contemporary experts back in June, for precious research and aircraft resources to find the night and foul weather, electronic assistance, bombing system, identified by the Oslo Report – the November 1939, intelligence report from inside the German government.

"We know the counter-measures developed by the scientists' work, essentially deflecting the beams for the German pilots. We have proven that element. However, we are struggling with consistently finding the specific frequency and location of the beams for any given night's operations. Dr. Jones, with the direct assistance of Doctor Watson-Watt's Bawdsey Station specialists, believes they have the science and engineering defined for a specific search-scanner that should rapidly determine the essential characteristics of the night's beams. They are working around the clock to produce a work-able search unit, and I might add more importantly, they are developing the procedures to utilize the device and counter-measures to protect our cities."

Doctor Robert Alexander Watson-Watt had been credited with pioneering the development in the use of radio waves to determine the direction, range and altitude of aircraft. By this day, he served as Superintendent, Bawdsey Research Station, Bawdsey Manor, near Felixstowe in Suffolk. The Chain Home network of radar sites so vital to the conduct of fighter defense operations during the Battle of Britain – the first leg of a three-legged stool, without it, the stool falls.

"I need to say this is our highest priority . . . to protect our citizens from the abuse inflicted upon us by the Germans," Churchill said. "We appear to have dominated the daylight hours, at least in good weather. We have not experienced a daylight raid in several days, now, while the night raids continue unabated." Since not all his guests were ULTRA-knowledgeable, Churchill could not refer to the secret information he had seen in the last two days, telling him the Germans had moved more than a few fighter squadrons back to Germany and granted extended leave to many pilots in the remaining fighter squadrons to rest up through the winter in preparation for spring operations of an unspecified nature. "We must find the means to defeat the night-bombers."

"Are you saying we have won the battle?" asked Morrison.

"I will make no such claim as long as bombs fall on our cities. We have a long way to go in this battle and in this war. The winter weather will offer some respite from the mortal combat we endured during the summer months, and until we can stop the night-bombers, the battle remains undecided. Beyond the air battle and the threat of invasion, we still have those dreadful U-boats to contend with on our supply lines."

"My bailiwick," interjected the First Lord. "We are making progress. The airborne microwave RDF units developed by Watson-Watt's lads have proven quite capable in finding surface objects as well . . . like a submarine conning tower or even a periscope, for instance."

Research and experimentation using radio waves reflected off objects had been carried out in numerous countries for several decades. Watson-Watt first demonstrated a practical system using radio waves to detect airborne objects – aircraft – bombers, fighters . . . anything metallic that flew – in the Daventry experiment of February 1935. It was Sir Henry Tizard, in his capacity as chairman of the Air Ministry's Committee for the Scientific Study of Air Defence in 1933, who saw the potential of Watson-Watt's radio technology. The British called the technology Radio Direction Finding (RDF). When the technical exchange program known as the Tizard Mission began in September, the British-American collaboration soon collectively adopted the American term for the technology – RADAR (RAdio Direction And Ranging) – that became so prevalent in usage the acronym soon became a noun. The combination of the British cavity magnetron, which enabled much higher power output, thus greater range capability, and the American duplexer switch, which allowed the same antenna to be utilized for both sending and receiving signals, enabled dramatic and rapid advances in the miniaturization of the electronics vital for use on space and weight constrained aircraft.

"Excellent," proclaimed Churchill. "The sooner we can find these vermin wolves of the sea the sooner we can eliminate them."

"The Intelligence Branch," Alexander added, "believes the Germans are producing U-boats faster than we are sinking them."

"Which is precisely why we should have landed a major force in Narvik last spring to take control of the iron ore mines in Northern Sweden." Churchill turned back to Sinclair. "Do you need any assistance from me to ensure the highest priority on these radio developments here or in America?"

Archie Sinclair shook his head and answered, "No. The joint team informed us last month they have formed a specific, secure, secret laboratory at the Massachusetts Institute of Technology in Cambridge, Massachusetts. They call it the Radiation Laboratory, or Rad Lab for short. The sole purpose of the Rad Lab is the rapid development, experimentation and productionization, as they say, of various classes of radar applications from gun direction, mapping, weather detection and of course interception of airborne targets. The scientists believe we will soon have a more advanced version of our airborne Mark-IV

Airborne Intercept unit that is being deployed with the Bristol Beaufighters, and I might add a tuned airborne surface search unit to aid Coastal Command in finding the bastards."

"We must keep the pressure on the scientists and engineers until we get ahead of this threat," Churchill pronounced. "Please pass along my sincerest appreciation for the extraordinary efforts of Doctors Jones and Watson-Watt, Sir Henry, and all the others working so diligently in defense of the realm. I sense that we are at or near the cusp of major changes that will enable us to attain the ultimate victory."

"We will," confirmed Eden for the other ministers.

"The Americans seem to have enthusiastically embraced Sir Henry's mission and our experts," said Lord Halifax.

"Excellent news," answered Churchill. "I, for one, held some trepidation as to what such an exchange might yield. It seems we may well be on our way to exceeding our wildest expectations or even imagination."

"You may well be correct, Winston," added Halifax. "The signs certainly point in that direction. Lord Lothian had repeatedly conveyed his admiration for the work of our experts on this Tizard exchange."

"Speaking of our cousins, the Americans, our featured film this evening is the latest Charlie Chaplin movie, "*The Great Dictator*," which I understand is an exceptionally funny and poignant parody of our adversary Herr Hitler."

Churchill led the others out of the study, down the hall, and into the room set-up as a small movie theater since Churchill's ascension to the premiership and his use of Chequers for his relaxation. The sound of the rain outside filled the silence until the movie began.

Despite the continuing ravages of war, a demonstrable air of vindication could be felt among the British leaders. They had survived the trial of their summer of vulnerability. The verdict had been rendered. They all knew there would be many more trials ahead before the Germans were defeated and subdued. Winston Churchill knew there would be difficult days ahead; yet, there was no denial, the verdict on the latest trial felt good. The German invasion had been thwarted at least for the difficult, inclement weather of autumn and winter. The British people remained free, and they were finally gaining strength. They lived to fight another day.

———

About the Author
Cap Parlier

—

Cap and his wife, Jeanne, live on the Great Plains of Kansas, along with two dogs and a cat. Their four grown children have begun their families as well as producing precious grandchildren – the rewards of long life. He is a graduate of the U.S. Naval Academy [USNA 1970], a retired Marine aviator, a Vietnam veteran and a former experimental test pilot. Cap has retired from the corporate world and can now fully indulge his passion for the story. His books are available from print and digital sources worldwide.   Cap has numerous other projects completed, but not yet published, and others in development, including screenplays, historical novels and a couple of history books.

—

Interested readers may wish to visit his website at http://www.parlier.com for his essays and other items, or subscribe to his weekly Blog: "Update from the Heartland."

Cap can be reached at: Cap@SaintGaudensPress.com.

—

Books by Cap Parlier:

Anod series
The Phoenix Seduction (1995)
Anod's Seduction (2004) [reprint of The Phoenix Seduction]
Anod's Redemption (2004)

Sacrifice (2000)
The Clarity of Hindsight (2016)

To So Few series
To So Few – In the Beginning (2014)
To So Few – The Prelude (2014)
To So Few – Explosion (2015)
To So Few – The Trial (2016)
To So Few – The Verdict (2017)

---

and with **Kevin E. Ready:**
TWA 800 - Accident or Incident? (1998)

Coming soon from Cap Parlier, **To So Few – Deflection**, the sixth book of the series novel of flight and a warrior's life.

These and other great books available from Saint Gaudens Press
Post Office Box 405
Solvang, CA 93463-0405
URL: http://www.SaintGaudensPress.com
Visit Cap Parlier's Web Site at: http://www.Parlier.com

Saint Gaudens, Saint Gaudens Press and the Winged Liberty colophon are trademarks of Saint Gaudens Press Inc.